CYBERCARTOGRAPHY IN
A RECONCILIATION COMMUNITY

CYBERCARTOGRAPHY IN A RECONCILIATION COMMUNITY

ENGAGING INTERSECTING PERSPECTIVES

Edited by

STEPHANIE PYNE
D. R. FRASER TAYLOR

ELSEVIER

Elsevier
Radarweg 29, PO Box 211, 1000 AE Amsterdam, Netherlands
The Boulevard, Langford Lane, Kidlington, Oxford OX5 1GB, United Kingdom
50 Hampshire Street, 5th Floor, Cambridge, MA 02139, United States

Notices
Knowledge and best practice in this field are constantly changing. As new research and experience broaden our understanding, changes in research methods, professional practices, or medical treatment may become necessary.

Practitioners and researchers must always rely on their own experience and knowledge in evaluating and using any information, methods, compounds, or experiments described herein. In using such information or methods they should be mindful of their own safety and the safety of others, including parties for whom they have a professional responsibility.

To the fullest extent of the law, neither the Publisher nor the authors, contributors, or editors, assume any liability for any injury and/or damage to persons or property as a matter of products liability, negligence or otherwise, or from any use or operation of any methods, products, instructions, or ideas contained in the material herein.

Library of Congress Cataloging-in-Publication Data
A catalog record for this book is available from the Library of Congress

British Library Cataloguing-in-Publication Data
A catalogue record for this book is available from the British Library

ISBN: 978-0-12-815343-7
ISSN: 1363-0814

For information on all Elsevier publications visit our website at
https://www.elsevier.com/books-and-journals

Publisher: Candice Janco
Acquisition Editor: Amy Shapiro
Editorial Project Manager: Susan Ikeda
Production Project Manager: James Selvam
Cover Designer: Mark Rogers
Cover Photographer: David Porter Jr.

Typeset by TNQ Technologies

In living memory of Joseph and Barbara Pyne

Contents

12. Conclusion: building awareness to bridge relationships

Stephanie Pyne and D.R. Fraser Taylor

Contributors

Melissa Castron MA Candidate, Department of History (Archival Studies), University of Manitoba, Winnipeg, MB, Canada

Trina Cooper-Bolam Doctoral Candidate, Cultural Mediations and Geomatics and Cartographic Research Centre (GCRC), Carleton University, Ottawa, ON, Canada

James Gerencser College Archivist, Dickinson College, Carlisle, PA, United States

Skylee-Storm Hogan MA Candidate, University of Western Ontario, London, ON, Canada; Shingwauk Residential Schools Centre, Algoma University, Sault Ste. Marie, ON, Canada

Tilly Laskey Curator, Maine Historical Society, Portland, Maine, United States

Krista McCracken Researcher/Curator, Shingwauk Residential Schools Centre, Algoma University, Sault Ste. Marie, ON, Canada

Kevin Palendat MA Candidate, Department of History (Archival Studies), University of Manitoba, Winnipeg, MB, Canada

Stephanie Pyne Postdoctoral Research Fellow, Geomatics and Cartographic Research Centre (GCRC), Carleton University, Ottawa, ON, Canada

Susan Rose Charles A. Dana Professor and Chair of Sociology, Dickinson College, Carlisle, PA, United States

D.R. Fraser Taylor Chancellor's Distinguished Research Professor of International Affairs, Geography and Environmental Studies, and Director, Geomatics and Cartographic Research Centre (GCRC), Carleton University, Ottawa, ON, Canada

Jeff Thomas Independent Curator and Photographer, Ottawa, ON, Canada

Andrew Woolford Professor, Department of Sociology & Criminology, University of Manitoba, Winnipeg, MB, Canada

Foreword

Stephanie Pyne

I was surprised in May 2008, at community launch presentation at Shawanosowe School in Wigwaasminising (Whitefish River First Nation) for a predecessor atlas to much of the work described in this volume. During this visit, I had planned to tell the grades fives and sixes the 'Rainbow Raven and the Fly in the Bottle' story below, which reflects the aims and methods of ongoing work related to intercultural reconciliation. I had brought five acrylic paintings on canvas, which were created and later digitized and included in the Lake Huron Treaty Atlas (https://lhta.ca/index.html). Before the discussion with the class began in the gym, I asked a group of children to take the paintings and arrange them in the order they thought they might go in. To my surprise, without first hearing the story, the children arranged the paintings in the right order (after having a great time running around the gym with them). When I told them they had got it right, and when I asked them how they knew the order of the paintings for a story they had not yet heard, the children replied that they had heard that story about 'the crow and the ant' in Ojibwe class. Although the story of Rainbow Raven and the Fly in the Bottle has different origins from the Ojibwe story recalled by the children, the themes and directions of the two stories were sufficiently similar to my story that the children were able to enact both stories by arranging a series of artworks, which — apparently — told a story in themselves.

The following story came to me in September 2004 when I was living in Banff, Alberta: all at once, with complete energy, form and life, in answer to questions I was pursuing in my master's studies in philosophy related to intercultural reconciliation and worldviews; epistemology, ontology and ethics. While there are many more stories about this story, it is offered here as a point of departure for further reflection.

Rainbow Raven and the Fly in the Bottle — The Fly

> What is your aim in philosophy? To show the fly the way out of the fly-bottle. *Wittgenstein, 1953, 2009, 110.*

There was once a happy little fly who buzzed along humming his little tune in the wind. The fly loved the wide open spaces and the many things there were to see in the world. When he flew in the wide open he felt like he was zooming through a garden of colour in the sky. One unfortunate day, however, the little fly flew right into the end of a tall clear glass bottle. The fly was confused and afraid. He was accustomed to flying wherever he wanted in his rainbow-coloured garden in the sky (Fig. F.1).

FIGURE F.1   Screenshot from the Nenboozhoo Map of the Lake Huron Treaty Atlas showing location 1 on Sleeping Buffalo Mountain in Banff, Alberta, Canada, associated with the Rainbow Raven and the Fly in the Bottle story.

Rainbow Raven

Rainbow Raven was not always the colours of the rainbow. She was once almost pure black. Although in just the right light, Raven's feathers glinted electric green and purple. Raven had a good eye from the sky for colourful objects that glittered from the ground below.

Each day Raven flew around looking for shiny treasures. She brought them back to her nest one by one and dropped them inside. At the end of the day, Raven returned to her nest to marvel at her treasures. When she tired of this, Raven arranged her bobbles in different patterns and tried to make something with a useful purpose. This is where she had difficulty. In the darkness of night, the objects no longer glittered. Even in the light of the full moon, Raven had no success in making anything that worked from her bobble collection. This went on every night for years and years until finally Raven became very tired of trying. It seemed that she had tried everything. Raven became very sad and eventually even lost interest in the bobbles. Still, she continued to scoop up new treasures every day because that is what ravens do. Raven began to cry out in despair and cried 'cawww, cawww' because she had lost her zest for life (Fig. F.2).

Raven Meets Fly

One day as Raven was out scouting, she spotted a crystal clear object with a little black speck darting inside in all directions. This sight interested her not because of its shiny glimmer, but because of its lively movement. As she drew nearer, she saw that it was the little fly trapped in a bottle, bouncing from side to side and buzzing frantically. The fly spotted Raven in the distance and called up to her, 'Raven, Raven, help me please!'

'Just fly up', cawed Raven.

But the fly was too confused in the confined space to tell the difference between up, down, sideways, frontwards or backwards and buzzed back, 'I'm trapped, I'm trapped!' Raven swooped down and tried to poke her beak into the end of the bottle to scoop the little fly out, but the hole in

FIGURE F.2 Screenshot from the Nenboozhoo Map of the Lake Huron Treaty Atlas showing location 2 on Sleeping Buffalo Mountain in Banff, Alberta, Canada, associated with the Rainbow Raven and the Fly in the Bottle story.

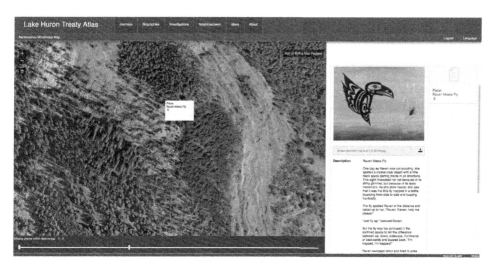

FIGURE F.3 Screenshot from the Nenboozhoo Map of the Lake Huron Treaty Atlas showing location 3 on Sleeping Buffalo Mountain in Banff, Alberta, Canada, associated with the Rainbow Raven and the Fly in the Bottle story.

the top was too small for her beak to fit into. She thought of her bobbles and how she had finally found a purpose for them. However, she didn't know how to make them work. The fly buzzed and buzzed. Raven thought and thought. She spent every day flying in circles around the bottle waiting for an answer to come to her and crying raindrops of pity for the trapped little fly (Fig. F.3).

Raven Helps Fly

Eventually Raven stopped crying and listened. Then the Wind spoke to her and

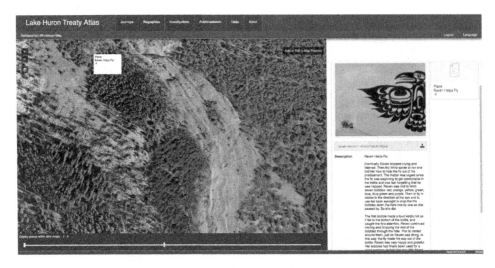

FIGURE F.4 Screenshot from the Nenboozhoo Map of the Lake Huron Treaty Atlas showing location 1 on Sleeping Buffalo Mountain in Banff, Alberta, Canada, associated with the Rainbow Raven and the Fly in the Bottle story.

FIGURE F.5 Screenshot from the Nenboozhoo Map of the Lake Huron Treaty Atlas showing location 5 on Sleeping Buffalo Mountain in Banff, Alberta, Canada, associated with the Rainbow Raven and the Fly in the Bottle story.

told her how to help the fly out of his predicament. The matter was urgent since the fly was beginning to get comfortable in the bottle and was fast forgetting that he was trapped. Raven was told to fetch seven bobbles: red, orange, yellow, green, blue, blue-green and purple. Then to fly in circles in the direction of the sun and to use her keen eyesight to drop the bobbles down the hole one by one as she passed by. So she did.

The first bobble made a loud kerplunck as it fell to the bottom of the bottle, and caught the fly's attention. Raven continued circling and dropping the rest of the bobbles through the hole. The fly circled around them, just as Raven was doing. In this way, the fly made his way out of the bottle. Raven was very happy and grateful. Her bobbles had finally been used for a good purpose, to help her new little friend in his great time of need. The fly was free, free at last to fly again in the garden in the sky (Fig. F.4).

Rainbow Raven

The Great Spirit on the Wind made a rope from the bobbles in the bottle and waved it over Raven's black wings. Then gently cast the rainbow rope across the sky to catch Raven's tears. Suddenly, Raven became a beautiful multicoloured bird, a Rainbow Raven. She cawwed thank you in awe, which made the Great Spirit realize that her voice did not match her colourful plumage. So he said to her in a serious voice, 'Raven, because you cared enough to help the little fly out of the bottle, and because you have tried for many years to make your bobbles work for some purpose, I will give you a voice to speak words. Instead of picking up bobbles on the ground, you will pick up words in the wind. Follow my directions and you may help the people of our world find a way out of the thinking traps created in their minds and hearts' (Fig. F.5).

Preface

The contributions to this volume were created in the midst of ongoing events, activities and projects related to reconciliation and residential schools (Canada)/boarding schools (United States), and broader reconciliation contexts. As such, they constitute a sample of thinking and practice — a dynamic cross-section of interrelated approaches that share a commitment to both social and spatial forms of justice. While many books concerning reconciliation in the case of residential schools and beyond are about the conditions requiring reconciliation, *Cybercartography in a Reconciliation Community: Engaging Intersecting Perspectives* is more concerned with the nature of approaches taken to reconciliation and education, especially in — but not limited to — the residential school context. As such, it is a gathering of perspectives on issues related to transdisciplinary research primarily in a residential school reconciliation context, and provides a demonstration of Cybercartography's unifying power when it comes to identifying intersections in disparate perspectives.

The 'cyber' in Cybercartography refers to more than simply the online environment. Equally important is the connotation of community, which can refer to many groups, including those involved in cybercartographic atlas development and use (Martínez and Reyes, 2005). Cybercartography is a transdisciplinary practice that involves many different teams engaging in the development of a variety of online, interactive, multimedia atlases. Cybercartography in a Reconciliation Community continues the approach taken in *Cybercartography: Theory and Practice* (Taylor, 2005) and *Developments in the Theory and Practice of Cybercartography* (Taylor and Lauriault, 2014), which include chapters by contributors from a diverse array of communities of knowledge discussing issues related to cybercartographic theory and practice in the context of an equally diverse array of cybercartographic atlas projects and related cartographic initiatives. In this case, we narrow the context to consider in more detail the workings of one specific cybercartographic atlas project, the Residential Schools Land Memory Mapping Project (funded by the Social Sciences and Humanities Research Council of Canada), which involves an innovative collaborative approach to mapping institutional material and a broadly construed notion of 'volunteered geographic information' (VGI). Exploring Cybercartography through the lens of this atlas project provides for a comprehensive understanding of both Cybercartography and transdisciplinary research, while informing the reader of education and reconciliation initiatives in Canada, the United States, the United Kingdom and Italy, in relation to broader historical geographies.

Most of the book's contributors have participated in some way in cybercartographic atlas development, most often contributing a perspective or a skill from within their particular community of practice and learning new skills and knowledge in the process. Their book contributions concern issues related to their particular approach to reconciliation, often in relation to residential schools, as well as issues that intersect with work in Cybercartography on the Residential Schools Land Memory Mapping Project. For example, Andrew Woolford (Winnipeg, Manitoba, Canada) is a sociologist and an internationally recognized genocide scholar; Susan Rose (Carlisle, Pennsylvania, USA) is also a sociologist who works on the

Carlisle Indian School Project, which has some shared features with both the Shinwauk Residential Schools Centre at Algoma University (SRSC, Sault Ste. Marie, Canada) and the Residential Schools Land Memory Mapping Project; Jeff Thomas (Ottawa, Ontario, Canada) is an urban Indigenous artist whose work engages with Cybercartography and who has curated residential school material; Trina Cooper-Bolam (Ottawa, Ontario, Canada) is a PhD candidate in Cultural Mediations (Carleton University) who has collaborated on residential schools projects with Jeff Thomas and Tilly Laskey is a museum curator specializing in Indigenous art and culture whose academic training includes art history and anthropology. Including their work in the book is intended to provide the reader with an understanding of the work and thought of the Residential Schools Land Memory Mapping Project community. This is important because – despite disciplinary differences – many common themes exist in issues discussed by the contributors. These commonalities become apparent in the chapters focused on aspects of the cybercartographic approach to the Residential Schools Land Memory Mapping Project.

Cybercartography in a Reconciliation Community includes both conceptual and applied dimensions and provides a good example of a reflexive approach to both research and knowledge dissemination. The positionality aspect of reflexivity is reflected in the chapter contributions made by project team members and others affiliated with the project and by the chapters concerning various aspects of cybercartographic atlas design and development research. The book aims to contribute to theoretical and practical knowledge of collaborative transdisciplinary research through its reflexive assessment of the relationships, processes and knowledge involved in cybercartographic research. Closely related goals include contributing to a broader ontology of cartography and providing insights into reconciliation and education processes. Its style is primarily narrative, which is consistent with the storytelling approach that characterizes many Indigenous approaches to knowledge dissemination and enhances its accessibility amongst a broad audience. It takes a high-speed tour approach that is inspired by Bourdieu (1992) in his Paris Workshop and is based on a recognition of common approaches and goals. As with many themes and concepts presented in this volume, collecting them together in a single volume serves to 'put them on the map' for further reflection and integration into ongoing theory and practice. In addition to shedding light on the processes, relationships and content and form of the Atlas being developed under this project, the book will serve in different respects as a design and development resource for the development of the atlas it describes and perhaps will serve in different ways as a model for other projects as well.

Stephanie Pyne and Fraser Taylor

References

Bourdieu, P., 1992. The practice of reflexive sociology. In: Bourdieu, P., Wacquant, L. (Eds.), An Invitation to Reflexive Sociology. University of Chicago Press, Chicago, pp. 217–253.

Martínez, E., Reyes, C., 2005. Cybercartography and society. In: Fraser Taylor, D.R. (Ed.), Cybercartography: Theory and Practice, Modern Cartography Series, vol. 4. Elsevier, Amsterdam, pp. 99–121 (Chapter 5).

Taylor, D.R.F., 2005. In: Cybercartography: Theory and Practice. Modern Cartography Series, vol. 4. Elsevier, Amsterdam.

Taylor, D.R.F., Lauriault, T. (Eds.), 2014. Developments in the Theory and Practice of Cybercartography: Applications and Indigenous Mapping. Elsevier, Amsterdam.

Acknowledgements

First a hearty and grateful acknowledgement of the contributing authors, many of whom are also contributing to the Residential Schools Land Memory Mapping Project (RSLMMP). This includes Jeff Thomas (Independent Artist, Curator and Writer, Ottawa, Ontario), Krista McCracken (Researcher/Curator, Arthur A. Wishart Library and Shingwauk Residential Schools Centre, Algoma University), Skylee-Storm Hogan (MA candidate, Public History, University of Western Ontario, see below), Susan Rose (Community Studies Center, Dickinson College), James Gerscener (Carlisle Indian School Digital Resource Centre), Andrew Woolford (Sociology, University of Manitoba); Trina-Cooper Bolam (PhD candidate, PhD Cultural Mediations, Carleton University, see below), Melissa Castron (MA, candidate, History/Archival Studies, University of Manitoba, see below), Kevin Palendat (MA, History/Archival Studies, University of Manitoba), Chris Calesso (MA, History/Archival Studies, University of Manitoba), Tilly Laskey (Independent scholar specializing in Native art and culture and a curator at the Maine Historical Society), Federica Burini (Foreign Languages, Literatures and Cultures, Planning and Management of Tourism Systems, University of Bergamo), and Greg Bak (History/Archival Studies, University of Manitoba) who were part of discussions and activities referred to in Chapters 8, 10, and 11. Special mention is in order for the contribution of Dr. D.R. Fraser Taylor, the book's coeditor, Director of the Geomatics and Cartographic Research Centre (GCRC) at Carleton University and principal investigator on the RSLMMP, and a superlative mentor; with thanks also extended to the technical and administrative staff of the GCRC, which is the major centre for research on Cybercartography.

Ongoing research and dialogue with RSLMMP research assistants have enriched both the project and the book. For this thanks to Trina Cooper-Bolam (see above); Chris Calesso (see above); Kevin Palendat (see above); Melissa Castron (see above); Noelle Dietrich (Sociology, University of Manitoba); Joanne Robertson, Meghan Caveen, Jamie McIntyre, and Alex Boston (Shingwauk Residential Schools Centre, Algoma University); Anja Novkovic and Annita Parish (Concordia University). A number of students have learnt about much of this book's contents and the ongoing work on the RSLMMP through coursework at both the undergraduate and graduate levels, which intersects with the project. Much of this work has yet to be reported on. Thank you in particular to the students of Framing Indigenous Studies (INDG, 2011; Carleton University, 2017); Territorial and Environment Studies (44136 −ENG, University of Bergamo, 2015−18; PhD seminar in Transcultural Humanistic Studies, University of Bergamo, 2017, 2018); Truth and Reconciliation Commissions, Museums and Archives (INF1005H and INF 1006H, University of Toronto, ISchool, 2019); and Public Administration in Nunavut (PADM 1005, Carleton University/Nunavut Sivuniksavut, 2019).

Several organizations are involved in ongoing work related to mapping and the Residential Schools Legacy. In this regard, the Children of Shingwauk Alumni

Association (CSAA), the Assiniboia Reunion and Commemorative Gathering Governing Council, the Legacy of Hope Foundation, in particular Jane Hubbard, and the National centre for Truth and Reconciliation deserve special mention for their ongoing contributions to the Residential Schools Land Memory Mapping Project and this book.

Acknowledgement is extended to the following past and present Atlas collaborators and coinvestigators not already mentioned above on the RSLMMP who have also participated in discussions of various aspects of the RSLMMP, including this book: Sebastien Caquard (Geography, Planning and Environment, Concordia University); Jonathan Dewar (former Director, Shingwauk Residential Schools Centre, Algoma University and Executive Director, First Nations Information Governance Centre); Sarah de Leeuw (Northern Medical Program, University of Northern British Columbia); Don Jackson (Founding Director, Shingwauk Residential Schools Centre, Algoma University and Special Advisor to the CSAA; Ry Moran, Director, National Centre for Truth and Reconciliation, University of Manitoba); Doris Young (Advisor to the President on Aboriginal Affairs, University College of the North); Lorena Fontaine (Indigenous Governance, University of Winnipeg; Jeff Barnes, Geography, Dawson College; Shirley Horn (CSAA, Algoma University); Alex Maass (Archeology, University of Southampton, UK); Nadia Myre (Visual Artist (Montreal, Quebec); Glen Lowry (Culture and Community, Emily Carr University of Art and Design); Janet McLeod (Mountain Therapy Psychological Services, Canmore, Alberta); Sharla Peltier (Education, University of Alberta); and Sarah Story (MA, History/ Archival Studies, University of Manitoba).

Gratitude is also extended to Rosa Orlandini, Romola Vasantha Thumbadoo, and Margaret Pearce, and to my family, for their ongoing support, including Jesse Pyne and Jason Pyne for their contributions.

With respect to those no longer with us, special thanks in memory of their ongoing wisdom to Jacob Wawatie, Louise Wawatie Wilmer Noganosh, Lewis Debassige — all former residential schools survivors—and to Blaine Belleau and William Commanda, David Allan Campbell, Stephen Valentine, Thomas Manyguns, Barbara George, Aline Pyne and Joanna van Loonen. Just as tossing a stone in a pond emanates concentric circles of energy outward, the circles of gratitude for discussions, research and other positive influences extends to Ginawaydaganuc, All My Relations.

Introduction

Stephanie Pyne*,[1], D.R. Fraser Taylor[2], Trina Cooper-Bolam[3]

[1]Postdoctoral Research Fellow, Geomatics and Cartographic Research Centre (GCRC), Carleton University, Ottawa, ON, Canada; [2]Chancellor's Distinguished Research Professor of International Affairs, Geography and Environmental Studies, and Director, Geomatics and Cartographic Research Centre (GCRC), Carleton University, Ottawa, ON, Canada; [3]Doctoral Candidate, Cultural Mediations and Geomatics and Cartographic Research Centre (GCRC), Carleton University, Ottawa, ON, Canada
*Corresponding author

1.1 Opening words

These are exciting and challenging times. According to the Seven Fires Prophecy of the Anishinabek (Anishinaabe people) of Turtle Island or North America, these are times when people are beginning to seek healing and greater understanding in their relationships with each other, the land and themselves. The Seven Fires Prophecy is an Anishinaabe teaching that has been carried

FIGURE 1.1.1   Grandfather William Commanda giving a reading of the Seven Fires Prophecy Belt. *Courtesy Romola Vasantha Thumbadoo.*

forward across many generations; it is encoded in a wampum belt, which was in the care of the North American Indigenous elder, Dr Grandfather William Commanda for about 40 years until the time of his passing at the age of 97 on 3 August 2011. Grandfather Commanda was 'a widely acknowledged public figure, guardian of three Algonquin wampum belts (sacred mnemonic record-keeping and governance devices including the Seven Fires Prophecy Belt, the Welcoming Belt, and Border Crossing Belt), Officer of the Order of Canada, and recipient of two honorary doctorate degrees' (Thumbadoo, 2017, ii) (Fig. 1.1.1).

Romola Vasantha Thumbadoo worked with Grandfather Commanda for many years to implement his vision for intercultural peace building and environmental well-being, which she continued — in part — through her studies toward a PhD Romola's thesis entitled, 'Gina-waydaganuc and the Circle of All Nations: The Remarkable Environmental Legacy of Elder William Commanda' (2017). Her thesis provides many inspiring examples of the ways Grandfather Commanda practiced 'his fully inclusive concept of Ginawaydagunuc that All is Related in the cosmic world' through his 'informal global eco-peace Circle of All Nations community'. According to Thumbadoo, 'the Circle of All Nations was conceptual-ized by William Commanda as a bridge-building mechanism and interface to speak to the incommensurability between Indigenous and non-Indigenous knowledge systems on envi-ronment, relationships and critical social justice and peace issues' (2017, ii).

Grandfather Commanda often read the belt at gatherings, and in his later years, sometimes referred to the written version prepared by Eddie Benton Banai (1988), which tells of the mes-sages of seven prophets at seven distinct fires. The prophecy is about choices and provides guidance regarding how to interpret and understand an unfolding 'reality'. The first three fires involve a prophet coming to the people to tell them about events that would transpire and the actions they should take. The people were told to continue moving west, from their initial home on the eastern edge of Turtle Island, and to continue, following the Mide (the spiritually wise people) and their interpretations of the signs indicated in the prophecies.

At the time of the fourth fire, a prophet came to the people in the form of twins, offering two sets of scenarios and choices:

> The Fourth Fire was originally given to the people by two prophets. They came as one. They told of the coming of the Lightskinned Race. One of the prophets said, 'You will know the future of our people by what face the Light-skinned Race wears. If they come wearing the face of nee-kon'-nis-i-win (brotherhood), then there will come a time of wonderful change for generations to come. They will bring new knowledge and articles that can be joined with the knowledge of this country. In this way two nations will join to make a mighty nation. This new nation will be joined by two more so that the four will form the mightiest nation of all. You will know the face of brotherhood if the Light-skinned Race comes carrying no weapons, if they come bearing only their knowledge and a handshake'.

> The other prophet said, 'Beware if the Light-skinned Race comes wearing the face of ni-boo-win' (death). You must be careful because the face of brotherhood and the face of death look very much alike. If they come carrying a weapon [...] beware. If they come in suffering [...] they could fool you. Their hearts may be filled with greed for the riches of this land. If they are indeed your brothers, let them prove it. Do not accept them in total trust. You shall know that the face they wear is one of death if the rivers run with poison and the fish become unfit to eat. You shall know them by these many things' (Benton Banai, 1988, 89–90).

The path taken at this juncture would affect the nature of the choices based on the messages of the future fires, at which the prophets foretold of the hardships the people would face over time as a result of broken relationships with the newcomers to Turtle Island (North America). At the time of the seventh fire, people would begin to emerge who would be concerned with healing the broken relationships born of the past; and to do so, they would seek the wisdom of Elders in order to learn how to re-engage in relationships at all scales in healthy and balanced ways. Many of the choices spoken of at the various fires involved choices being made not only by the Anishinaabek, but by the newcomers as well, making the Seven Fires Prophecy a valuable interpretative framework for understanding how to reconcile relationships between the Anishinaabek and the newcomers to Turtle Island.

There is increasing agreement that Indigenous peoples worldwide have much to contribute to informing the change in attitudes that needs to take place. For example, Peterson (1999) recommends adopting a both-and attitude rather than the either-or attitudes that predominate in Western cultures, following the lessons he learnt from Central African peoples. According to a 'both-and' view, the individual and the community exist in mutual tension, yet strive for a balance between needs for autonomy and security in an intertwined web of reflexively related cooperative relationships. This same pattern of reflexivity extends to 'both-and' relations between humans and the environment – a pattern that is contrasted with dualistic and reductionist, Western scientific approaches to knowledge that oppose individual and community; and humans and the environment (95). A 'both-and' ethical epistemology and ontology underlies a holistic approach to life that emphasizes processes and values balance in all relationships. Peterson goes further by recommending that the Western 'either-or' worldview be informed by the Central African worldview in order to transcend its dualistic and reductionist ways. Adopting a 'both-and' approach to their two-pronged approach to contemporary Indigenous cartography, Johnson et al. (2006) follow the Hawaiian practice of looking backwards to face the future in their work:

> 'The concepts of "past" and "future" are explained by Hawaiians using bodily directions, the front of the body faces the "past" while the back faces "future". Hawaiians "face" their "future" with their backs because the future is an unknown. On the other hand, "past" is knowable; it can be "seen" in front of each of us, shaping our character and consciousness. Hawaiians believe that knowing who they are, genealogically, and

where they came from, geographically and metaphysically, makes them capable of making more informed decisions about the direction to move in the future' (82).

The 730-page United Nations Environment Program (UNEP) publication, *Cultural and Spiritual Values of Biodiversity* (1999), which contains Peterson's reflections, is based on knowledge that was shared at two major sets of meetings between scholars and individuals representing Indigenous communities worldwide. The document recommends not only including, but also following Indigenous ways in order to mitigate the extreme and growing crisis with respect to the loss of both species diversity and traditional Indigenous languages. Denzin et al. (2008) echo this view in their opening words to the *Handbook of Critical and Indigenous Methodologies*:

> We seek a productive dialogue between indigenous and critical scholars. This involves a re-visioning of critical pedagogy, a re-grounding of Paulo Freire's (2000) pedagogy of the oppressed in local, indigenous contexts. We call this merger of indigenous and critical methodologies critical indigenous pedagogy (CIP). It understands that all inquiry is both political and moral. It uses methods critically, for explicit social justice purposes. It values the transformative power of indigenous, subjugated knowledges. It values the pedagogical practices that produce these knowledges (Semali and Kincheloe, 1999, 15), and it seeks forms of praxis and inquiry that are emancipatory and empowering. It embraces the commitment by indigenous scholars to decolonize Western methodologies, to criticize and demystify the ways in which Western science and the modern academy have been a part of the colonial apparatus. This revisioning of critical pedagogy understands with Paulo Freire and Antonio Faundez (1989, 46) that 'indigenous knowledge is a rich social resource for any justice related attempt to bring about social change' (Semali and Kincheloe, 1999, 15) (2).

The need for all people to seek the guidance of the Elders is reflected clearly in the words of the seventh prophet to the Anishinaabek who was said to have a strange light in his eyes:

> In the time of the Seventh Fire, New People will emerge. They will retrace their steps to find what was left by the trail. Their steps will take them to the Elders who they will ask to guide them on their journey. But many of the Elders will have fallen asleep. They will awaken to this new time with nothing to offer. Some of the Elders will be silent out of fear. Some of the Elders will be silent because no one will ask anything of them. The New People will have to be careful in how they approach the Elders. The task of the New People will not be easy. If the New People will remain strong in their Quest, the Water Drum of the Midewiwin Lodge will again sound its voice. There will be a Rebirth of the Anishinabe Nation and a rekindling of old flames. The Sacred Fire will again be lit.

> It is at this time that the Light Skinned race will be given a choice between two roads. If they choose the right road, then the Seventh Fire will light the Eighth and final Fire, an eternal Fire of peace, love, brotherhood and sisterhood. If the light skinned race makes the wrong choice of roads, the destruction which they brought with them in coming to this country will come back at them and cause much suffering and death to all the Earth's people (Benton-Banai, 1988, 91).

Many have said this is the era of the seventh fire. Some even say that the eighth fire has already been lit. If this is the case, it has taken many years to gather enough kindling to get the eighth fire lit, and it could take many more to get it burning brightly. In the academic world, the kindling that will light the eighth fire of renewed relationships is a metaphor for the thoughts, ideas, concepts and approaches of people like Ludwig Wittgenstein, who stood up against the essentialism of the logical positivist movement by advocating a family resemblance approach to meaning and a therapeutic role for philosophy as conceptual clarification (Pyne, 2006, 2019). It was Wittgenstein who said:

Getting hold of the difficulty deep down is what is hard. Because it is grasped near the surface it simply remains the difficulty it was. It has to be pulled out by the roots; and that involves our beginning to think in a new way. The change is as decisive as, for example, that from the alchemical to the chemical way of thinking. The new way of thinking is what is so hard to establish.

Once the new way of thinking has been established, the old problems vanish; indeed they become hard to recapture. For they go with our way of expressing ourselves and, if we clothe ourselves in a new form of expression, the old problems are discarded along with the old garment (1980, 48).

1.2 Multidisciplinary research in a reconciliation context

It is becoming more and more clear that attitudes associated with colonialism have led to actions that have not only inflicted damage and hardship on the Indigenous Peoples of the World, but on others as well. The words spoken at the fourth fire referred to in Section 1.1 have come to pass: The rivers have started to 'run with poison', and the fish have 'become unfit to eat' (Benton Banai, 1988, 90). There is a need for reconciliation not only 'between peoples' (intercultural reconciliation), but also between people and the land, which could be referred to as 'environmental reconciliation'. At a deeper supporting level, there seems to be a growing movement toward what might be termed 'epistemological and ontological reconciliation'. This 'deeper' level is often concomitant with intercultural and environmental reconciliation, and achieving epistemological and ontological reconciliation is often necessary for these other forms of reconciliation to be successful (Wittgenstein, 1980). This is the space where the assumptions underlying colonialism exist, the space where critical academic thinkers spend much of their time.

The need to reconcile current and future thinking with past approaches in Western thought began with a renewed and growing interest in social justice. Precipitated by the writing of thinkers like philosopher and political theorist John Rawls (1957, 1963, 1968, 1971), academics in both philosophy and the social sciences began to consider moral and ethical concerns in contrast to the strict empiricism of the logical positivism that dominated the first half of 20th-century thought. In geography, David Harvey (1973) was a leader in bringing social justice issues to the fore. Since then, critical geography has worked through the birth and development of feminist thinking and finds expression today through such concepts as performativity (del Casino and Hanna, 2006; Thrift, 1996, 2000, 2004a, 2004b; Turnbull, 2000, 2007) and relational space (Allen, 2004; Amin, 2004; Massey, 2004; Thrift, 2004a) — all in an aggregate attempt to reconcile our current and future attitudes and worldviews with those we have inherited from the past and to a great extent carry with us, unnoticed. In general, the 20th century saw a shift from positivistic approaches to social thought based on notions of neutral scientific rationality to efforts concerned with social justice, decolonization that acknowledged the socially constructed and value-laden nature of knowledge. Since then, efforts toward decolonization worldwide have included initiatives aimed at reconciliation in response to atrocities and wrongdoings, from enforced disappearances to the forcible removal of children from their homes.

Cartography has a colonial history, which critical cartographers are working to overcome by adopting new attitudes and attempting to use cartography's power in a decolonizing and socially just manner. Critical cartographers seek to rethink and redo mapping by acknowledging and transcending a colonial past (Kitchin and Dodge, 2007; Pickles, 2004; Turnbull, 2007) in which maps were used by nation-states primarily to assert and reinforce their territorial claims. Today, mapping practices are increasingly being engaged in as central components of the solutions to complex social and economic challenges (Taylor and Pyne, 2010; Taylor and Caquard, 2006). The movement began with an emphasis on decolonizing history by Brian Harley (1988, 1989, 1990, 2001), who focused on deconstructing colonial maps to understand the ways they were used to exercise and promote colonial authority. Over time, critical cartographers began to look more deeply into the ontological status of maps themselves. They came to view maps as processes that give rise to emergent, 'ontogenetic' knowledge and are never complete (Kitchin and Dodge, 2007). They considered epistemological issues as well, such as overcoming incommensurability in mapping Indigenous perspectives and knowledge (Turnbull, 2007).

Maps are increasingly thought of in performative terms and are far more inclusive and reflective of multiple agendas than they were in their positivist and colonial past. This is largely due to conceptual advances in critical cartography, which are beginning to see the applied fruits of their labour in this first quarter of the 21st century. While the emergence of maps and atlases reflecting critical perspectives is encouraging, there remains a need for more work in this area. This is especially true in the area of digital mapping in the information age, which presents a vast range of new possibilities for mediating reconciliation and social justice initiatives, in addition to challenges.

The information age is seeing a proliferation of knowledge and knowledge dissemination, and an increase in inter-, multi- and transdisciplinary research networks focused on acquiring and sharing knowledge on related or common issues. Cartography can be useful in mediating both networked research processes and knowledge dissemination to the broader public, in addition to being an excellent educational vehicle. As concerns with technology advancement have progressed from the past century into the current century, so have critiques of status quo approaches to knowledge acquisition, and we find ourselves today in a world with increased emphasis on reflective and reflexive approaches to 'information'.

Critical cartographers challenge conventional status quo assumptions about space through reflexive deconstructive and reconstructive mapping practices (Caquard et al., 2009; Harley, 1988, 1989, 2001; Pyne and Taylor, 2012; Pearce, 2008), which have the potential to be consistent with Indigenous perspectives and understandings. This is good news for approaches to reconciliation that aim to be consistent with Indigenous ways of knowing and doing, which are often described as being holistic (Louis, 2007).

Critical approaches to cartography continue to participate in the collective quest to decolonize knowledge and knowledge-gathering processes. Deconstructive approaches to map interpretation question the motives and context of the map-maker (Harley, 1988, 1989; Crampton, 2001, 2003); reconstructive approaches include making maps through participatory processes and community collaboration (Caquard, 2011; Corner, 1999; Cosgrove, 2005; Crampton, 2009; Crampton and Krieger, 2006; Elwood and Ghose, 2004; Fox et al., 2005; Kitchin, 2008; Kitchin and Dodge, 2007; Pearce, 2008; Pearce and Louis, 2007; Pyne, 2013,

2014; Turnbull, 2007; Taylor and Lauriault, 2014). Fraser Taylor recognized the new digital context of maps in his definition of cartography as '[t]he organisation, presentation, communication and utilisation of geo-information in graphic, digital or tactile form. [Cartography] can include all stages from data presentation to end use in the creation of maps and related spatial information products' (1991, 214).

Today's cartography extends to participatory collaborations with individuals from a variety of knowledge communities. It can be 'on the ground', and involve art, experience and immediacy, on one hand; and involve technologies, on the other (Carver, 2003; Craig et al., 2002; Irwin et al., 2009; Parker, 2006). There is growing attention to mapping experience, emotions and Indigenous perspectives (Caquard, 2011; Harmon, 2003; Hirt, 2012; Louis, 2007; Louis et al., 2012; Pyne, 2013; Taylor and Lauriault, 2014). Collaborative projects to map traditional knowledge and place names provide one example of the potential for these initiatives to contribute to the empowerment of Indigenous communities (Laidler et al., 2010; Pyne, 2013; Pyne and Taylor, 2012; Tobias, 2000).

The turn to technology taken by cartography has not lessened its focus on history, instead offering new possibilities for organizing digital archival collections according to geospatial principles. Research in education and curriculum development highlights the need for approaches to teaching and learning that incorporate increased self-reflection and are more responsive to students' needs in an increasingly complex, information-rich and multicultural world (Basit, 2013; Torres, 2011; Tilley and Taylor, 2013). When it comes to integrating geographic information systems into curricula, '[r]ecent work regarding the use of geographical information systems (GIS) in schools has suggested that it supports both higher level thinking processes and increased intrinsic motivation' (West, 2003, 267). Sociology and education and curriculum studies share a common thread: the significant theoretical and practical elaborations of the concept of 'reflexivity' that have come out of research on qualitative methods in the field of sociology (Bourdieu and Wacquant, 1992; Denzin and Lincoln, 2005; Denzin et al., 2008; Ellingson, 2009; Finlay and Gough, 2003). This concept is most commonly reflected in discussions of 'positionality' in both theory and practice. Accounting for positionality in research involves acknowledging the significance of researcher and participant perspectives on research processes and outcomes (Clandinin and Connelly, 2000; Davies, 2012; Gergen and Gergen, 1991; Holmes, 2010); an acknowledgement that extends to critical self-reflections on the research processes themselves, especially when they involve collaborative research efforts with team members from different knowledge environments (Chia, 1996; Holland, 1999). Cartography offers new possibilities for organizing digital archival collections according to geospatial principles. Critical archival studies take an inclusive approach to digital archives database management; it seeks ways to engage the broader community in the creation and use of these databases (Christen, 2011; Flinn et al., 2009; Shilton and Srinivasan, 2007), in addition to employing ethical, decolonizing approaches to understanding and interpreting these processes (Bastian, 2013; Christen, 2011; Nesmith, 2006).

Ongoing disciplinary debate in archaeology continues around the various meanings and ethical dimensions of 'public archaeology', and there is a shift toward community-based research (Chambers, 2004; Deetz, 1998; Faulkner, 1999, 2000; Harrison et al., 2012; Herscher and McManamon, 2000; Smith and Wobst, 2005; Zimmerman, 2000). Community archaeology involves participatory research reflecting multiple perspectives in its research results

(Cressey et al., 2003; Derry, 2003; Derry and Malloy, 2003; Marshall, 2002, 2004; Shackel, 2004). Robbins and Robbins (2014) speak of 'building a polyvocal public memory' (29). Rosenzweig and Dissard (2013) view archaeological sites, as 'living, social landscapes where local power, authority, and tradition are enacted' (Peterson, 2013, 1). Community archaeologists work well with geographers and cartographers who share a common interest in place-based research (Rippon, 2013). The interest in knowledge dissemination and educational outreach is shared as well and extends to the fields of sociology and curriculum development studies, two disciplines that have tended to work well together for quite some time (Brighton, 2011).

Trina Cooper-Bolam's (2020) work at the intersection of critical museology and critical heritage studies seeks to develop self-reflexive and recursive praxiological methods toward the reconciliatory musealization of sites of difficult history. In the following five paragraphs, Cooper-Bolam provides a brief description of the emergence and development of critical museal concerns and discourses, which share synergies and critical alignments with the other counter-hegemonic and decolonizing movements mentioned in this chapter.

New approaches to museology in critical museology and critical heritage studies began initially with a critique of pre-20th-century museum orthodoxy and involved integration of the discourses and practices of public history (previously considered folklore, local or amateur history) and resulted in the development of a new museum ethics and set of practices (Marstine, 2011). This critique challenged the very notion of the museum's place and function in society, including the previously unquestioned authority of museums, which conveyed hegemonic historical interpretations to the public through master narratives. In contrast, the new museology promoted shared authority (Conrad, 2008; Frisch, 1990) and multivocality (Black, 2011; Franco, 1997; Misztal, 2003; Phillips, 2005, 2011); embraced controversy and complexity (Conrad, 2008) through strategies of interpretive dissonance and resistance (Dahl and Stade, 2002; Roberts, 1997); and concomitantly promoted a shift toward public history-making, communicating cultural continuities and enabling cultural empowerment (Conrad, 2008; Rosenzweig and Thelan, 1998; Allen and Anson, 2005; Carson, 2008). The museum was theorized as moving from an outdated 'temple' to a 'forum' model (Cameron, 1971) intended to democratize the institution (Pruulmann-Vengerfeldt, 2014), to situate and mobilize the museum in the political, and to create opportunities for affective, empathetic and meaningful engagements with history (Rosenzweig and Thelan, 1998). This shift has (1) enabled identity work (Trofanenko, 2006; Rounds, 2006), including national identity and memory (Friesen et al. (2009); (2) contributed new tools for building historical literacy and fostering participatory historical culture (Seixas, 2012) and (3) articulated a role for museums in bridging cultural divides and addressing difficult histories and knowledge (Britzmann, 1998; Lehrer et al., 2011; Rosenberg, 2011; Simon, 2011).

Anthony Shelton (2013) distinguishes between three museologies: critical, praxiological and operational; and outlines the following four epistemological positions upon which critical museology is predicated: (1) history is constructed and demands sceptical scrutiny, (2) museums are collecting institutions that must be rescued from both the dead hand of an objectivist history and from psychological reductionism or 'cold passion', (3) museums 'reproduce a teleological circle' (11), which must be interrogated and opened to new knowledge claims, generate 'new heterologies and explode the limited range of exhibition genres' (12) and (4) museums are fundamentally 'more heterotopic than the societies in which they operate and are therefore potentially disruptive of them' (13). These positions demand corresponding ethical positions and practices. Shelton's methodological interdictions suggest new avenues for situating the museum, which also have implications for practice. Stressing the hypercomplexity of the networks in which museums participate, and viewing museums as Bennett's (1995) 'distinctive exhibitionary complexes', Shelton urges us to better understand them 'as hubs within hypercomplex, though not necessarily cohesive networked fields'(19). This can be seen as a call to recognize other fields with which museums are networked, and to analyze and better understand their entanglement. The moral obligations of museums and their associated pedagogies include pedagogies of witness (Simon, 2005, 2011, 2014), together with their spatial performativities (Bonder, 2009; Milozs, 2015; Rankin and Schmidt, 2009; Razack, 2007).

Posited by Laurajane Smith (2006) and expanded by Rodney Harrison (2013), critical heritage studies nudged the critical study of heritage in the direction of renewed ethics and counterhegemonic critique, emancipatory politics, intangible heritage (Smith and Akawaga, 2008), new materialist critique and new, 'unofficial' and public, heritage practice. Discourse on 'difficult heritage' (Tunbridge and Ashworth, 1996) and 'difficult pasts' (Kertzer, 1989; Vinitzky-Seroussi, 2002; Wagner-Pacifici and Schwartz, 1991) focuses on dissonances associated with histories of atrocities, adding 'dark side' (Yiftachel, 1998), shadow values (Tunbridge and Ashworth, 1996), antivalues (Gamboni, 1997) and 'discord value' (Dolff-Bonekämper, 2008) to values-based criteria, and strategies of multivocality and fragmentation within heritage practice (Vinitzky-Seroussi, 2002). Smith (2006) situates heritage in the performative, where 'performance of heritage and the understanding or idea of heritage are mutually constitutive and reinforcing', (3) but that heritage is also used as a resource for subaltern groups to contest and redefine received values and identities (4).

A focus on sites, and in particular the former sites of Indian industrial and residential schools, yields yet other relevant and provocative literature, particularly those of Christina Cameron (2010) who sees them as potential sites of conscience; Geoffrey Carr (2009) who sees them as 'Atopoi of the Modern'; Anna Brace (2014), who considers them as potential sites of continuous creation and layermaking (8–9); and Corntassel et al. (2009), who consider sites of remediation through ceremonial destruction. Simon's (2006) statement, 'there is no futurity (no break from the endless repetition of a violent past) without memories that are not your own but nevertheless claim you to a responsible memorial kinship and the corresponding thought such a problematic inheritance evokes' (203) offers motivation and context for his pedagogy of the 'terrible gift', which 'centers the demands of testament and supports the corresponding work of inheritance' (203) – sites of trauma being one such inheritance and bequest. James E. Young (1992) interrogates site of memory and forgetting, cautioning against site uses and monuments that have the potential to effect, 'sealing off memory from awareness' (272).

Literature on representing trauma in the museum and at sites of trauma also informs this research, encompassing works by Cathy Caruth (1995) and Jenny Edkins (2003) who complicate the role of primary and secondary witness in trauma and implicate survival and witnessing with the originating trauma, Marita Sturkin (1991), who with Caruth stresses the limits to the communicability of trauma and its context, and Paul Antze and Michael Lambek (1996) who situate healing in interpretation and translation. Trauma as spectacle, explored through Krzysztof Wodiczko's projections of trauma interpreted by Mark Jarzombek (2006) together with Roger Simon's critiques of the spectacle of trauma staged by and through the TRC process in Canada, similarly informs Trina Cooper-Bolam's (2014, 2018) critique of representations of trauma; in addition to the concept of 'historic trauma' (Wesley-Esquimaux and Smolewski, 2004; Chansonneuve, 2005), which is comprised of the pervasive trauma caused and perpetuated by colonialism, and specific traumas incurred through the Indian residential school system and interactions with other instruments of colonial subjugation, experienced and passed intergenerationally through Indigenous Peoples, and its critique (Kirmayer et al., 2014).

A growing number of reconciliation initiatives are gathering steam in a combined effort to reconcile relationships shaped by the Residential Schools legacy in Canada. This includes work begun by the Truth and Reconciliation Commission (TRC), whose aims include acknowledging Residential School experiences, impacts and consequences and promoting awareness and public education of Canadians about the Indian Residential Schools legacy and its impacts (see http://www.trc.ca/). The Commission was mandated to create a National Research Centre, which was established in 2015 to provide broad access to information and ongoing interactions with research and education partners, including the broader public and survivors. In addition, it released its final report in 2015 which included a vast number of calls to action (TRC, 2015a). Despite a growing number of critiques concerning the TRC investigation process (Chung, 2016; James, 2012, 2017; Niezen, 2017), these calls to action have been supporting a growing number of initiatives that reflect the broader shift toward decolonizing policy, education, art and theoretical and applied research and that began before the creation of the TRC in 2009. For example, the new approach to museology described above gained greater impetus following controversies sparked by the Glenbow Museums' The Spirit Sings and the Royal Ontario Museum's Into the Heart of Africa exhibitions, which led to reconsiderations of museum practice and new ways of engaging with source communities, Indigenous Peoples and collections; the 1992 Task Force on Museums and Indigenous Peoples contributed recommendations, since taken up by museums subjectively and to varying degrees, on ethics, consultation and representation, human resourcing and management (power structures), and collecting practices, access and repatriation (Wilson et al., 1992); and TRC Call to Action #67 urges the Canadian Museums Association in collaboration with 'Aboriginal Peoples' to undertake a 'national review of museum policies and best practices to determine the level of compliance with the United Nations Declaration on the Rights of Indigenous Peoples and to make recommendations' (TRC, 2015b, 290) (Cooper-Bolam, 2020).

In the post-TRC Calls to Action period, educators are looking at ways to foster critical thinking skills, reflexivity and ethical awareness in students; and academics are beginning to blend research with education, applying holistic approaches that include community-generated knowledge and knowledge from other sources. Today's academics, educators and policymakers are mindful of the new age of digital technology, which presents a vast range of new possibilities for analyzing, interpreting, expressing and disseminating research results, in addition to introducing new challenges (Crampton, 2003). Concomitant with the relatively recent shift to more aware and engaged research processes is the increase in inter-, multi- and transdisciplinary approaches (Rescher, 2001), as the value of multiple perspectives begins to settle into the research Zeitgeist. However, bringing multiple perspectives to the understanding of an issue also breeds the possibility for increased complexity and fragmentation of knowledge (Brauen et al., 2011; Rescher, 2001). Cartographic approaches are especially useful when it comes to organizing this knowledge, and documenting knowledge gathering processes in intuitively appealing and epistemologically effectual ways (Brauen et al., 2011).

1.3 Nature and purpose

Cybercartography in a Reconciliation Community: Engaging Intersecting Perspectives is concerned with presenting an intersecting set of approaches taken to reconciliation and education. In this regard it focuses especially on reconciliation in relation to cases involving residential and boarding schools in North America and workhouses in the United Kingdom and Ireland. The book could be looked at as a gathering of perspectives that intersect in a transdisciplinary manner in either the residential school reconciliation context or in the broader and more general context of critical studies and decolonization. At the same time, it is a demonstration of the unifying power of Cybercartography when it comes to identifying and creating research intersections between different types of knowledge and practice. As will become apparent in the reading of many chapters in this book, Cybercartography is an umbrella framework for a variety of contributions that can lead to a common output: In this case, The Residential Schools Land Memory Atlas (RSLMA), which is in turn the major output of the Residential Schools Land Memory Mapping Project (RSLMMP; funded by the Social Sciences and Humanities Research Council (SSHRC) of Canada, 2015–20). While the scope of this book goes beyond this project in some respects, it is nevertheless intrinsically related to it, as we hope to make apparent through its reading (Pyne, 2019; Taylor et al., 2019).

Cybercartography is a transdisciplinary practice that involves many different teams engaging in the development of a variety of online, interactive, multimedia atlases (see Chapter 2 for a more detailed summary of Cybercartography). *Cybercartography: Theory and Practice* (Taylor, 2005), *Developments in the Theory and Practice of Cybercartography* (Taylor and Lauriault, 2014), and *Further Developments in the Theory and Practice of Cybercartography: International Dimensions and Language Mapping* (Taylor et al., 2019) include chapters by contributors from a diverse array of communities of knowledge that discuss a variety of issues related to cybercartographic theory and practice from the perspective of their own distinctive positions and interests. The 'cyber' in Cybercartography refers to more than simply the online environment. Equally important is the connotation of community, which can refer to many individuals and groups, including those involved in cybercartographic atlas development and use (Martínez and Reyes, 2005; Pyne, 2013, 2019; Pyne and Taylor, 2015; Taylor and Pyne, 2010). *Cybercartography in a Reconciliation Community* continues this approach within the context of one specific cybercartographic atlas project, the RSLMMP (funded by the SSHRC of Canada, 2015–20), which involves an innovative collaborative approach to mapping institutional material, volunteered geographic information and innovative strategies for integrating archival, cartographic, critical heritage, artistic and participatory dimensions. Exploring Cybercartography through the lens of this cybercartographic atlas project sheds light on both Cybercartography and transdisciplinary research, while informing the reader of education and reconciliation initiatives relating to Residential Schools in Canada, boarding schools in the United States and workhouses in the United Kingdom and Ireland.

Following a section outlining the Residential Schools reconciliation context in Canada (Section 1.3.1), which draws heavily on Cooper-Bolam (2014), this chapter briefly discusses the following frameworks, which generally guide this book: David Crocker's (1999, 2015) understanding of 'reconciliation' in a transitional justice framework (Section 1.3.2); reflexivity, including the reflexive presentation style of Pierre Bourdieu (1992) in his Paris Workshop (Section 1.3.3); David Turnbull's (2000) 'talk, template, traditions' model as an interpretative framework for understanding the nature of cybercartographic atlas development (Section 1.3.4); and the distinction between implicit and explicit forms of cartography as important conceptual tools for understanding transdisciplinary intersections (Section 1.3.5).

1.3.1 A residential schools reconciliation context

In 1996, the Canadian Royal Commission on Aboriginal Peoples released its report acknowledging the negative effects of residential schools policy in Canada in addition to a host of related injustices. This was followed by the Indian Residential Schools Settlement Agreement (IRSSA), the 2008 Canadian government apology to residential schools survivors in the House of Commons on 11 June 2008 (Harper, 2011) and establishment of the 2008–15 TRC, which was tasked to investigate the government-funded, church-run schools. According to the Commission, more than 150,000 First Nations, Métis and Inuit children in Canada were forcibly taken to residential schools and denied many freedoms, including their right to speak their language, practice their culture and communicate with family. Moreover, many former students have reported being victims of physical and/or sexual abuse, and several generations of survivors' descendants have been plagued with the negative intergenerational effects of their residential schools experiences. In addition to gathering survivors' testimonies and hosting story-sharing meetings, the TRC was tasked with setting up the National Centre for Truth and Reconciliation as a place to store residential schools–related materials and issued a set of recommendations that have had a huge impact on Canadian research and education policy. As this Canadian example illustrates, reconciliation is a process that takes place over time and involves many actors. This process has included a shift toward decolonizing education policies to foster critical thinking skills, reflexivity and ethical awareness. At the same time, critical approaches to reconciliation around residential schools reveal issues with implementation, for example 'the contested definition of "residential schools"' (Cooper-Bolam, 2014), which limited the number of schools to those considered under the IRSSA (http://www.residentialschoolsettlement.ca/english.html) and its mandated TRC (http://www.trc.ca/).

The Residential School system operated for over a century, with the first IRSSA-recognized school being the Mohawk Institute in Brantford, Ontario, which opened its doors in 1831, and the last being Gordon Indian Residential School in Punnichy, Saskatchewan, which closed in 1996. The Aboriginal Healing Foundation (AHF), a not-for-profit organization that involved a focus on including survivor voices, held a more expansive view of the Indian Residential School system in Canada. For example, it considered a broader set of schools than the 139 schools considered by the IRSSA and the TRC, and extended the definition to include 'industrial schools, boarding schools, homes for students, hostels, billets, residential schools, residential schools with a majority of day students, or a combination of the above' (AHF, 2001, 5). It is indeed a contentious issue that many survivors were excluded from the IRSSA and its compensation and commemoration programs due to the failure of their schools or the circumstances of their residence to meet IRSSA criteria.

Most of the residential schools were situated away from the Indian reserves and other Indigenous settlements in Canada and in many cases were established in remote locations, inaccessible to the parents of their pupils. This policy may have in part been the result of a visit to the United States in 1879 by Nicholas Flood Davin, a bureaucrat who was sent to learn from the American Indian boarding school practices by Prime Minister John A. Macdonald. Davin, who was concerned with government obligations to its 'Indian wards' pursuant to the British North America Act, learnt about the American 'aggressive assimilation' policy and later recommended that the Canadian government establish industrial boarding schools far from reserves (Miller, 1996; Milloy, 1999; Titley, 1986) (see Chapter 5 for a consideration of American boarding schools and Chapter 6, which considers both American boarding schools and Canadian Indian residential Schools).

Two earlier recommendations influenced the Canadian government mandated design of residential schools: those of the 1842 Bagot Commission calling for the establishment of agricultural or industrial-based boarding schools; and Egerton Ryerson's, 1847 report, which made similar recommendations but also exhorted the benefits to be gained from religious instruction (Miller, 1996). And so in 1892, the Federal Government and Catholic, Anglican, Methodist and Presbyterian churches entered into a formal partnership to establish and administer a system of Indian industrial boarding schools (later called residential schools). The agreement stipulated that while the Federal Government would fund the schools, the churches would manage their operations. Although industrial training programs were gradually replaced in favour of less expensive 'formal' education, students continued on a half-day system with half of the day spent performing manual labour (often to support the operation of the school and its farms), and the other half in religious and rudimentary 'educational' instruction.

Varied archival sources indicate that, almost from the onset, parents and Indigenous leaders had unaddressed concerns regarding the competency of many teachers (Barman et al., 1986, 11) and instances of abuse (10). According to Educator Celia Haig-Brown, 'abuse at the schools was widespread: emotional and psychological abuse was constant, physical abuse was meted out as punishment, and sexual abuse was also common. Survivors recall being beaten and strapped; some students were shackled to their beds; some had needles shoved in their tongues for speaking their native languages' (Haig-Brown, 1998, 16). In 1907, Indian Affairs' Chief Medical Officer, Dr P.H. Bryce, who, along with several other doctors, had surveyed the conditions of the schools, noted a 15 to 24% death rate among the children, which rose to 42% for sick children sent home to die (Milloy, 1999, 75). In some individual institutions, for example File Hills IRS in Saskatchewan, 75% of the students had died over the 16 years of the school's operation (Milloy, 1999, 91; Fournier and Crey, 1997, 49). The Department of Indian Affairs, under the leadership of Deputy-Superintendent Duncan Campbell Scott deliberately ignored recommendations by Bryce and failed to implement a standard of care that could have prevented the ongoing spread of tuberculosis and other diseases in the schools (Milloy, 1999, 97). Instead, he terminated Bryce and expanded the IRS system through compulsory attendance (Titley, 1986, 87, 90). The government's willful disregard for the basic needs of the children in its care, and the perpetuation of the conditions that were known to cause death, appears to fall under Article II, subsection (c) of the United Nations Convention on the Prevention and Punishment of the

Crime of Genocide: Deliberately inflicting on the group conditions of life calculated to bring about its physical destruction in whole or in part.

In the Epilogue of *A National Crime: The Canadian Government and the Residential School System — 1879 to 1986*, John Milloy (1999) discusses the schools' deepest secret, 'persistent, widespread sexual abuse', which came to light through media reports and court cases in the 1980s (xvii). Convictions in the late 1990s against former Alberni IRS boys supervisor Arthur Henry Plint 'on 36 counts of sexual assault committed between 1948 and 1968 and in 2003, against dorm supervisor Donald Bruce Haddock, who pleaded guilty to four counts of indecent assault which occurred at Alberni IRS between 1948 and 1954 [...] led BC Supreme Court Justice Douglas Hogarth to declare "the Indian residential school system was nothing more than institutionalized pedophilia"' (United Church of Canada, 2014).

As of 2014, adjudicators in the Independent Assessment Process (IAP), a component of the IRSSA, the Federal Government and the TRC were battling each other over what to do with the '800,000 audio recordings, transcripts and other documents associated with 38,000 claims of sexual abuse, physical abuse and other heinous acts' (Wittmeier, 2014). Access to these valuable records is essential to understanding the nature of the abuses suffered in the schools. Despite the possibility of IAP records becoming available to researchers, because of the incalculable number of survivors who experienced acts of abuse but who did not participate in the IAP, the true extent of abuses suffered by students in Canadian residential schools will remain unknown. While some of these survivors were excluded from the IAP process because their schools did not fall under the IRSSA, others whose schools were included were reluctant or unable to endure the vicarious trauma of communicating their experiences of abuse. Given that suicide was one of the documented negative impacts of the Common Experience Payment (CEP), a lump sum issued to every qualifying survivor (Reimer and Bombay, 2010, 54), one can imagine the multiple negative impacts of recounting serious childhood traumas and their ongoing impacts. In addition, consider the toll on survivors of the emotionally taxing labour involved in researching and gathering the records and resources required to meet the evidentiary standards of the IAP.

One common feature shared by all survivors, including those reporting positive aspects of their school experiences, is the experience of being taken from their home and community and brought to a system designed to 'kill the Indian in the child' for the purpose of indoctrination into settler society, as Prime Minister Stephen Harper affirmed on the occasion of his historic apology to survivors (Harper, 2011). Until extant oral histories and archival records collected by the TRC and delivered to the National Research Centre for Truth and Reconciliation are catalogued, aggregated and analyzed across multiple vectors, our picture of residential schools will remain limited. Given that only those survivors sufficiently advanced in their healing journey braved IRSSA processes and entrusted their mechanisms of witness with their stories, we will never fully know what survivors experienced in the schools and will never fully grasp their legacy on survivors, their families and communities, on Indigenous cultures, and on Canada as a whole. With that caveat, enough is presently known about the Residential Schools legacy to compel, on the part of Canada and Canadians, an ethic of devoir de mémoire. In its interim report, the TRC asserted: 'Canadians have been denied a full and proper education as to the nature of Aboriginal societies, and the history of the relationship between Aboriginal and non-Aboriginal peoples' (2012, 86) — a denial that has contributed to racism and hostility. Such an education is but one of the many ongoing calls on the part of Aboriginal peoples and their non-Aboriginal allies for recognition of the IRS system — and indeed the violence of settler colonialism — as a genocide. Yet, the drive to inscribe histories of violence and trauma associated with the Residential Schools legacy upon our annals — committing them to our technologies of memory and in doing so realizing our duty to remember — is, at every pass, drawn into contention with Canada's equally motivated drive to contain and control its historiography.

In this book, we focus on reconciliation with respect to the legacies of Indian residential schools in Canada, Native boarding schools in the United States and workhouses in the United Kingdom and Ireland. In addition to the significance of reconciliation in the post-conflict examples mentioned above, the legacies associated with these schools and workhouses are important examples of problematic intercultural legacies requiring sustained, multiperspective attention in order to redress the intergenerational and intercultural trauma associated with them.

1.3.2 Reconciliation in a transitional justice framework

'What should be meant by "reconciliation" as a social goal after human rights violations and violent conflict within a nation? What are the main obstacles to achieving reconciliation and some means to overcome them?' (Crocker, 2015, 1).

Reconciliation can be looked at as a process of 'reckoning with past wrongs' (Crocker, 1999, 2015; Pyne and Taylor, 2015). There are many national contexts to which reconciliation applies in a manner that reaches to both the individual and communal experience:

> Countries often reckon with past wrongs. Among these wrongs are an insurgent group's human rights violations of government officials or innocent citizens, a government's illegal and armed response to subversives or innocent civilians, a despotic government's abuses of its own citizens, and crimes that adversaries commit against each other in a civil war. Argentina (1976–83), Chile (1973–90), and South Africa (before 1995) are examples of state-sanctioned human rights violations of (suspected) dissidents and insurgents; Nicaragua (1979–90), Rwanda (1998), and Sudan (after 2002) illustrate human rights violations in the context of civil wars. In the breakup of Yugoslavia (1991–99), members of different ethnic and religious groups committed atrocities against those who had been fellow citizens. In each type of case, but especially in the last two, foreign nations also bear responsibility for things done and left undone (Crocker, 2015, 1–2).

In Pyne and Taylor (2015) we appealed to the transitional justice framework put forward by David Crocker (1999) as a good way to interpret 'reconciliation', and began to explore the potential for Cybercartography as a reconciliation tool. This section re-presents this framework as a useful way to assess reconciliation processes such as those in Canada with respect to its Indian Residential Schools legacy. Reconciliation is 'a widely accepted objective and guiding principle in attempts to deal with the aftermath of painfully repressive regimes around the world' (Bhandar, 2004, 834). It is also a process or a set of processes geared toward achieving this objective, making it an inherently reflexive concept — bending back on itself and demanding a holistic approach to its implementation. Reconciliation initiatives such as commissions, investigations and hearings often include projects aimed at producing an 'official truth' or 'official history' (Bhandar, 2004, 834). However, the 'official' nature of fact-finding commissions with their Crown-appointed commissioners and their rather formal protocol for providing testimony has been critiqued as a further manifestation of colonial attitudes in the guise of a quest for reconciliation. In the context of constitutional reconciliation, Brenna Bhandar expresses difficulties with the extent to which a 'single universe of comprehensibility' can be created through the official production of such knowledge (834); and, she asks some very relevant questions: 'Who actually produces the knowledge? Who collects and interprets the data that come to form the official history? Are the data rendered through testimonials of the victims? Are they rendered through the testimony of those in power? What is left out?' (834). Echoing the views of Residential Schools survivors in Canada, and others concerned with this reconciliation context, Bhandar questions whether or not these circumstances will lead to a merely 'partial' reconciliation.

The jury is still out on a commonly accepted definition of 'reconciliation'. In the meantime, David Crocker's (1999, 2015) conceptualization of the term in the context 'transitional justice' provides a useful way to approach 'reconciliation'. Going beyond a unidimensional approach, which favours either criminal or social justice measures or 'tools' such as trials or truth commissions, Crocker's account includes both of these; in addition to other measures, including international criminal tribunals; social shaming and banning of perpetrators from public office; public access to police records; public apology or memorials to victims; reburial of victims; compensation to victims or their families; literary and historical writing; and blanket or individualized amnesty or legal immunity from prosecution, depending on the context (1999).

While Crocker's earlier work on the nature of transitional justice, which includes 'truth and reconciliation', is based on his observations of a comprehensive set of cases, his 2015 considerations of transitional justice revolve around the case of Peru in its 'efforts to transition from 30 years of conflict and authoritarianism — in which as many as 70,000 people died or disappeared — to social justice, democracy, and peace' (2015, 1). Crocker (1999, 2015) presents the following eight multicultural goals of transitional justice: truth; providing a public platform for victims; accountability and punishment; rule of law; compensation to victims; institutional reform; long-term development; reconciliation; and public deliberation. Achieving transitional justice involves working toward these goals, which are mutually interdependent and actually become means for each other. For example, inclusive public deliberation contributes to effective long-term development.

Crocker (2015) addresses the issue of precedence in considering these means and goals, placing reconciliation in the lead in the case of Peru's situation, in addition to providing a more nuanced account in which there are 'five ideals of reconciliation that have emerged in the theory and practice of transitional justice — ways in which nations reckon with past wrongs'(1). Crocker illustrates these ideals with a favourable analysis of the comprehensive concept of reconciliation employed by Peru's Truth and Reconciliation Committee, which released its Final Report in 2003. (1). Despite his favourable analysis, given the immeasurable challenges associated with reconciliation, Crocker also identifies 'some serious obstacles that impede Peru's reaching the CVR's ideal of reconciliation, and [suggests] some ways to overcome them' (1).

Referring to reconciliation as the 'preeminent goal of reconciliation' Crocker introduces five interrelated meanings of reconciliation in a transitional justice context. Consistent with his 1999 article, Crocker distinguishes between thin and thick conceptions of reconciliation; however, with the benefit of the Peruvian Truth and Reconciliation case, he extends his initial formulation of 'reconciliation' to admit cases that reflect degrees of thickness.

Minimal reconciliation marks the low end of the reconciliation spectrum, as it is 'nothing more nor less than "nonlethal coexistence" in the sense that former enemies — whether groups or individuals — cease tyrannizing and killing each other' (5). Whereas maximum reconciliation is exemplified by Bishop Desmond Tutu's idea of reconciliation 'as a person-to-person relation characterized by the African concept of ubuntu: forgiveness, mercy (rather than justice), a shared moral vision, mutual healing and social harmony' (2015, 5). Noting the difficulties associated with achieving maximum reconciliation and the desire to surpass minimal reconciliation, Crocker offers three intermediary degrees of reconciliation, which offer some hope for getting past 'the depth of hostility between past opponents, as well as both moral and practical objections to coercing mutuality, contrition or the granting of forgiveness' (5): 'mutual security', 'deliberative reciprocity' and 'historical reconciliation'.

The first phase beyond minimal reconciliation involves the norms of tolerance, avoidance of stereotypical thinking, trust and restraint and leads to 'mutual security', where 'former enemies may reconcile in the sense of going beyond nonlethal coexistence to establish and comply with workable "rules of the game" (rule of law) that mitigate conflict and guide cooperation on such matters as security, economic opportunities, and taxation' (5). The next phase, 'deliberative reciprocity', builds on the 'mutual security' developed in the previous phase and involves 'exchange of proposals, reasons and criticism — the parties try to solve a concrete problem, make a policy choice together or establish a rule under which they agree to live […]. They enter into a rational give and take in order to define problems, build on areas of common concern, and forge principled compromises with which most (if not all) can live'(6). Finally, Crocker presents 'historical reconciliation', which involves people reconciling with their common past via

> multiethnic (or multinational) and multidisciplinary historical investigation concerning the causes, nature, and consequences of past wrongdoing. Former enemies can settle accounts with the past and to some extent with each other in and through forging at least the general outline of a common narrative about what happened and why. Although complete consensus—unless pitched at a very abstract level—will be impossible and sometimes undesirable and although agreement on details may be elusive, we can hope for convergence on the main framework and identify those disagreements that call for further investigation and dialogue or informed tolerance (7).

In the case of reconciliation in the context of the Residential Schools legacy in Canada, there has been some attention to the goals of truth in terms of new approaches to including previously excluded perspectives, reconciliation in terms of new approaches to relationships, and in terms of compensation. Each of these areas has problematic aspects, which are being addressed in situ by a variety of intersecting approaches reflecting common themes and interests. It would seem that — although contested in many respects — the processes that have led to the work of Canada's TRC on Indian Residential Schools and the publication of its final report have contributed to progress at various nodes of Crocker's 'reconciliation continuum'. This includes work reported on in this volume, which strives toward both 'historical reconciliation' and ubuntu.

1.3.3 Reflexivity

Increasing scepticism of the neutral, objective, 'scientific gaze' has led critical cartographers and scholars from other disciplines to adopt 'a reflexive stance in which we recognize all social activity, including research itself, as an ongoing endogenous accomplishment' (Cunliffe, 2003, 983). Although there remains 'some uncertainty and confusion about how reflexivity should be defined and practiced' (Finlay and Gough, 2003, 1), the most widespread interpretation of 'reflexivity' focuses on its application to the individual, an interpretation that has resulted in a focus on the researcher's evolving identity — their 'positionality' in relation to others. Reflexivity is a dynamic and multidimensional concept that invites holistic, pan-disciplinary interpretation. Resisting reduction and binary opposition, 'reflexivity' has broad application at multiple scales. Depending on the context, 'reflexivity' can be considered 'a philosophy, a research method or a technique' (Cunliffe, 2003, 984). At a general level, Melvin Pollner conceived 'radical reflexivity' as 'an "unsettling", [...] regarding the basic assumptions, discourse and practices used in describing reality' (1991, 370). At an even more general level of applicability, 'reflexivity' means 'to bend back upon oneself' (Finlay and Gough, 2003, ix), which indicates a cyclical as opposed to a linear perspective. Linda Finlay and Brendan Gough identify some of the main strands of investigation involving reflexivity, including 'the humanistic-phenomenological and psychoanalytic emphasis on self-knowledge, "critical" traditions such as feminism, which prioritize sociopolitical positions, and social constructionist and "postmodern" approaches, which attend to discourse and rhetoric in the production of research texts'(1).

According to Ann Cunliffe, reflexivity 'unsettles' representation by suggesting that we are constantly constructing meaning and social realities as we interact with others and talk about our experience. We therefore cannot separate ontology and epistemology, nor can we ignore the situated nature of that experience and the cultural, historical and linguistic traditions that permeate our work (Cunliffe, 2003, 985). Cunliffe provides a definition of 'reflexivity' that situates 'reflexivity' between the 'two root metaphors [of] otherness and betweeness' involving deconstruction and reconstruction respectively: 'Texts grounded in otherness build on postmodern and poststructuralist commitments by incorporating deconstructionist or contradiction-centered approaches [...] Texts grounded in betweeness draw on ethnographic work, specifically constructionist approaches, proposing that social realities are constructed between us in our conversations' (Cunliffe, 2003, 986). Although both perspectives define reflexivity as a 'turning back', deconstructionist approaches focus on identifying assumptions and presuppositions by 'problematizing explanations and revealing philosophical, ideological, linguistic and textual uncertainties'; while constructionist approaches 'focus on our ways of being and acting in the world, how we make sense of our experience' (Cunliffe, 2003, 989—990). Deconstructionist accounts focus on critically tracking the past, and often 'call for narrative circularity — tracing the situated and partial nature of our accounts' (989); and constructionists consider 'ontological issues of who we are and how we interact and create our realities with others' (Cunliffe, 2003, 989—990).

Following the two-pronged approach put forward in Johnson et al. (2006) and Pyne (2013), preference is given to critical academic approaches that emphasize holistic methods and are somewhat consistent with Indigenous ways. This book expresses holism via the concept of reflexivity. As such the book is both a reflexive work and a compendium of reflexive works in a way that incorporates and goes beyond positionality and reflection on research processes. In doing so, it follows Pierre Bourdieu's approach in his Paris Workshop (1992), which includes (1) transparency, (2) emergence and (3) a high speed tour approach.

With respect to transparency, this book and many of its chapters are akin to Pierre Bourdieu's idea of 'a research presentation', which 'is in every respect the very opposite of an exhibition, of a show in which you seek to show off and to impress others' (1992, 219). Taking a reflexive approach to presentation allows for the exposure of details that might otherwise remain hidden and increases the 'chances of benefiting from the discussion and the more constructive and good-willed I am sure the criticisms and advice you will receive' (219). Transparency is also useful for tracking the emergent nature of research. The emergent nature of the intersecting work related to residential/boarding schools is consistent with a popular conception of 'maps as becoming' in critical cartography. Bourdieu illustrates both the significance and the vulnerability of 'the emergent approach' in his critique of the perfectionist 'homo academicus':

> I will [...] present the research work that I am presently conducting. You will then see in a state that one may call "becoming", that is muddled, cloudy, work that you usually see only in a *finished* [author's emphasis] state. Homo academicus relishes the finished. Like the *pompier* [author's emphasis] (academic) painters, he or she likes to make the strokes of the brush, the touching and retouching disappear from his works. I have at times felt a great anguish after I discovered that painters such as Couture, who was Manet's master, had left behind magnificent sketches, very close to impressionist painting — which constructed itself against pompier painting — and that they had often "spoiled", in a sense, these works by putting the finishing touches stipulated by the ethic of work well done and well polished whose expression can be found in the academic aesthetic. I will try to present this research work in progress in its fermenting confusion (1992, 219).

Finally, Bourdieu's high-speed tour approach is adopted in many cases in order to provide a picture of the ongoing processes involved midstream in the 5-year transdisciplinary RSLMMP — involving theory and practice, research and education. Bourdieu summarizes this approach in the following manner:

> One of the functions of a seminar such as this one is to give you an opportunity to see how research work is actually carried out. You will not get a complete recording of all the mishaps and misfirings, of all the repetitions that proved necessary to produce the final transcript, which annuls them. But the high-speed picture that will be shown to you should allow you to acquire an idea of what goes on in the privacy of the 'laboratory' or, to speak more modestly, the workshop — in the sense of the workshop of the artisan or of the Quattrocento painter (Bourdieu, 1992, 220).

A reflexive perspective is more than causal understanding, which most often reflects a linear approach. Reflexivity acknowledges mutual effects and implies a holistic need for balance. It is a dynamic and multidimensional concept that invites holistic, pan-disciplinary interpretation. There are many examples of reflexivity in the various contributions of the book, which have multiple functions, including (1) knowledge dissemination of ongoing work and conceptual progress on reconciliation-related issues and (2) knowledge generation in the context of the ongoing design and development of the cybercartographic RSLMA, which is the main output of a 2015–10 SSHRC-funded project by a similar name. A good example is the processes involved in the writing of Chapter 3, which (1) included continual transitioning between written text and map in the development of both the chapter and the map, (2) resulted in the creation of a new map module for the RSLMA, (3) catalyzed a series of atlas design and development decisions and (4) led to the development of new research relationships.

1.3.4 Talk, templates and tradition: An iterative approach to research and education

David Turnbull's arational, performative approach to understanding the way Gothic cathedrals were constructed involves viewing them 'literally as "laboratories"' (2000, 54), in which iterative interactions occurred between the three dimensions of talk, templates and tradition. As people from different knowledge traditions talked with one another about various issues related to the construction process, transformations in knowledge occurred. One important manifestation of these transformations was templates: Guides that other people could use to continue patterns that worked. The people incorporated the templates into their practices in a way that in turn contributed to the emergence of new traditions, and the performative cycle of collaborative construction continued over several generations until the cathedral was completed (Turnbull, 2000). This is contrasted with a view of design and development as a hierarchical process, governed from above by the all-seeing eye of rationality and technoscientific logic. It acknowledges the synchronicity and rhizomatic growth of both social relations and design and development knowledge (Caquard et al., 2009; Hess, 2004). Talk is a key ingredient of the iterative processes in the design and development of the RSLMA. In contrast with the construction process in the making of the Chartres, 'talk' in the collaborative creation of the RSLMA extends to text and is often mediated by such technologies as the telephone and the Internet. Like mortar in bricklaying, talk is the real glue that holds the construction process together. The category of 'templates' is a broad category, including but not limited to concept and design drawings, geonarrative concepts and Anishinaabe stories. 'Tradition' is a likewise broad category that includes this work being a part of the reconciliation context, as well as being part of the cybercartographic and critical cartographic traditions.

Most of the book's contributors have participated (and continue to participate) in the talk, template and tradition approach to the transdisciplinary and iterative development of a series of cybercartographic map modules that will comprise the RSLMA, the central cybercartographic output of the RSLMMP. Their contributions reflect distinctive yet complementary perspectives, and collaborations include the transdisciplinary teaching and learning of skills across communities of practice. The contributors to this book discuss issues related to their particular approach to reconciliation in relation to residential/boarding schools, as well as issues that intersect with work in Cybercartography on the RSLMMP.

1.3.5 Implicit and explicit cartography

The distinction between implicit and explicit cartography is a working conceptual approach that involves 'playing with' the 'epi-map', 'para-map' concepts of Denis Wood and Martin Fels (2008). Along these lines, the entire book could be looked at as being in an epi-map relationship with the in-development RSLMA in the sense that it both comments on the Atlas and contributes to its ongoing development. Viewing the book and the Atlas in this light introduces a performative and relational lens that is at the same time reflexive (Wood and Fels, 2008; Pyne, 2013). In this way the book is an implicit approach to cartography, while the Atlas can be viewed as an explicit approach to cartography. Consistent with mainstream thinking in critical cartography, both the book and the RSLMMP, which is discussed throughout, assume a broad approach to mapping that can include both implicit and explicit approaches to cartography. According to this understanding, all of the approaches presented in this book constitute one or both of these approaches. For example, Chapter 5, which discusses ongoing reconciliation work around the Carlisle Indian Industrial School, participates in implicit mapping insofar as it 'map[s] out pathways' for teaching the history of the school in a written narrative format; while it could be said to participate in explicit mapping insofar as it reports on various map projects associated with its work.

1.4 Book contents

This book's contributions are broadly consistent with qualitative inquiry in its overriding concern with social and related forms of justice (Denzin and Lincoln, 2005; Denzin et al., 2008); and in its commitment to participating in the broad trend toward the democratization of research, which involves aspects such as intersectionality and inclusion (Rescher, 2001 Pyne, 2013). In this spirit, the book includes chapters discussing the work and thought of members of the RSLMMP research community. This is important because — despite disciplinary differences — many common themes exist in issues discussed by the contributors.

Chapter 2 gives an historiological account of the RSLMMP. It extends the theory and practice of Cybercartography with respect to key concepts such as transdisciplinarity, iterative processes, reflexivity and emergent knowledge and expands on a new geonarrative method, 'geo-transcription', first introduced in Pyne (2013), in addition to providing a summary of David Turnbull (2000) 'talk, template and tradition' approach as a useful interpretative framework for describing and explaining iterative development processes in Cybercartography (Pyne, 2013, 2019).

Chapter 3 is the first in a series of chapters contributed by authors who work independently of the RSLMMP on residential/boarding schools research, and whose work intersects in some way with the project. Reflecting an approach akin to Laura Ellingson's (2009) 'multigenre crystallization', this chapter provides an example of a reflexive geonarrative approach — sometimes a conversation, sometimes a report, sometimes a discourse — that 'zooms in and out' in implicit cartographic fashion from general discussion to the more particular, intersecting cartography with art broadly in the context of residential schools reconciliation. The collaborations discussed in this chapter provide a good example of transdisciplinary research at the intersection of art (including curatorial studies) and cartography that reflect Turnbull's 'talk, templates, tradition' approach.

Chapter 4 discusses ongoing research processes at the Shingwauk Residential Schools Centre linked to the production of the Shingwauk Residential Schools Maps in RSLMA, with a particular focus on critical archival practice and the concept of place-based history.

Chapter 5 discusses work to create mapping resources with respect to the important American case of the Carlisle Indian Industrial School, which is 'a major site of memory for Indian nations across the country and for those interested in the history of American education and colonization'. It provides a glimpse into the Carlisle School Digitization Project, which includes archives development, mapping and relationship building. Whereas Chapter 4 deals with a Canadian 'residential school' and Chapter 5 deals with an American 'boarding school', Chapter 6 compares schools from each country in a case study analysis that presents a useful spatial concept, 'the settler-colonial mesh'. It addresses the important issue of diversity in historical geographies of schools and is an example of implicit cartography in its concern with the spatiality of the residential schools legacy in North America, with phrases such as 'the space of destructive assimilative education' and 'the space of Indigenous boarding schools'; in addition to its use of a key spatially oriented concept: 'the settler-colonial mesh', which is a multiscalar and multidimensional concept that responds to the complexity of 'settler colonial practices of assimilative education over time'. While the chapter focuses on 'assimilative boarding schools', it is also interested in 'broader processes of settler colonial domination', reflecting a similar zooming in—zooming out narrative strategy to this book and its other chapters, in addition to constituting another example of implicit cartography, where the most zoomed-out level is an account of broader historical geographical issues going beyond the particular stories associated with residential schools. With respect to intersectionality, Chapter 6 covers some of the same ground covered in Chapter 5; and especially relevant to the intersection with cartography is Woolford's appeal to Cybercartography as an ideal vehicle for addressing the micro-level of settler colonial practices of assimilative education.

Chapter 7 broadens the historical geographical scope in establishing lines of ideological, operational, and eventually museological, continuity between 19th-century workhouses in England and Ireland and Indian residential/boarding schools in North America. Surveying workhouse museums in Portumna, Ireland and Southwell, United Kingdom, comparative and related sites of difficult history, this chapter investigates emergent reconciliatory museologies. Motivated by a desire to contribute to the development of a postrupture museological ethics and repertoire in Canada, Cooper-Bolam visited these sites to explore how they might inform ongoing museological inventions at the Shingwauk Indian Residential School site in Sault Ste. Marie, Ontario, Canada (discussed in Chapter 4). Framed within overlapping categories of site intervention and use, interpretation design, and animation and programming, Cooper-Bolam investigates the Portumna and Southwell sites for evidence of critical historical recovery, reclamation and transformation toward social and spatial justice. Cooper-Bolam's approach is consistent with the talk, templates, tradition approach discussed further in this book and acknowledges the importance of in-person interactions, relationships and reciprocities and of experiential knowledge and memory, derived in part, from the affective and embodied phenomena that only arises from visiting and dwelling in sites of research. Chapters 3—7 are interrelated in terms of their historical geography. The American model influenced the Canadian approach to residential schools in the late 19th century and after, and both were influenced by the UK workhouse and orphanage models. While there were many aspects of modelling, architecture played a significant role in implementing assimilationist policies and in affecting student experience (deLeeuw, 2007).

Chapters 8 and 9 provide insight into the in-development Residential Schools Land Memory Mapping Atlas with reflexive accounts of the processes involved in the emergence of the RSLMA. Chapter 9 discusses an interesting approach to community-based research involving volunteered geographic information generated by survivors of Assiniboia Indian Residential School in Winnipeg, Manitoba, Canada at their 2017 reunion; and Chapter 8 charts aspects of roughly the first half of the iterative development RSLMA, including work with research assistants from universities across Canada. This chapter sheds light on the role of Cybercartography as both a research and a presentation framework at the intersection of community-based approaches to archival studies, cartography and museology, for example. It became apparent early on in the project — and even before — that we would be involving broad concepts of community and participation that involved the research team itself as a community. Combined with the concept of emergence in the context of project design and development, is not only research but also extensive talking between team members, which has resulted in a series of templates with the aim of contributing to both the broader reconciliation tradition and the more specific cybercartographic tradition.

Chapter 10 comments on ethical issues such as trust, and introduces Sara Ruddick's maternal thinking framework as a useful ethics of care approach to apply to both the institutional and the participatory aspects of research projects, especially those such as the RSLMMP, which involve 'difficult knowledge' and are focused explicitly on reconciliation between Indigenous peoples and others. Chapter 11 discusses ongoing work at the intersection of research and education that is inspired by the personal geography of Giacomo Beltrami (1779—1855), who travelled amidst Dakota and Ojibwe people in the United States in the early 1800s, before the entrenchment of residential/boarding school policy in North America. It is concerned with the broader colonial historical geography in which the residential/boarding schools are enmeshed. Finally, Chapter 12 concludes the book with reflections on research intersections, transitional justice, reflexivity, talk templates, iterative processes, transdisciplinarity and community; and a summary of the main issues, trends and concepts emerging from the book.

1.5 Conclusion

Cybercartography in a Reconciliation Community includes both conceptual and applied dimensions and provides a good example of a reflexive approach to both research and knowledge dissemination. The positionality aspect of reflexivity is reflected in the chapter contributions made by project team members and others affiliated with the project and by the chapters concerning various aspects of cybercartographic atlas design and development research. The book aims to contribute to theoretical and practical knowledge of collaborative transdisciplinary research through its reflexive assessment of the relationships, processes and knowledge involved in cybercartographic research. Closely related goals include contributing to a broader ontology of cartography and providing insights into reconciliation and education processes. Its style is primarily narrative, which is consistent with the storytelling approach that characterizes many Indigenous approaches to knowledge dissemination and enhances its accessibility amongst a broad audience.

All of the chapters in this book touch on cartographic concepts — either implicitly, explicitly or both — and feed nicely into the book's intersectional approach. Many authors have included discussion of the broader historical geographical context of Residential Schools and more particular stories reflecting students' memories and experiences. As such, they contribute to the truth component identified by David Crocker (1999, 2015) as one of the eight important goals of transitional justice. Bourdieu (1992) reflexive approach to conducting workshops, which is useful for tracking emergent mapping practices (Kitchin, 2008; Kitchin and Dodge, 2007; Turnbull, 2007), allows us to track the emergence — the various stages, shifts and intersecting dimensions — of the RSLMMP and related work. So, without further adieu, we welcome you into our transdisciplinary processual laboratory, to give you a picture of the engine room of our collaborative, diversified, yet still interlinked research.

References

AHF, 2001. Aboriginal Healing Foundation Program Handbook, third ed. Aboriginal Healing Foundation, Ottawa, ON.

Allen, J., 2004. The whereabouts of power: politics, government and space. Geografiska Annaler 86B, 19—32.

Allen, G., Anson, C., 2005. The Role of the Museum in Creating Multi-Cultural Identities: Lessons from Canada, vol. 33. Edwin Mellen Press, Lewiston.

Amin, A., 2004. Regions abound: towards a new politics of place. Geografiska Annaler 86B, 33—44.

Antze, P., Lambek, M. (Eds.), 1996. Tense Past: Cultural Essays in Trauma and Memory. Routledge, New York.

Barman, J., Hébert, Y., McCaskill, D., 1986. Indian Education in Canada, first ed. University of British Columbia, Vancouver.

Basit, T., 2013. Ethics, reflexivity and access in educational research: issues in intergenerational investigation. Research Papers in Education 28 (4), 506—517.

Bastian, J., 2013. The records of memory, the archives of identity: celebrations, texts and archival sensibilities. Archival Science 13 (2—3), 121—131.

Bennett, T., 1995. The Birth of the Museum: History, Theory, Politics. Routledge, London.

Benton-Banai, E., 1988. The Mishomis Book: The Voice of the Ojibway. Indian Country Press, St. Paul, Minnesota.

Bhandar, B., 2004. Anxious reconciliation(s): unsettling foundations and spatializing history. Environment and Planning D: Society and Space 22, 831—845.

Black, G., 2011. Museums, memory and history. Cultural and Social History 8 (3), 415—427.

Bonder, J., 2009. On Memory, trauma, public space, monuments, and memorials. Places: Forum of Design for the Public Realm 21 (1), 62—69.

Bourdieu, P., 1992. The practice of reflexive sociology. In: Bourdieu, P., Wacquant, L. (Eds.), An Invitation to Reflexive Sociology. University of Chicago Press, Chicago, pp. 217—253.

Bourdieu, P., Wacquant, L. (Eds.), 1992. An Invitation to Reflexive Sociology. University of Chicago Press, Chicago, pp. 217—253.

Brace, A., 2014. Heritage Alternatives at Sites of Trauma: Examples of the Indian Residential Schools of Canada (Ph.D. thesis). University of York.

Brauen, G., Pyne, S., Hayes, A., Fiset, J.P., Taylor, D.R.F., 2011. Transdisciplinary participation using an open source cybercartographic toolkit: the atlas of the lake Huron Treaty relationship process. Geomatica 65 (1), 27—45.

Brighton, S., 2011. Applied archaeology and community collaboration: uncovering the past and empowering the present. Human Organization 70 (4), 344—354.

Britzman, D.P., 1998. Lost Subjects, Contested Objects: Toward a Psychoanalytic Inquiry of Learning. State University of New York Press, Albany.

Cameron, D.F., 1971. The museum, a temple or the forum. Curator: The Museum Journal: Museum 14 (1), 11—24.

Cameron, C., 2010. World heritage sites of conscience and memory. In: Offenhäuser, C.D., Zimmerli, W., Albert, M.-T. (Eds.), World Heritage and Cultural Diversity. German Commission for UNESCO, Bonn, pp. 112—119.

Caquard, S., 2011. Cartography 1: mapping narrative cartography. Progress in Human Geography 37 (1), 135—144.

Caquard, S., Pyne, S., Igloliorte, H., Mierins, K., Hayes, A., Taylor, D.R.F., 2009. A "living" atlas for geospatial storytelling: the cybercartographic atlas of indigenous perspectives and knowledge of the great lakes region. Cartographica 44 (2), 83—100.

Carr, G., 2009. Atopoi of the modern: revisiting the place of the Indian residential school. ESC: English Studies in Canada 35 (1), 109—135.

Carson, C., 2008. The end of history museums: what's plan B? The Public Historian 30 (4), 9—27.

Caruth, C. (Ed.), 1995. Trauma: Explorations in Memory. Johns Hopkins University Press, Baltimore.

Carver, S., 2003. The future of participatory approaches using geographic information: developing a research agenda for the 21st century. Urban and Regional Information Systems Association (URISA) Journal 15. APA I 61—72.

Chambers, J., 2004. Epilogue: archaeology, heritage, and public endeavor. In: Shackel, P., Chambers, E. (Eds.), Places in Mind: Public Archaeology as Applied Anthropology. Routledge, New York, London, pp. 193—208.

Chansonneuve, D., 2005. Reclaiming Connections: Understanding Residential School Trauma Among Aboriginal People. Aboriginal Healing Foundation, AHF Research Series, Ottawa.

Chia, R., 1996. The problem of reflexivity in organizational research: towards a postmodern science of organization. Organization 3 (1), 31—59.

Christen, K., 2011. Opening archives: respectful repatriation. American Archivist 74 (1), 185—210.

Chung, S., 2016. The morning after Canada's Truth and Reconciliation Commission report: decolonisation through hybridity, ambivalence and alliance. Intercultural Education 27 (5), 399—408.

Clandinin, D.J., Connelly, F.M., 2000. Narrative Inquiry: Experience and Story in Qualitative Research. Jossey-Bass, San Francisco.

Conrad, M., 2008. Towards a participatory historical culture. Canadian Issues October, 66—99.

Cooper-Bolam, T., 2014. Healing Heritage: New Approaches to Commemorating Canada's Indian Residential School System (Master's thesis). Carleton University, Ottawa. https://curve.carleton.ca/search?s=cooper+bolam.

Cooper-Bolam, T., 2018. On the call for a residential schools national monument. Journal of Canadian Studies 52 (1), 57—81. https://www.worldcat.org/title/heritage-reader/oclc/460822647.

Cooper-Bolam, T. 2020. Gifts of the Terrible Gift — A Post-TRC Investigation in Praxiological Museology (Unpublished PhD Dissertation, Carleton University, Ottawa (forthcoming).

Corner, J., 1999. The agency of mapping: speculation, critique and invention. In: Cosgrove, D. (Ed.), Mappings. Reaktion Books, London, pp. 213—252.

Corntassel, J., Chaw-win-is, T'lakwadzi, 2009. Indigenous storytelling, truth-telling, and community approaches to reconciliation. ESC: English Studies in Canada 35 (1), 137—159.

Cosgrove, D., 2005. Maps, mapping, modernity: art and cartography in the twentieth century. Imago Mundi 57 (1), 35—54.

Craig, W., Harris, T., Wiener, D., 2002. Community Participation and Geographic Information Systems. CRC Press, Boca Raton, FL.

Crampton, J., 2001. Maps as social constructions: power, communication and visualization. Progress in Human Geography 25 (2), 235—252.

Crampton, J., 2003. The Political Mapping of Cyberspace. University of Chicago Press, Chicago.

Crampton, J.W., 2009. Cartography: performative, participatory, political. Progress in Human Geography 33 (6), 840—848.

Crampton, J., Krieger, J., 2006. An introduction to critical cartography. ACME: An International E-Journal for Critical Geographies 4 (1), 11—33.

Cressey, P., Reeder, R., Bryson, J., 2003. Held in trust: community archaeology in Alexandria, Virginia. In: Derry, L., Malloy, M. (Eds.), Archaeologists and Local Communities: Partners in Exploring the Past. Society for American Archaeology, Washington DC, pp. 1—17.

Crocker, D., 1999. Reckoning with past wrongs: a normative framework. Ethics and International Affairs 13, 43—64.

Crocker, D., 2015. Obstacles to reconciliation in Peru: an ethical analysis. Unpublished English translation of Obtsaculos para la reconciliacion en el Peru. In: Giusti, M., Gutiérrez, G., Salmón, E. (Eds.), La Verdad nos Hace Libres. Sobre las Relaciones entre Filosofía, derechos Humanos, Religión y Universidad. Fondo Editorial de la Pontificia Universidad Católica del Perú, Lima, Peru.

Cunliffe, A., 2003. Reflexive inquiry in organizational research: questions and possibilities. Human Relations 56 (8), 983—1003.

Dahl, G.B., Stade, R., 2002. Anthropology, museums, and contemporary cultural processes: an introduction. Ethos: Journal of Anthropology 65 (2), 168–169.

Davies, P., 2012. 'Me', 'me', 'me': the use of the first person in academic writing and some reflections on subjective analyses of personal experiences. Sociology 46 (4), 744–752.

Deetz, J., 1998. Discussion: archaeologists as storytellers. Historical Archaeology 32 (1), 94–96.

Del Casino, V.J., Hanna, S.P., 2006. 'Beyond the 'binaries': a methodological intervention for interrogating maps as representational practices. ACME: An International E-Journal for Critical Geographies 4 (1), 34–56.

de Leeuw, S., 2007. Intimate colonialisms: the material and experienced places of British Columbia's residential schools. Canadian Geographer 51 (3), 339–359.

Denzin, N.K., Lincoln, Y.S., 2005. The SAGE Handbook of Qualitative Research, third ed. Sage Publications, Thousand Oaks, California.

Denzin, N.K., Lincoln, Y.S., Tuhiwai Smith, L., 2008. Handbook of Critical and Indigenous Methodologies. Sage, Thousand Oaks, California.

Derry, L., 2003. Consequences of involving archaeology in contemporary community issues. In: Derry, L., Malloy, M. (Eds.), Archaeologists and Local Communities: Partners in Exploring the Past. Society for American Archaeology, Washington DC, pp. 19–29.

Derry, L., Malloy, M. (Eds.), 2003. Archaeologists and Local Communities: Partners in Exploring the Past. Society for American Archaeology, Washington DC.

Dolff-Bonekämper, G., 2008. Sites of memory and sites of discord: historic monuments as a medium for discussing conflict in Europe. In: Fairclough, G.H., Harrison, R., Jameson, J.H., Schofield, J. (Eds.), The Heritage Reader. Routledge, London & New York. https://www.worldcat.org/title/heritage-reader/oclc/460822647.

Edkins, J., 2009. Trauma and the Memory of Politics. Cambridge University Press, Cambridge, UK.

Egerton, R., 1847. Statistics respecting Indian schools. Department of Indian Affairs 73–77. https://commons.wikimedia.org/wiki/File:Egerton_Ryerson_on_Residential_Schools.pdf.

Ellingson, L., 2009. Engaging Crystallization in Qualitative Research. Sage, Thousand Oaks, CA.

Elwood, S., Ghose, R., 2004. PPGIS in community development planning: framing the organizational context. Cartographica 38 (3–4), 19–33.

Faulkner, N., 1999. Postcolonial archaeology: issues of culture, identity, and knowledge. In: Hodder, I. (Ed.), Archaeological Theory Today. Polity Press, Cambridge, pp. 241–261.

Faulkner, N., 2000. Archaeology from below. Public Archaeology 1 (1), 21–33.

Finlay, L., Gough, B. (Eds.), 2003. Reflexivity: A Practical Guide for Researchers in Health and Social Science. Blackwell Publishing, Oxford.

Flinn, A., Stevens, M., Shepherd, E., 2009. Whose memories, Whose archives? Independent community archives, autonomy and the mainstream. Archival Science 9 (1–2), 71–86.

Fournier, S., Crey, E., 1997. Stolen from Our Embrace, first ed. Douglas & McIntyre, Vancouver.

Fox, J., Suryanata, K., Hershock, P., 2005. Mapping Communities: Ethics, Values, Practice. East-West Center, Honolulu, Hawaii.

Franco, B., 1997. Public history and memory: a museum perspective. The Public Historian 19 (2), 65–67. https://www.jstor.org/stable/3379145?seq=1#page_scan_tab_contents.

Freire, P., 2000. Pedagogy of the Oppressed. The Continuum International Publishing Group, New York.

Friere, P., Faundez, A., 1989. Learning to Question. A Pedagogy of Liberation. Continuum, New York.

Friesen, G., Muise, D., Northrup, D., 2009. Variations on the theme of remembering: a national survey of how Canadians use the past. Journal of the Canadian Historical Association 20 (1), 221–248.

Frisch, M., 1990. A Shared Authority: Essays on the Craft and Meaning of Oral and Public History. State University of New York Press, Albany.

Gamboni, D., 1997. "The Destruction of Art." Iconoclasm and Vandalism since the French Revolution. Reaktion Books, London.

Gergen, K.J., Gergen, M.M., 1991. Toward reflexive methodologies. In: Steier, F. (Ed.), Research and Reflexivity. Sage, Beverly Hills, CA, pp. 76–95.

Haig-Brown, C., 2001. Resistance and Renewal, first ed. Arsenal Pulp Press, Vancouver, BC.

Harley, B., 1988. Secrecy and silences: the hidden agenda of state cartography in early modern Europe. Imago Mundi 40, 57–76.

Harley, B., 1989. Deconstructing the map. Cartographica 26, 1–20.

Harley, B., 1990. Cartography, ethics and social theory. Cartographica 27, 1–23.

Harley, J.B., 2001. The New Nature of Maps: Essays in the History of Cartography. Johns Hopkins University Press, Baltimore.

Harmon, K., 2003. You Are Here: Personal Geographies and Other Maps of the Imagination. Princeton Architectural Press, New York, NY.

Harper, S., 2011. Statement of Apology – to Former Students of Indian Residential Schools on Behalf of the Government of Canada. June 11. Ottawa. https://www.aadnc-aandc.gc.ca/DAM/DAM-INTER-HQ/STAGING/textetext/rqpi_apo_pdf_1322167347706_eng.pdf.

Harrison, R., 2013. Heritage: Critical Approaches. Routledge, New York.

Harrison, R., Ferris, N., Wilcox, M., 2012. The Archaeology of the Colonized and its Contribution to Global Archaeological Theory. Oxford University Press, Oxford.

Harvey, D., 1973. Social Justice and the City. Johns Hopkins University Press, Baltimore.

Herscher, E., McManamon, F., 2000. Public education and outreach: the obligation to educate. In: Zimmerman, L., Vitelli, K., Hollowell-Zimmer, J. (Eds.), Ethical Issues in Archaeology. Altamira Press, Walnut Creek, pp. 49–51.

Hess, M., 2004. Spatial relationships? Towards a reconceptualization of embeddedness. Progress in Human Geography 28 (2), 165–186.

Hirt, I., 2012. Mapping dreams/dreaming maps: bridging Indigenous and western geographical knowledge. Cartographica, Special issue on Indigenous Cartography and Counter Mapping 47 (2), 105–120.

Holland, R., 1999. Reflexivity. Human Relations 52 (4), 463–484.

Holmes, M., 2010. The emotionalization of reflexivity. Sociology 44 (1), 139–154.

Irwin, R., Bickel, B., Triggs, V., Springgay, S., Beer, R., Grauer, K., Xiong, G., Sameshima, P., 2009. The City of Richgate: A/r/tographic cartography as public pedagogy. Jade 28 (1), 61–70.

James, M., 2012. A carnival of truth? Knowledge, ignorance and the Canadian truth and reconciliation commission. The International Journal of Transitional Justice 6, 182–204.

James, M., 2017. Changing the subject: the TRC, its national events, and the displacement of substantive reconciliation in Canadian media representations. Journal of Canadian Studies 51 (2), 362–397.

Jarzombek, M., 2006. The post-traumatic turn and the art of Walid Ra'ad and Krzystof Wodiczko: from theory to trope and beyond. In: Saltzman, L., Rosenberg, E. (Eds.), Trauma and Visuality in Modernity. Dartmouth College Press, University Press of New England, Hanover, N.H, pp. 249–271.

Johnson, J., Louis, R., Pramono, H., 2006. Facing the future: encouraging critical cartographic literacies in Indigenous communities. ACME: An International E-Journal for Critical Geographies 4 (1), 80–98.

Kertzer, D.I., 1989. Ritual, Politics, and Power. Yale University Press, New Haven.

Kirmayer, L.J., Gone, J.P., Moses, J., 2014. Rethinking historical trauma. Transcultural Psychology 51 (3), 299–319.

Kitchin, R., 2008. The practices of mapping. Cartographica 43 (3), 211–216.

Kitchin, R., Dodge, M., 2007. Rethinking maps. Progress in Human Geography 31 (3), 331–344.

Laidler, G.J., Elee, P., Ikummaq, T., Joamie, E., Aporta, C., 2010. Mapping sea-ice knowledge, use, and change in Nunavut, Canada (Cape Dorset, Igloolik, Pangnirtung). In: Krupnik, I., Aporta, C., Gearheard, S., Laidler, G.J., Kielsen-Holm, L. (Eds.), SIKU: Knowing Our Ice, Documenting Inuit Sea-Ice Knowledge and Use. Springer, Dordrecht.

Lehrer, E., Milton, C.E., Patterson, M.E. (Eds.), 2011. Curating Difficult Knowledge. Palgrave Macmillan, UK.

Louis, R.P., 2007. Can you hear us now? Voices from the margin: using Indigenous methodologies in geographic research. Geographical Research 45 (2), 130–139.

Louis, R.P., Johnson, J.T., Pramono, A.H., 2012. Introduction: indigenous cartographies and counter-mapping. Cartographica, Special issue on Indigenous Cartography and Counter Mapping 44, 77–79.

Marshall, Y., 2002. What is community archaeology? World Archaeology 34 (2), 211–219.

Marshall, Y., 2004. "To have and enjoy the liberty of conscience": community responsive museum outreach education at the Bowne House. In: Shackel, P., Chambers, E. (Eds.), Places in Mind: Public Archaeology as Applied Anthropology. Routledge, New York, pp. 85–100.

Marstine, J., 2011. Routledge Companion to Museum Ethics. Routledge, Milton Park, Abingdon, Oxon.

Martínez, E., Reyes, C., 2005. Cybercartography and society. In: Fraser Taylor, D.R. (Ed.), Cybercartography: Theory and Practice, Volume 4 in Modern Cartography Series. Elsevier, Amsterdam, pp. 99–121 (Chapter 5).

Massey, D., 2004. Geographies of responsibility. Geografiska Annaler 86B, 5–18.

Miller, J.R., 1996. Shingwauk's Vision. University of Toronto Press, Toronto.

Milloy, J.S., 1999. A National Crime. University of Manitoba Press, Winnipeg.

Miłosz, M., 2015. 'Don't Let Fear Take over': The Space and Memory of Indian Residential Schools (Masters thesis (Architecture)). University of Waterloo. Waterloo, Ontario.

Misztal, B., 2003. Theories of Social Remembering. Maidenhead, Philadelphia.

Niezen, R., 2017. Truth and Indignation: Canada's Truth and Reconciliation Commission on Indian Residential Schools. University of Toronto Press, North York, Ontario.

Nesmith, T., 2006. The concept of societal provenance an records of nineteenth-century Aboriginal-European relations in Western Canada: implications for archival theory and practice. Archival Science 6 (3–4), 351–360.

Parker, B., 2006. Constructing community through maps? Power and praxis in community Mapping. The Professional Geographer 58 (4), 470–484.

Pearce, M.W., 2008. Framing the days: place and narrative in cartography. Cartography and Geographic Information Science 35, 17–32.

Pearce, M.W., Louis, R., 2007. Mapping indigenous depth of place. American Indian Culture & Research Journal 32 (3), 107–126.

Peterson, J., 2013. From the guest editor. Near Eastern Archaeology 76 (3), 1.

Peterson, R., 1999. Central African voices on the human-environment relationship. Cultural and spiritual values of biodoversity. In: Cultural and Spiritual Values of Biodiversity, UNEP, London, England. Intermediate Technology Publications, pp. 95–98.

Phillips, R.B., 2005. CHR Forum: re-placing objects: historical practices for the second museum age. The Canadian Historical Review 86 (1), 83–110.

Phillips, R.B., 2011. Museum Pieces: Toward the Indigenization of Canadian Museums, vol. 7. McGill-Queen's Press, Montreal and Kingston.

Pollner, M., 1991. Left of ethnomethodology: the rise and decline of radical reflexivity. American Sociological Review 56, 370–380.

Pickles, J., 2004. A History of Spaces: Cartographic Reason, Mapping and the Geocoded World. Routledge, London, New York.

Pruulmann-Vengerfeldt, P., Runnel, P., 2014. Democratising the museum: reflections of participatory technologies. Participations. Journal of Audience and Reception Studies 11 (1), 332–335.

Pyne, S., 2006. Full Responsible Reason and Good Development (Master's thesis, unpublished). Carleton University, Ottawa. https://curve.carleton.ca/904fd230-bd1a-4164-b21d-d613ac2dd66c.

Pyne, S., 2013. Sound of the Drum, Energy of the Dance: Making the Lake Huron Treaty Atlas the Anishinaabe Way (Unpublished Ph.D. dissertation). Carleton University, Ottawa. https://curve.carleton.ca/392a68ac-086c-4470-976d-101d4e96f9f7.

Pyne, S., 2014. The role of experience in the iterative development of the lake Huron Treaty atlas. In: Taylor, D.R.F., Lauriault, T. (Eds.), Developments in the Theory and Practice of Cybercartography: Applications and Indigenous Mapping. Elsevier, Amsterdam (Chapter 17).

Pyne, S., 2019. Cybercartography and the critical cartography clan. In: Taylor, F., Anonby, E., Murasugi, K. (Eds.), Further Developments in the Theory and Practice of Cybercartography: International Dimensions and Language Mapping. Elsevier, London, Forthcoming.

Pyne, S., Taylor, D.R.F., 2012. Mapping indigenous perspectives in the making of the cybercartographic atlas of the lake Huron Treaty relationship process. Cartographica, Special Issue on Indigenous Cartography and Counter Mapping 47 (2), 92–104.

Pyne, S., Taylor, D.R.F., 2015. Cybercartography, transitional justice and the residential schools legacy. Geomatica 69 (1), 173–187.

Rankin, E., Schmidt, L., 2009. The apartheid museum: performing a spatial dialectics. Journal of Visual Culture 8 (1), 76–102.

Rawls, J., 1957. Justice as fairness. Journal of Philosophy 54 (22), 653–662.

Rawls, J., 1963. The sense of justice. Philosophical Review 72 (3), 281–305.

Rawls, J., 1968. Distributive justice: some addenda. Natural Law Forum 13, 51–71.

Rawls, J., 1971. A Theory of Justice. Belknap Press, Cambridge, Massachusetts.

Razack, S., 2007. When place becomes race. Race and Racialization: Essential Readings 54, 74–82.

Reimer, G., Bombay, A., 2010. The Indian Residential Schools Settlement Agreement's Common Experience Payment and Healing, first ed. Aboriginal Healing Foundation, Ottawa, Ont.

Rescher, N., 2001. Philosophical Reasoning: A Study in the Methodology of Philosophizing. Blackwell Publishing Inc., Malden, Massachusetts.

Rippon, S., 2013. Historic landscape character and sense of place. Landscape Research 38 (2), 179–202. https://www.tandfonline.com/doi/abs/10.1080/01426397.2012.672642.

Robbins, C., Robbins, M., 2014. Engaging the contested memory of the public square: community collaboration, archaeology, and oral history at Corpus Christi's Artesian Park. The Public Historian 36 (2), 26–50.

Roberts, L., 1997. From Knowledge to Narrative. Smithsonian Institution Press, London.

Rosenberg, T.J., 2011. History museums and social cohesion: building identity, bridging communities, and addressing difficult issues. Peabody Journal of Education 86, 115–128.

Rosenzweig, M., Dissard, L., 2013. Common ground: archaeological practice and local communities in southeastern Turkey. Near Eastern Archaeology 76 (3), 152–158.

Rosenzweig, R., Thelan, D., 1998. The Presence of the Past: Popular Uses of History in American Life. Columbia University Press, New York.

Rounds, J., 2006. Doing identity work in museums. Curator: The Museum Journal 49 (2), 133–150.

Semali, L.M., Kincheloe, J.L., 1999. What Is Indigenous Knowledge: Voices from the Academy. Falmer Press, New York.

Seixas, P., Morton, T., 2012. The Big Six Historical Thinking Concepts. Nelson Education Ltd, Toronto.

Shackel, P., 2004. Working with communities: heritage development and archaeology. In: Shackel, P., Chambers, E. (Eds.), Places in Mind: Public Archaeology as Applied Anthropology. Routledge, New York, pp. 1–16.

Shelton, A., 2013. Critical museology: a manifesto. Museum Worlds 1 (1), 7–23.

Shilton, K., Srinivasan, R., 2007. Participatory appraisal and arrangement for multicultural collections. Archivaria 63, 87–101.

Simon, R.I., 2005. The Touch of the Past. Palgrave Macmillan, New York.

Simon, R.I., 2006. The terrible gift: museums and the possibility of hope without consolation. Museum Management and Curatorship 21 (3), 187–204.

Simon, R.I., 2011. A shock to thought: curatorial judgment and the public exhibition of 'difficult knowledge'. Memory Studies 4 (4), 432–449.

Simon, R.I., 2014. A Pedagogy of Witnessing. State University of New York Press, Albany.

Smith, L., 2006. Uses of Heritage. Routledge, New York.

Smith, C., Wobst, H., 2005. Decolonizing archaeological practice and theory. In: Smith, C., Wobs, H.M. (Eds.), Indigenous Archaeologies: Decolonizing Theory and Practice. Routledge, London, pp. 5–16.

Smith, L., Akagawa, N. (Eds.), 2008. Intangible Heritage. Routledge, New York.

Sturken, M., 1991. The wall, the screen, and the image: the Vietnam Veterans Memorial. Representation, Special Issue: Monumental Histories 35 (Summer), 118–142.

Taylor, D.R.F., 1991. A conceptual basis for cartography/new directions for the information era. Cartographica 28 (4), 1–8.

Taylor, D.R.F., 2005. Cybercartography: theory and practice. In: Modern Cartography Series, vol. 4. Elsevier, Amsterdam.

Taylor, D.R.F., Caquard, S. (Guest Eds.), 2006. Special Issue on Cybercartography. Cartographica 41 (1) 1–5.

Taylor, D.R.F., Pyne, S., 2010. The history and development of the theory and practice cybercartography. International Journal of Digital Earth 3 (1), 1–14.

Taylor, D.R.F., Lauriault, T. (Eds.), 2014. Developments in the Theory and Practice of Cybercartography: Applications and Indigenous Mapping. Elsevier, Amsterdam.

Taylor, F., Anonby, E., Murasugi, K. (Eds.), 2019. Further Developments in the Theory and Practice of Cybercartography: International Dimensions and Language Mapping. Elsevier, London, Forthcoming.

Thumbadoo, R., 2017. Ginawaydaganuc and the Circle of All Nations: The Remarkable Environmental Legacy of Elder William Commanda (Ph.D. dissertation). Carleton University. https://curve.carleton.ca/aa4e3cbb-5b83-464d-8286-a901fcd77b06.

Thrift, N., 1996. Spatial Formations. Sage Publications, London.

Thrift, N., 2000. Afterword. Environment and Planning D: Society and Space 18, 213–255.

Thrift, N., 2004a. Intensities of feeling: towards a spatial politics of affect. Geografiska Annaler 86B, 57–78.

Thrift, N., 2004b. Performance and performativity: a geography of unknown lands. In: Duncan, J., Johnson, N., Schein, R. (Eds.), A Companion to Cultural Geography. Blackwell, Oxford, pp. 121–136.

Tilley, S., Taylor, L., 2013. Understanding curriculum as lived: teaching for social justice and equity goals. Race, Ethnicity and Education 16 (3), 406–429.

Titley, E.B., 1986. A Narrow Vision: Duncan Campbell Scott and the Administration of Indian Affairs in Canada. UBC Press, Vancouver. https://www.ubcpress.ca/a-narrow-vision.

Tobias, T.N., 2000. Chief Kerry's Moose: A Guidebook to Land Use and Occupancy Mapping, Research Design, and Data Collection. Union of BC Indian Chiefs, Ecotrust Canada, Vancouver.

Torres, J., 2011. Participation as a pedagogy of complexity: lessons from two design projects with children. Urban Design International 17 (1), 62–75.

Trofanenko, B., 2006. Interrupting the gaze: on reconsidering authority in the museum. Curriculum Studies 38 (1), 49–65.

TRC, 2012. Truth and Reconciliation Commission of Canada: Interim Report. Ottawa. http://www.myrobust.com/websites/trcinstitution/File/Interim%20report%20English%20electronic.pdf.

TRC, 2015a. Report of the Truth and Reconciliation Commission of Canada. McGill-Queen's University Press, Montreal, Kingston, London, Chicago. http://caid.ca/DTRC.html.

TRC, 2015b. Canada's residential schools: the legacy. In: The Final Report of the Truth and Reconciliation Commission of Canada, vol. 5. McGill-Queen's University Press, Montreal, Kingston, London, Chicago. http://caid.ca/TRCFinVol52015.pdf.

Tunbridge, J.E., Ashworth, G.J., 1996. Dissonant Heritage. J. Wiley, Chichester.

Turnbull, D., 2000. Masons, Tricksters, and Cartographers: Comparative Studies in the Sociology of Scientific and Indigenous Knowledge. Harwood Academic, Amsterdam.

Turnbull, D., 2007. Maps, narratives, and trails: performativity, hodology, and distributed knowledges in complex adaptive systems: an approach to emergent mapping. Geographical Research 45, 140–149.

UNEP, 1999. Cultural and Spiritual Values of Biodiversity. Intermediate Technology Publications, London, England.

United Church of Canada, 2014. Alberni Indian Residential School - the Children Remembered. https://thechildrenremembered.ca/school-locations/alberni/.

Vinitzky-Seroussi, V., 2002. Commemorating a difficult past: Yitzhak Rabin's memorials. American Sociological Review 67 (1), 30–51.

Wagner-Pacifici, R., Schwartz, B., 1991. The Vietnam Veterans memorial: commemorating a difficult past. American Journal of Sociology 97 (2), 376–420.

Wesley-Esquimaux, C.C., Smolewski, M., 2004. Historic Trauma and Aboriginal Healing, first ed. Aboriginal Healing Foundation, Ottawa.

West, B., 2003. Student attitudes and the impact of GIS on thinking skills and motivation. Journal of Geography 102, 267–274.

Wilson, T.H., Erasmus, G., Penney, D.W., 1992. Museums and first peoples in Canada. Museum Anthropology 16 (2), 6–11.

Wittgenstein, L., 1980. Culture and Value. Translated by Peter Winch. The University of Chicago Press, Chicago.

Wittmeier, B., 2014. Residential school abuse-claim documents should Be destroyed adjudicator argues. Edmonton Journal. http://www.bishop-accountability.org/news2014/05_06/2014_06_19_Wittmeier_ResidentialSchool.htm.

Wood, D., Fels, J., 2008. The Natures of Maps: Cartographic Constructions of the Natural World. University of Chicago Press, Chicago.

Yiftachel, O., 1998. Planning and social control: exploring the dark side. Journal of Planning Literature 12 (4), 395–406.

Young, J.E., 1992. The counter-monument: memory against itself in Germany today. Critical Inquiry 18 (2), 267–296.

Zimmerman, L., 2000. Regaining Our nerve: ethics, values, and the transformation of archaeology. In: Lynott, M., Wylie, A. (Eds.), Ethics in American Archaeology. Society for American Archaeology, Washington, pp. 71–74.

Cybercartography, emergence and iterative development: The Residential Schools Land Memory Project (RSLMMP)

*Stephanie Pyne**

Postdoctoral Research Fellow, Geomatics and Cartographic Research Centre (GCRC), Carleton University, Ottawa, ON, Canada

*Corresponding author

2.1 Introduction

In an age when reconciliation is practically a household word, public engagement with mapping is increasing in a variety of ways. Although in some cases, maps are still being created and used for exploitative purposes, in an increasing number of instances, map creation and use is contributing to the enrichment of democratic deliberation processes and the enhancement of social and related forms of justice. With a focus on inclusion, transformation and equality, critical approaches to cartography, including participatory, collaborative—and even experimental—approaches, have the potential to enhance reconciliation initiatives and provide a new vehicle for storytelling and sharing (Pyne, 2019; Pyne and Taylor, 2015). The cybercartographic approach to the Residential Schools Land Memory Mapping Project (RSLMMP) provides a good example of a reflexive and emergent approach to collaborative cartography that reflects the central characteristics of Cybercartography, while at the same time maintaining its own distinctive traits. This chapter includes a brief overview of Cybercartography in order to provide the necessary background for the subsequent tracing of the iterative development of the RSLMMP, which has been developing in an iterative manner especially since the emergence of the Residential Schools Map Module (https://lhta.ca/residentialschools/index.html?module=module.residential_schools) in the cybercartographic Lake Huron Treaty Atlas (LHTA; https://lhta.ca/index.html) in 2011. While there is much to discuss when it comes to describing and explaining the relationship between the Residential Schools legacy and the broader historical geographical context reflected in the LHTA, including 'treaty history', this chapter focuses mostly on the iterative development of the Residential Schools component of the LHTA in order to provide a better understanding of historical and methodological foundations of the RSLMMP.

Iterative processes are essential to a performative understanding of cartography. For example, the design and development of the LHTA and its progeny, the Residential Schools Land Memory Atlas, have both involved iterative processes comprised generally of conception, creation, review and revision phases. Many of the iterative processes that have come into play in the making of both Atlases are features of the broader cybercartographic framework and are in some ways like the processes involved in the development of other cybercartographic atlases. Performativity is central to all dimensions of these iterative processes, which involve active collaborations and include diversity of Atlas contributors (Brauen et al., 2011; Pulsifer et al., 2008). David Turnbull's (2000) 'talk, template and tradition' approach to understanding the collaborative social processes responsible for construction processes (introduced in Chapter 1) provides a handy framework to understand the performative nature of collaborative map and atlas development.

This chapter adopts the reflexive approach of Bourdieu (1992) referred to in Chapter 1, which involves a "high-speed tour" of over 10 years of ongoing work that is transparent, related to emergent knowledge and consistent with Indigenous approaches to research. This approach is appropriate and useful for describing research processes and project development details from previous work, which has laid the foundation for the development of the central output of the RSLMMP: Residential Schools Land Memory Atlas. Some of these details are presented elsewhere (Pyne, 2013; Pyne and Taylor, 2015). However, in order to convey the fact that the RSLMMP came from somewhere and did not just pop out of thin air, I present details that may not at first glance seem to be directly related to the thematic content of the RSLMMP, but which nevertheless are related in a number of significant ways. In addition, it is important to realize that with many dimensions along which cybercartographic atlas development can occur over time, any discussion of iterative evolution is relative, and can be engaged in across a series of shifting conceptual scales. In this chapter, I present only a brief sketch reflecting the project's iterative evolution. In terms of implicit cartography, this strategy is akin to a zooming in on various moments in the project's processes, and to zooming out in terms of going back in time to consider the theoretical and practical origins of the project.

2.2 Cybercartography: Evolution of a theoretical and practical framework

Cybercartography began in the 1960s as an approach taken by Fraser Taylor to integrating mapping into community economic and social development in Kenya, although he did not use this term at the time. By incorporating participatory mapping practices into his approach to community development, Taylor was taking a multidimensional approach to 'development' (Pyne, 2013, 2019). In contrast with the increasingly popular 'development from below' approach, Taylor agreed with the grassroots people that depending on how it was applied, such an approach could continue to reinforce an established hegemony and fail to deliver on the goals of development—goals that included (and continue to include) enhanced agency and empowerment, and improved community and individual cultural and economic well-being. Despite changes in technology and an increase in critical theoretical approaches to cartography over the years since then, many of the people-related aspects of Taylor's early work remain the same: action, awareness, care, effort, initiative, knowledge-sharing, transformation and production of alternative findings.

In 1997, Taylor provided an initial characterization of Cybercartography:

> Cybercartography will see cartography applied to a much wider range of topics than has traditionally been the case [...] It will also utilize an increasing range of emerging media forms and telecommunication networks such as the Internet and the World Wide Web. It will be a multidimensional cartography using multimedia formats and is more likely to be an integral part of an information package than a stand-alone product [...] Cybercartography will also be highly interactive and engage the user in new ways. In organizational terms, it will see new partnerships being created between national mapping organizations, the private sector and educational institutions and the products of Cybercartography are likely to be compiled by teams of individuals from very different disciplines and professional perspectives working together (1997, 4).

Taylor's expanded view of cartography was in sync with the critical cartography movement that began with Harley (1988, 1989) and the increase in cartography's public popularity, which has seemed to go in lockstep with the birth and maturing of personal computing, the Internet and Geographic Information Systems (MacEachren, 2013). Despite its sympathetic relationship with other critical approaches to cartography, Cybercartography remained unique in terms of its history, both in terms of theoretical approach and practice (Taylor, 2005; Taylor and Lauriault, 2014).

While at first glance, a cybercartographic atlas is an interactive multimedia website with maps, it is also the collaborative processes that go into making the atlas maps. Although they are often referred to simply as 'maps', these web pages are more accurately described as 'map modules'; they have generous space for an interactive digital map, and side panel, which displays information and media (content) that is relevant to particular points on the map. The cybercartographic Nunaliit Atlas Framework, which underpins this type of website, is Free and Open Source Software (FOSS) and available at http://nunaliit.org/. As a holistic approach to online atlas development, Cybercartography is consistent with Indigenous worldviews. As a relationship-focused approach, it involves reciprocity and engaging people in the production of maps to tell the stories they wish to tell, and giving these stories back to communities for education and further input (Taylor, 1997, 2003, 2005; Taylor and Caquard, 2006; Taylor and Pyne, 2010; Pyne, 2013). Moreover, it takes an inclusive, transdisciplinary theoretical and methodological approach that involves subject specialists sharing knowledge across disciplinary barriers (Lauriault and Taylor, 2005; Pyne, 2013; Taylor and Lauriault, 2014; Fig. 2.2.1).

It is no doubt due to his 'on the ground' experience with the multiple dimensions of social, political and economic development processes (and their interrelationships) through his early development work in Kenya that Taylor realized the atlas context was required in order to 'reflect' multiple dimensions through cartographic practices; a single map would just not do. What was needed was a vehicle to tell different stories cartographically, one that reflected the multiple dimensions of 'social and economic processes' and the interrelationships between them. What was needed was an atlas. The most obvious characteristic of Cybercartography is that it is about atlases. The significance of Cybercartography as an atlas-making — rather than a mere map-making — framework cannot be understated. An atlas ties together a collection of maps with a narrative logic that contributes to understanding and interpretation of both itself and the maps (Akerman, 1991, 1993; Wood and Fels, 2008; Pyne, 2013). Akerman (1991) presents Mercator's founding vision for the 'Atlas' as narrative with a broad scope that covers a variety of dimensions. Mercator's inspiration and guiding metaphor for his proposed project to 'mirror reality' was none other than the god-king himself, Atlas. Mercator's secondary stated purpose, to 'set before your eyes, the whole world … and … by this meanes

FIGURE 2.2.1   Screenshot of the cybercartographic Residential Schools Map Module, showing the central map display and side panels as one example of a cybercartographic atlas display. This screenshot is also presented in Chapter 3 (see Fig. 3.3.6.1).

leade the Reader to higher speculations' (Mercator quoted in Akerman, 1991, p. 1–2) corresponds to the central purpose of Cybercartography, which includes ongoing work applying a cybercartographic approach to Residential Schools: To contribute to an enhanced understanding and improved relationships or reconciliation (Crocker, 1999, 2008, 2015).

A primary concern for cartography is whether or not cartography is capable of meaningfully conveying such things as experience, Indigenous perspectives and knowledge, and critical academic approaches to the status quo (Johnson et al., 2006; Pearce, 2008; Pearce and Louis, 2007; Turnbull, 2007). The multimedia, multisensory, multimodal, interactive and transdisciplinary nature of the cybercartographic approach to atlas-making is geared to conveying multiple perspectives and synthesizing them in the form of cybercartographic output such that the whole is greater than the sum of its parts (Taylor, 1997, 2005; Taylor and Caquard, 2006; Taylor and Pyne, 2010; Pyne, 2019). In addition to being consistent with the holistic nature of Indigenous approaches to knowledge and knowledge creation, this approach is consistent with Howard Gardner's Multiple Intelligence Theory (Taylor, 2014). In addition, Cybercartography takes a broad, inclusive approach to both science and art by acknowledging the holistic relationship between science, which includes geospatial technologies, and art, which extends to storytelling. Alluding to the holistic nature of Cybercartography, Taylor emphasized how intuition and practice preceded the theoretical development of Cybercartography; and in the process of the iterative development of Cybercartography, Taylor (2003, 2014) further specified seven key elements, and six related features (Taylor, 2014), which emerged primarily through the interplay of theory and practice across a number of collaborative cybercartographic projects with Indigenous individuals and communities.

In 2003, a Social Sciences and Humanities Research Council (SSHRC) major grant helped fund the Cybercartography and the New Economy Project, which focused on use and usability in a transdisciplinary approach to the design and the development of two educational atlases, and involved the creation of the Geomatics and Cartographic Research Centre (GCRC) at Carleton University in Ottawa, Canada. Completed in 2007, this research resulted in many publications, including the publication of *Cybercartography: Theory and Practice* (Taylor, 2005), with a special issue of *Cartographica* devoted to describing and summarizing the implications of this work. From this point, Cybercartography began to involve increased practice and theoretical rigor, combined with a focus on mapping with Indigenous peoples and more attention to outreach and education (Taylor and Lauriault, 2014). The flexibility of Cybercartography's Nunaliit Atlas Framework (Hayes et al., 2014) allows a range of atlas projects to share technological innovations with other projects. Examples include Views from the North (Payne et al., 2014); the Inuit Siku (sea ice) Atlas (Ljubicic et al., 2014); the Kitikmeot Place Names Atlas (Keith et al., 2014); the Gwich'in Atlas (Aporta et al., 2014); the Arctic Bay Atlas (Taylor et al., 2014); and the LHTA project (Pyne, 2013, 2014; Pyne and Taylor, 2012; Taylor and Pyne, 2010). Each cybercartographic atlas project evolves uniquely. The cybercartographic research framework guides all projects and is further developed through them. Collaborative relationships are essential to cybercartographic atlas development. Atlas contributors come from a variety of interrelated communities of practice (Pyne and Taylor, 2012; Pulsifer et al., 2011). In case of the LHTA, these contributions have resulted in a diverse range of map themes reflecting a complex multidimensional approach to 'treaty history' that extends to such themes as the Indian Residential/Boarding School legacy in North America (Pyne, 2013; Pyne and Taylor, 2015).

A holistic and reflexive understanding of the evolving RSLMMP requires an excursion into its prehistory. This includes its foundation, the LHTA project, which involved the expansion of the Treaties Module in the pilot cybercartographic Atlas of Indigenous Perspectives and Knowledge (Great Lakes — St. Lawrence Region (GLSL Atlas (Caquard et al., 2009); into a new Atlas with 27 new working maps—including the Residential Schools Map—and conversion to an improved Nunaliit Atlas II Framework. Work under the Public Outreach Grant supported further technological improvements, new maps and the creation of new collaborative research relationships with a broad community, such as academics, students, subject specialists, Indigenous community members, artists, elders and educators, some of which are being continued under the present RSLMMP). Fig. 2.2.2 traces the iterative

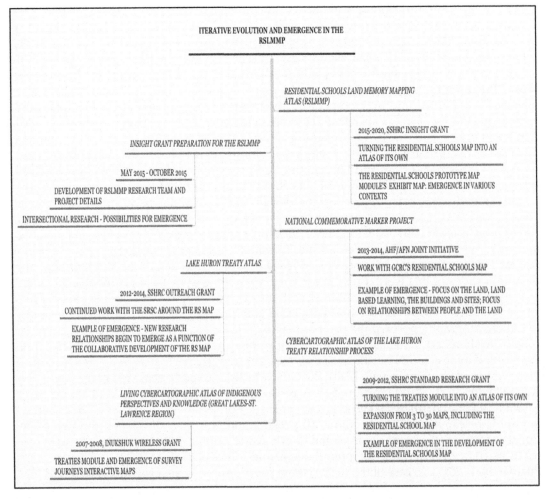

FIGURE 2.2.2   Time-lined tree diagram showing the evolution of project foundations leading to the Residential Schools Land Memory Mapping Project.

evolution of the RSLMMP, beginning with the pilot GLSL Atlas and including projects related to the development of the cybercartographic LHTA.

Learning, emergent knowledge and transformation are central aspects of the types of iterative processes that give rise to cybercartographic atlases; and these iterative processes occur along at least four interrelated dimensions: conceptual, financial, technological and content (Brauen et al., 2011; Pyne, 2013, 2014, 2019; Pyne and Taylor, 2012). Each of these categories is construed broadly. For example, the conceptual dimension includes epistemological and ethical aspects, in addition to ontological and methodological considerations, and is apparent in design decisions that contribute atlas development. The 'financial' dimension refers primarily to the various grants that have supported this ongoing atlas work, which began with a focus on treaties and shifted to a focus on residential schools. Grants have interrelated aims, guidelines and an overarching ethics, which applicants must essentially 'comply with' in order to even qualify for being considered in a funding competition. Moreover, should an application be successful, the funded project will be somewhat constrained by specifications indicated in the proposal. The 'technological' dimension is manifest primarily in computer programming and geospatial technologies development activities and is heavily contingent upon the funding and content dimensions. All cybercartographic atlas projects involve FOSS in the context of a sharing community of software developers. Brauen et al. (2011) describe the nature and benefits of FOSS in the following manner:

> Enabling a diversity of groups to participate in atlas projects is a desirable outcome of our research, related also to the potential for multiple perspectives to be accommodated within the narrative of a single atlas (Taylor and Pyne, 2010), and is best achieved if the cost barriers to adoption and use of the technologies we produce are kept as low as possible. In this way, our participation as a FOSS producer and user inserts our project into what Kelty (2008, p. 3) has called a recursive public: 'a public that is vitally concerned with the material and practical maintenance and modification of the technical, legal, practical, and conceptual means of its own existence as a public'. Kelty argued that participation in the creation and use of FOSS is one way in which such a public asserts itself as 'capable of speaking to existing forms of power through the production of actually existing alternatives'(3). We would argue that enabling the creation of spatial narratives through the production of open source technologies is consistent with the vision for such a public. By choosing to develop a team around FOSS, we chose, possibly without knowing it, to develop a core set of capabilities related to web mapping, geospatial technologies and open geospatial standards within our project that would not have developed if we would have had a vendor-provided set of tools (32).

The 'recursive public' focus of the FOSS movement combined with its interest in nonproprietary information and creating an equitable community for software knowledge sharing makes it a good choice for atlas design and development along the technology dimension.

Finally, the 'content' dimension relates broadly to the multimedia 'stories' identified for transformation into geonarratives. 'Content contributions to this cartographic telling of the Robinson Huron Treaty story come from a variety of interrelated, and sometimes overlapping knowledge perspectives, including Anishinaabe, historical, archival, survey, archaeological, legal, geographical, political, and critical academic understandings. In keeping with the multimedia orientation of the project, contributions can be text, images, videos, or audio' (Brauen et al., 2011, p. 33). The iterative development of geonarratives involves interactions between these four dimensions in a holistic interplay of theory and practice. For example, along the technological dimension, the Nunaliit Atlas Software Framework is continually updated and customized in relation to the conceptual and/or content dimensions of particular cybercartographic atlas projects. Nunaliit is thus developed in an iterative manner and released as 'FOSS' (Brauen et al., 2011, p. 29; Taylor and Pyne, 2010; Hayes et al., 2014). Interproject iterative processes involve individual cybercartographic Atlas projects building on, borrowing from or contributing to one another along any one, or a combination of dimensions. For example, the introduction of the 'timeline' feature along the technology dimension (in response to the demands of the narrative being mapped) in the first phase of the LHTA provided a basis for further technological research on timelines in Nunaliit, including the customized development of a timeline for the Thule Atlas in response to its unique content (https://thuleatlas.org/index.html). There are also intraproject iterative processes that occur within the life of an individual cybercartographic atlas project over its successive research and funding phases. The evolution of the RSLMMP provides a good example, with its prehistory in the LHTA project and its predecessor the Treaties Module of the GLSL Atlas, which began with the design and development of the Survey Journeys maps, and included the emergence of the concept and method of 'geo-transcription' (Pyne in Caquard et al., 2009; Pyne, 2013, 2014) along the conceptual dimension.

In addition to these atlas product development dimensions are individual and sociocultural-political development dimensions. Individual development relates to enhancement of (personal) empowerment via collaborative research and learning in cybercartographic atlas production (self-esteem, enriched knowledge, satisfaction of contributing to a social justice and education tool, and economic gain). The capacity for Cybercartography to engage individuals in an immersive learning situation is related to cartography's broader capacity to engage (Cartwright, 2013, 149). Sociocultural-political development refers to the enhanced interpersonal awareness and understanding that can result both through participation in and exposure to an interactive cybercartographic atlas. A rich or thick form of democracy referred to as deliberative democracy underlies this dimension. Communication and expression of multiple perspectives is a key feature of deliberative democracy, in addition to inclusivity, the ability to be heard/understood, multiple forms of expression (a specific aspect of inclusivity, engagement (comes from cartographic reflections), reflexivity, education, understanding and awareness versus knowledge as accumulation of facts, and cultivation of moral capabilities (Crocker, 2008; Van Staveren, 2001; Young, 2000).

The multidimensional iterative development of cybercartographic atlases can be understood in terms of the talk-templates-tradition framework invoked by Turnbull (2000) to interpret the multigenerational construction of Gothic cathedrals referred to in Chapter 1 (Section 1.3.4) combined with the reflexive approach to presentation inspired by Bourdieu (1992), which is also referred to in Chapter 1 (Section 1.3.3) and involves transparency, emergence and a 'high-speed tour' approach.

2.3 The Lake Huron Treaty Atlas: A high-speed tour

The LHTA is built in an iterative fashion on the Treaties Module of the GLSL Atlas, which aimed to present previously excluded knowledge and to gather dispersed historical artefacts from many digital sources such as museums, archives and private collections in order to

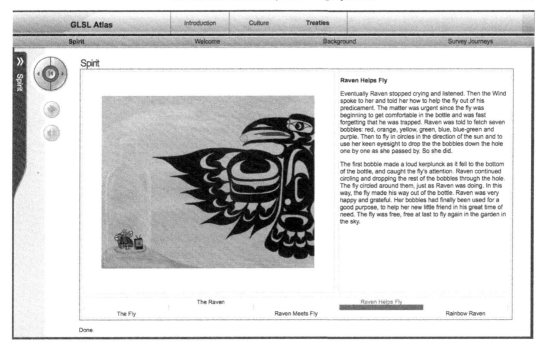

FIGURE 2.3.1 Screenshot showing fourth pane of spirit section of Treaties Module (GLSL Atlas) with the theme 'building awareness to bridge relationships', which was produced with early Nunaliit technology (Pyne, 2013, Foreword and 189).

provide social, geographical and cultural context, and to provide a digital geospatial gathering place for this material that could enable the contribution of additional knowledge and perspectives, including Indigenous community members and others (Caquard et al., 2009). The preserved though not currently online GLSL Atlas was built with the original Nunaliit I Cybercartographic Atlas technology and featured a Culture Module and a Treaties Module (Fig. 2.3.1).

The Survey Journeys Map Module was based on the survey diary of Dennis (1851), which was transcribed and mapped out with the aid of information recorded in the diary entries, survey plans, current and historical maps and satellite imagery. Each overnight stop in the journeys was mapped onto one of two significant historical background maps: a standard settlement map, which tells a story in itself and illustrates the potential for maps in general as effective narrative vehicles, and a sketch map made by Commissioner Alexander Vidal (Vidal and Anderson, 1849) reflecting his investigative process and including his impressions of Anishinaabe territories in the Lake Huron region, along with mining lots within the officially demarcated 'Indian Territories' that had been applied for by prospectors prior to the treaty-making process in 1850. The diary entries associated with the stops are included to the right of the map, together with an audio clip of a voice representing Dennis. The diary entries were transcribed from Dennis's original handwritten report by the first author, and read and recorded by my brother, Jason Pyne, who also included sound effects for features

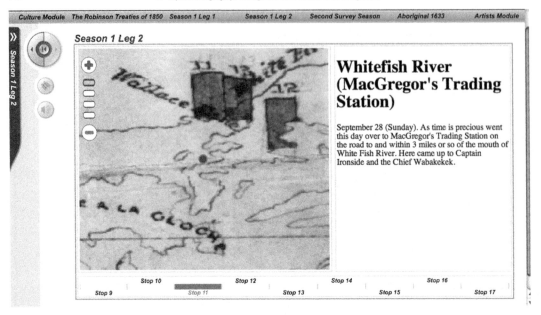

FIGURE 2.3.2   Screenshot showing the point representing one camp stop on the survey route of J.S. Dennis in the Treaties Module (GLSL Atlas), which was produced with early Nunaliit I technology.

such as the sound of horses clopping to match the entries referring to stagecoach travel and rain, and wind sounds for entries referring to stormy weather (Fig. 2.3.2).

A variety of iterative processes contributed to the conception of the Survey Journeys maps in the GLSL Atlas. The Treaties Module was not always focused on the Lake Huron Treaty region. At the beginning of the project, the context was a fairly broad definition of the Great Lakes region. Because this was a project to develop sample content, it was clear that after a certain period for preliminary background research, a decision would have to be made with respect to historical and geographical focus. After 5 weeks of initial research, with no specific focus yet identified, I received a forwarded email from Alan Corbiere, then-Director of the Ojibwe Cultural Foundation (currently a doctoral candidate in History at York University, Toronto), through Heidi Bohaker, a University of Toronto historian and Atlas-making team member. Alan had learned about the Atlas project through Heidi and Ruth Phillips, Director of the Great Lakes Research Alliance for the Study of Aboriginal Arts and Cultures (GRA-SAC) and a coinvestigator on the GLSL Atlas project. Alan's email referred to a chapter by Michael Marlatt, 'The Calamity of the Initial Reserve Surveys under the Robinson Treaty' (2004), which includes a relatively detailed description of the extended survey process for the Lake Huron and Lake Superior region treaties of 1850. After reading the paper, I was sure that the story of the Lake Huron Treaty signing and survey process would serve as an ideal sample content for the Treaties Module: a perfect example for a cybercartographic map module designed to include previously excluded worldviews. Marlatt's chapter is particularly instructive for a number of reasons. Based on land claims research conducted by the author, it provides a critical examination of the survey process from a surveyor's perspective.

Marlatt describes the political and economic context of the treaty signing and survey processes and, drawing on the original surveyor's reports, includes a nation-by-nation narrative tabulation of the various errors and omissions in the surveys themselves.

Before receiving the reference from Alan, I had emerging knowledge of (1) the parameters of the project; (2) Cybercartography as a theoretical and methodological framework, including some familiarity with the first two atlases produced by the GCRC and (3) treaty-related history; nevertheless, I was still not sure exactly 'what' to map, or how. After this point, I was sure. Amazed at how much more there was to the Robinson Huron Treaty than the actual treaty document and impressed by the research in the Marlatt chapter, I now had a working vision for a map and a story to focus on. The experience associated with this significant decision point, which resulted in the initial map conception, can be referred to as a 'mapping epiphany' (Pyne, 2013, 2014), a holistic concept that speaks to the significant role of emotion in decision-making processes (Van Staveren, 2001; Ruddick, 2009; Pyne, 2006) both in general and also in the particular approach to Cybercartography I was beginning to develop via theoretical and practical engagement with others.

Then, in January 2008, another iterative transformation occurred in the development of the Treaties Module when Alan visited the GCRC to map the traditional story of Nenboozhoo and the creation of Mindemoya Island for the 'Culture' section of the GLSL Atlas. I was fortunate to participate with Amos Hayes (the GCRC's technical manager) and Alan in this process. Although I already had a map idea with the decision to render the Marlatt chapter into a map, my participation in the creation of the Nenboozhoo Mindemoye map gave rise to a second epiphany in the iterative design and development of the Treaties Module. At this point, I had already engaged in a series of phone conversations with Michael Marlatt (who had sent me digital copies of archival records including the Dennis survey report, diary and field notes (Dennis, 1851, 1853), and sketch maps of reserve boundaries prepared by Dennis). I had also ordered a high-resolution digital copy of the Vidal—Anderson Commission 1849 Sketch Map (Vidal and Anderson, 1849) upon Alan's suggestion, and serendipitously received a hard copy of the sketch map from Library and Archives Canada on the day of Alan's visit. After a great day of mapping and talking, I arrived home that evening, got out of my car and began to walk toward my front door, when at that moment the idea was born to transcribe and map out—or geo-transcribe—the Dennis survey diaries, thereby supplanting the working plan to geo-transcribe Marlatt's (2004) summary (Caquard et al., 2009; Pyne, 2013, 2014).

Although the initial working plan had already been generated and discussed, the experience of mapping out Nenboozhoo's journey to Mindemoye had inspired a new yet related map idea: To transcribe and map out J.S. Dennis survey diaries, the primary document upon which the Marlatt chapter was based, in addition to mapping out the Marlatt chapter as a critical geonarrative summary of the survey diaries at a later time. The diary referred to more than 100 distinct campsites, provided additional details to the Marlatt chapter and was far easier to follow than the archived handwritten diary when translated into geonarrative form. This iterative shift in focus has not only created new possibilities for analysis and critical understanding but also helped shape ideas for additional geonarratives for inclusion in the later LHTA (Brauen et al., 2011, p. 34).

During the period following the launch of the GLSL Atlas on 23 April 2008, work along the conceptual dimension occurred with the writing of a paper about the project (Caquard et al.,

2009) and several conference presentations concerning the Treaties Module, including a presentation at the Canadian Association of Geographers annual conference entitled, 'Linking Cybercartography and Indigenous Tourism: The Reconciliation Trail', in which I sketched out the rationale for a new cybercartographic atlas project, inspired by the work to date on the Treaties Module, and which coincidentally coincided with the formal Canadian government apology to former Indian Residential Schools students in Canada, and the announcement of the Truth and Reconciliation Commission (TRC) of Canada (Pyne, 2008).

In addition, Dr Taylor (my then-doctoral supervisor) suggested to me that if I were interested in continuing the work I had started on the Treaties Module under a new project, he would serve as principal investigator and be available to me as a consultant-mentor in my work on the funding application to SSHRC for a project that would serve concurrently as the focus of my dissertation and as an additional GCRC atlas project involving Indigenous peoples and others. The preparation for the paper, the conference presentations and proposal writing included research into critical cartography along the conceptual dimension and provided an opportunity for me to both situate the Atlas project in the ongoing dialogue and to incorporate insights from those dialogues into my interpretation of the Atlas work. At that time I was paying attention to observations from critical cartography that included ideas relating to (1) maps as social constructions (Crampton, 2001); (2) the exclusion of certain perspectives from historical maps (Harley, 1988, 1989; 2001; Perkins, 2003) and (3) how such exclusion has been systemically guided by imperialistic and colonial goals (Craib and Burnett, 1998; Edney, 1997, 2005; Given, 2002; Godlewska, 1995). It was proposed that insights from critical cartographic thought and practice would continue to be incorporated into the theoretical grounding of the proposed project to expand the Treaties Module into a cybercartographic atlas, with maps being viewed first, as processes (Taylor, 1991, 1997; Del Casino and Hanna, 2006; Kitchin, 2008; Turnbull, 2007) and second, as media of expression that have the potential to rectify previous exclusionary practices (Corner, 1999; Sparke, 1998, 2005) through alternative approaches that are not overly prescribed, while remaining open to emergent possibilities (Lewis, 1998; Fox et al., 2005; Fig. 2.3.3).

Expansion of the Treaties Module of the GLSL Atlas into the LHTA occurred between 2009 and 2012 under an SSHRC of Canada Standard Research Grant, with the addition of 27 new working maps—including the Residential Schools Map—and conversion to an improved Nunaliit Atlas Framework. An SSHRC Public Outreach Grant (2012—14) supported further technological improvements, new maps and the creation of new collaborative research relationships. As a virtual geospatial public outreach and education interface, the Atlas was designed and developed to shed light on a variety of themes over time and across space, including treaties, institutional processes, biographies and Residential Schools. Including these components together in one atlas space acknowledges the interrelationships between them and provides a broad context to consider each individually. In addition to the inaugural Survey Journeys Maps, which provide the basis for critically tracking the survey portion of the Robinson Huron Treaty process, the Atlas includes biography maps and other maps intended to reflect contextual details relevant to understanding the treaty story. As part of a broader trend in critical cartography to engage in mapping in new decolonizing ways, the atlas maps deal with relationships between scales, spheres or domains, most notably the relationship between legal and regulatory institutions and people, their lives and

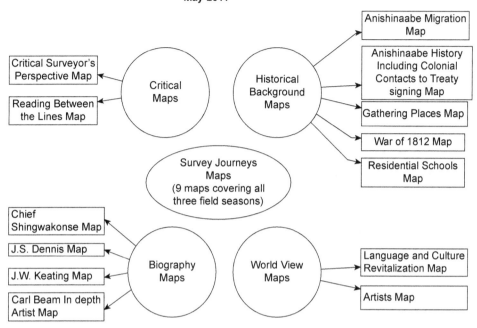

**Potential Geonarratives for the Lake Huron Treaty Atlas
May 2011**

FIGURE 2.3.3 Diagram reflecting conceptual-narrative vision for expanding the Treaties Module into a cyber-cartographic atlas, which was presented and discussed with indigenous collaborators and others (Pyne, 2013, p. 159).

experiences (Crampton, 2001; Harley, 1989, 1990, 2001; Harvey, 2000; Kitchin and Dodge, 2007; Turnbull, 2007; Wood and Fels, 2008; Fig. 2.3.4).

On 27 May 2009, the first version of the LHTA (then referred to as the Cybercartographic Atlas of the Lake Huron Treaty Relationship Process or CALTRP) was uploaded to the GCRC's server. This atlas-in-development was initially comprised of six map sections, three for season one and three for season two of the Robinson Huron Treaty survey journeys. The first iterative transformation occurred along the technology dimension with the decision to convert the Treaties Module content to the Nunaliit II open source atlas framework. Around this time and following, the technological, conceptual and narrative dimensions began to develop iteratively as the GCRC technical team and I began to have regular meetings concerning the form and functions of the maps envisaged for the Atlas. It was decided that the initial priority would be getting the envisaged maps up and running in the Nunaliit II framework. Once the survey journeys maps were working properly, it was thought that attention would turn to the creation of a map template for some of the additional maps planned for the Atlas.

Following the creation of the Survey Journeys Maps in Nunaliit II, up to 30 maps emerged over the next 3 years (see Pyne, 2013 for a summary of this process). This included the Residential Schools Map, which began to emerge in April 2011 when I visited Shingwauk University at Algoma University in Sault Ste. Marie, Ontario. I had been invited to meet

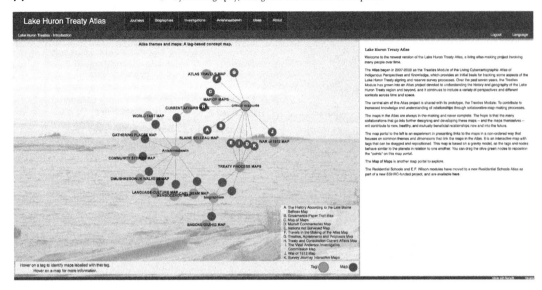

FIGURE 2.3.4   Screenshot of Lake Huron Treaty Atlas homepage, which was created with customized open source code from the Data Drive Documents (D3) website (https://d3js.org/) by Glenn Brauen, and which was developed in the Nunaliit II Atlas software Framework.

the teachers and students for their final week of classes and related celebrations to learn about Shingwauk University and to tell them about the atlas project. During this visit, I walked with a group of students and others from the Shingwauk University office (located in the former home of Shingwauk Industrial Home's first principal, EF Wilson) across the Algoma University front lawn to the main university building (the former Shingwauk Residential School) to attend one of these classes. As we crossed the lawn, it was suggested that I meet the people working at the Shingwauk Residential Schools Centre (SRSC), which was also located in the former Shingwauk Residential School. I proceeded immediately to the SRSC, just across the hall from the class we were headed to. As it turned out, the time was ripe for beginning discussions and actions toward mapping Residential Schools. That day, I met Tina Priest, an undergraduate intern working with the SRSC. I mentioned that if the Centre was interested, it could have a virtual map to supplement the large map on the wall in front of us, full of thumbtacks representing schools. Tina connected me with the then-Director and cofounder of the SRSC, Don Jackson, who I met with soon after. These initial conversations evolved into a working partnership between the GCRC and the SRSC, which has continued to the present and involves development across several dimensions.

By June, Tina had been issued a username and password for the working atlas and was able to make a visit in person to the GCRC in Ottawa, along with others from the SRSC. During the year that followed, Tina worked from the SRSC in collaboration with me and other GCRC team members, talking over issues and concerns on the phone and in person from time to time, exchanging email correspondence and interfacing through the online cybercartographic Nunaliit Atlas Framework itself. Through her work contributing to the development of the Residential Schools Map, Tina was able to learn through practice and interaction with individuals from a variety of knowledge spaces, including others in the Residential Schools education, research and healing community. The map development process also included talks and email correspondence with the SRSC's then-director, Don Jackson, concerning the potential for the eventual development of a cybercartographic Residential Schools Atlas to grow from this pilot work. The envisioned atlas would complement the work of the residential schools research, education and reconciliation community. In addition to drawing attention to the geospatial dimensions of the Residential Schools Legacy, such an atlas could aid in information dissemination, network building and public participation.

By April 2012, the Residential Schools Map included points accurate to the community level for most of the schools in North America, going beyond the 139 schools acknowledged in the Indian Residential Schools Settlement Agreement (IRSSA) discussed briefly in Chapter 1. An initial 'scour for any and all information' approach was taken to the Internet search portion of the map-making project. This approach allowed for the discovery of often inconsistent information concerning opening and closing dates for schools, for example; another important inconsistency across information sources was the number of buildings associated with each school over time (Fig. 2.3.5).

Soon after the launch of the CALTRP in August 2012, work began under an SSHRC Outreach Grant in a variety of areas including further developments to the Residential Schools map components of the Atlas. For example, Don Jackson gave several presentations on the Residential Schools component of the Atlas to a variety of individuals in the Residential Schools education and reconciliation community, including Trina Cooper-Bolam, then-Executive Director of the Legacy of Hope Foundation (LHF, referred to in Chapter 3) and Director

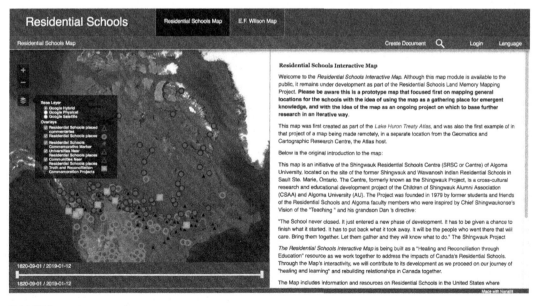

FIGURE 2.3.5 Screenshot of Residential Schools Map showing all layers in the central map window reflecting research to January 2019, and introductory text in the right side panel, and developed with Nunaliit II technology.

of Commemorative Projects for the Aboriginal Healing Foundation (AHF), two affiliated NGOs with programming related to the mandate of the TRC (referred to in Chapter 1).

In April 2013, Trina contacted us at the GCRC to inquire about collaborating on the map component of the National Indian Residential Schools Commemorative Marker Project with the Assembly of First Nations as part of the commemorative phase of the Indian Residential School Settlement Agreement (IRSSA, 2007). After several meetings, we (the GCRC team members) learned of the general desire for a comprehensive set of site-specific Residential School locations amongst the reconciliation research and education community. Seeing this as a good opportunity to improve the school locations in the Residential Schools Map of the LHTA and to explore a public participatory approach to gathering school location information, we initiated a pilot participatory outreach activity for gathering school location-related information, and reviewed and refined some of the school coordinates in the Residential Schools Map (in addition to creating a layer for universities and Indigenous communities in proximity to the schools). Through this process, we became more aware of the diversity of school sites and the rich potential for education and outreach associated with them. Despite our awareness of the need for a more in-depth and long-term participatory mapping project to implement this potential, we began pilot outreach activities with some communities to engage people in visiting the sites to obtain accurate site readings and media. We then used models from archaeology to generate a site survey toolkit. Emerging from this process, the vision for a new project began to crystallize, largely due to conversations and collaborative mapping activities with the survivors who participated in the pilot outreach activities; and throughout the year, there were ongoing discussions with various people working in the area of Residential Schools' reconciliation. The need to find more accurate locations for the schools provided an opportunity to both use and improve the Residential Schools Map, which inspired deeper questions regarding the role of Residential School sites in reconciliation processes and opportunities for education and collaborations (Pyne and Taylor, 2015).

In May 2014, I began an SSHRC postdoctoral research fellowship to further develop the Residential Schools map component of the LHTA and spent the first 6 months of the fellowship working in transdisciplinary collaboration with Fraser Taylor and up to 40 potential research team partners in the preparation of an application for an SSHRC Insight Grant to engage in the RSLMMP, which would further develop the Residential Schools Map in the context of a new cybercartographic atlas project. We were successful in our application, and in May 2015, we began the 5-year project to transform the Residential Schools Map into an Atlas of its own. The RSLMMP was proposed to build on and contribute to (1) the further development of the Residential Schools Map in the LHTA, (2) work begun with the Assembly of First Nations and the AHF on the National Commemorative Marker Project (Assembly of First Nations, 2014), (3) emerging work with the National Research Centre on Residential Schools at the University of Manitoba (nctr.ca), (4) intersecting work with the Embodying Empathy: Fostering Historical Knowledge and Caring Through a Virtual Indian Residential School project (http://www.katherinestarzyk.com/empathy/) and (5) work with survivor and other community groups, religious organizations and among Residential Schools and sites across Canada.

Some schools are still standing but abandoned, some have been demolished, while others are still standing and have been repurposed. Many schools have a history of rebuilding over time, yet memories are sometimes acknowledged primarily in relation to the most recent buildings. The proposed methodology would combine two types of knowledge about the schools and sites—archival and experiential—and involve a variety of participants in the collaborative creation of a series of cybercartographic maps. The online mapping of digital archival documents would involve researchers and students focusing on applying decolonizing theoretical approaches to map development; the 'on the ground' mapping would involve survivors sharing site-specific memories with students who will gather this knowledge and document their experiences through creative, artistic and cartographic media. Along with the standard knowledge dissemination outputs, including peer-reviewed articles and presentations, the RSLMMP was proposed to include a new Residential Schools Land Memory Atlas comprised of a new map series with a cybercartographic digital archives map, nine cybercartographic incorporating experience and memory, and a participant network map. Although the proposed list of project collaborators, coinvestigators and participants is quite long, the core research has been completed by myself in collaboration with university and community research assistants who have been hired under the project since 2015.

In keeping with a central aim of the project, to contribute to student training, it was proposed that as the postdoctoral fellow, I would gain experience in national research project design, development and project management, and providing guidance and assistance as required to all project participants, including the university research assistants we would hire under the project. In addition to increased technical knowledge and ability, I would also have the opportunity to make advancements in conceptual and theoretical approaches and understandings and develop enhanced research relationships. The proposed PhD research assistant, Trina Cooper-Bolam, had collaborated with us previously under the National Commemorative Marker Project (referred to above), and had a decade of experience working in Residential Schools' commemoration, heritage, and education and a proposed thesis project that aligned closely with our project. Through her participation on the RSLMMP, Trina would have the opportunity to contribute her background knowledge and experience and enhance existing theoretical and practical skills. The research assistants would increase their technical knowledge, including an improved understanding and awareness the social and cultural dimensions of technical work, and knowledge of Cybercartography, which includes other forms of geographic information processing (Taylor, 2005, 2014). Through their participation in the cybercartographic mapping of both 'document-based' and 'on the ground' data, students would also learn about the ethics and application of participatory research.

2.4 Residential Schools Land Memory Mapping Project, SSHRC Insight Grant, 2015—20

As the project was beginning to get rolling, initial work had already begun as a result of the work to get the research team of coinvestigators and collaborators together in the process of preparing the SSHRC Insight Grant application. I had just completed the first year of a 2-year SSHRC postdoctoral fellowship, which involved moving to Winnipeg to work under the supervision of Dr Greg Bak at the University of Manitoba learning about critical approaches to archival studies, and sharing knowledge related to my work in Cybercartography in a series of presentations to different audiences. During this time, I also gave presentations related to the developing cybercartographic approach to the residential schools legacy at the university (for example, Pyne, 2015, 2016a,b); and had meetings with Ry Moran, Director of the National Centre for Truth and Reconciliation (also at University of Manitoba) and Andrew Woolford who was managing two related projects, the SSHRC funded Embodying Empathy project (referred to above) and 'Remembering Assiniboia: Commemorative Justice for an Indian Residential School in Winnipeg's River Heights Neighbourhood', a local project, which is discussed further in Chapter 9. This initial work involved dialogue as the main vehicle of communication and early development in the project, in addition to research into archival studies literature as part of my participation in the Archival Studies Master's core courses and literature in participatory, story and tour-related mapping (see Chapter 8).

The first task of the day after beginning the project was to adjust the proposed project deliverables to match the actual funding received for the project, which was about 20% less than requested in the application. This involved reducing the promised nine school-related modules to six, with corresponding reductions in project costs, mostly related to participatory events related to the module development. We also required a project statement for ethics board review and approval to go forward with the project. Considering that we did not yet have any plans in place for participatory events and intended to first get our RAs going and involved in learning about Cybercartography and preliminary research necessary for later atlas design and development, we were granted initial ethics clearance to necessary to initiate the project (see Chapter 10).

In keeping with the proposed plans to include a reflexive aspect to shed light on the atlas-making process and the atlas development collaborators, we made some initial headway on the technical, conceptual and content dimensions with the creation of the project planning and network map, which is an ongoing experiment in reflexive mapping that uses the cybercartographic atlas framework for tracking and reporting on project progress details in a way that makes use of 'place' as an organizing factor, and which has changed names over the course of the development process. This map module involves tracking both ongoing project development and the 'travels' of research team members, especially those with a history of working in residential schools reconciliation. The metadata categories were designed to track such things as meeting minutes and project milestones and deliverables. Through the process of 'testing' the module by making entries, we decided to have separate layers for research bases, which were often universities, and research trips, which involved both knowledge gathering and knowledge dissemination. This change has made improvements to the organizational capacity of the map, which ties both the project and the participants to the land in important ways. However, certain aspects of the map remain cumbersome. Future development directions for this map in order to enhance its usability include updating to new cybercartographic possibilities available in atlases such as the Languages Of Iran Atlas (http://iranatlas.net/index.html) which makes extensive use of filters, and AXON, the Atlas of Ontario Neuroscience (http://axon.braininstitute.ca/index.html0, whose maps integrates customized open source code from the Data Drive Documents (D3) Website (https://d3js.org/), a practice that was pioneered in the development of the opening portal to the LHTA in collaboration with Glenn Brauen in 2012 (see Figs. 2.3.4 and 2.4.1).

At the same time, we began to develop the Travels Map, which was essentially intended to track and compare the career paths of the atlas collaborators. So far this working map

FIGURE 2.4.1 Screenshot showing the Project Journal Map of the in-development Residential Schools Land Memory Atlas with the red triangle in the central map panel representing my visit to the Guy Hill Residential School reunion in August 2015, with photographs taken at the reunion in the right side panel.

includes points for journeys of myself, Trina Cooper-Bolam, including her trip to Southwell and the Irish Workhouse (see Chapter 7), points associated with Jeff Thomas' exhibitions over time (see Chapter 3), in addition to other relevant points. We have also discussed mapping the public outreach and related events attended by the NCTR and have begun by mapping publicly available media (Fig. 2.4.2).

The selection of research assistants from the University of Manitoba was a central item of discussion in the early stages of the project. I prepared a job description for Greg to circulate among his students, and Greg got to know his new students better. Yet again, talk in the context of social relationship building came into play. During the first class at the beginning of September, Greg announced that all students in the Archival Studies program were invited to a nearby restaurant for a social gathering after class. This enjoyable exchange helped immeasurably in the research assistant selection process. After the meeting, I drove two senior students home; and it turned out that they were neighbours, not only to each other but to me as well. Closer to the end of term, Greg recommended a first-year MA student, Chris Calesso, who was interested in issues at the intersection of museum and archival studies and who was also curious about cartography. Chris was set to start in January 2016. While I was happy to have the selection of the University of Manitoba RA established, I was also wondering about the students I had met in September and thought would be appropriate as research assistants for the project. Then, in a serendipitous encounter I ran into one of these students in a local store on my birthday. And so, another University of Manitoba RA, Kevin Palendat, joined the project. Kevin was in the process of moving to Ottawa to complete an internship with Library and Archives Canada, which was a perfect

FIGURE 2.4.2 Screenshot showing the Travels Map of the in-development Residential Schools Land Memory Atlas with the purple circle representing Trina Cooper-Bolam's visit to Southwell Workhouse in the central map window, with a photograph of the Southwell Workhouse deadroom, courtesy of Graham Iddon and explanatory text by Trina Cooper-Bolam (see Chapter 7).

location for him to be close to the GCRC, and I would be moving back to Ottawa in May. This would mean that Kevin, Trina and I could have in person meetings about project development.

In addition to Kevin and Chris, Andrew Woolford suggested that an RA working on archival research related to the Remembering Assiniboia project, also be funded by our project, since her research informed it as well; and in a related process, Krista McCracken at the SRSC (Algoma University) was in the process of selecting an RA to work on the project through the SRSC (see Chapter 4 for an example of some of the work that ensued). In addition, I had a meeting with then-coinvestigator and Director of the SRSC, Jonathan Dewar, in Sault Ste. Marie, Ontario, about a variety of issues. For example, Jonathan mentioned the changes in research funding for Residential Schools initiatives with the dissolving of the wind-up of the TRC of Canada; I shared insights from the literature I had been reviewing on mapping and touring (for example, Sprake, 2012); we discussed the idea I had discussed with Don Jackson (the previous Director of the SRSC and collaborator on our original Residential Schools Map) in 2011 about creating a prototype geospatial portal for SRSC archives through this project; we confirmed that the SRSC would hire an undergraduate Algoma University RA (funded by the RSLMMP) to work with Krista McCracken, in addition to an Indigenous community mapping consultant, who I had worked with before and who I also met with when during my visit to Sault Ste. Marie in 2015. In our discussion about the planned case study module for the Shingwauk Residential School, Jonathan mentioned the SRSC's involvement in an adaptive reuse project under the Museum Assistance Program

(MAP), Reclaiming Shingwauk Hall (http://reclaimingshingwaukhall.ca/), which I had heard about previously from Trina Cooper-Bolam (the project's PhD RA) who was also working on this project. Trina and I had numerous discussions over the first 2 years of the project regarding intersections between the Reclaiming Shingwauk Hall project. For example, a significant portion of the research team's archival and cartographic research on the Shingwauk Residential Schools informs the cartographic dimensions of the reclaiming Shingwauk Hall project.

Between May 2015 and January 2016, there was technological progress with respect to the migration of Residential Schools Map and EF Wilson Map Modules from the LHTA to a new Residential Schools Atlas space; the creation and initial use of the project planning and travels maps; and the creation of a prototype Shingwauk Residential School case study module based on the prototype Exhibit Module (discussed in Chapter 11), with a focus producing a map background layer through research on official school blueprints, floor plans, site plans etc. Along the conceptual (including ethical) dimension, we were shifting from a conventional to a relational view of archival studies and considering the effects of this shift on technological design decisions; we had received preliminary ethics clearance and were in the process of preparing the full ethics clearance application, and ensuring that the RAs coming on board completed the mandatory SSHRC CORE ethics training, and we planned to have initial informal meetings with potential participants when we were ready to do so. With respect to the narrative dimension, we made the decision to begin by researching and incorporating archival records into the case study school map modules.

In addition, over the course of the first 6 months of the project, a shift in the project occurred on the conceptual and technological dimensions where we originally conceived the Atlas to be organized into one archival sources module and six case study school modules reflecting a participatory approach to mapping. After attending MA in Archival Studies classes and moving toward a broader decolonized approach to archives, I realized the broader archival nature of the project; and discussed the following idea with the RAs: to 'house' archival records in the school-specific modules in addition to any media we might gather reflecting the participatory dimension. We also discussed the ideal case of being able to convey comprehensive historical geographies of the schools and sites, which was related to our research to scope out all forms of record associated with schools (e.g., theses, newspaper articles, other media).

By May 2016, the end of year one, I noted in a project summary that nothing was yet planned for the participatory/outreach component of the project, and that we would focus on the atlas prototype development. Also, some progress had been made with respect to (1) the research direction and focus and (2) cybercartographic atlas prototype development.

With respect to prototype development, I was able to list 11 prototype Residential Schools Land Memory Atlas Modules.

i. six case study school modules to act as interactive geospatial portals to archival and other sources of information regarding these specific schools, where our initial research to create the prototype Shingwauk Industrial Home module was noted to have led to some productive ideas with respect to educational outreach, in addition to leading to the realization that extensive archival research is required in order to complete this module and construct the prototype modules for the other five case study schools;

ii. one prototype biographies map module with a series of map layers featuring work and travels of team participants;

iii. three orientation and support modules (mapping archival materials module, witness trauma education module, atlas orientation module) and

iv. one prototype project planning module.

In terms of outreach, given the large number of team members and associated research assistants, an outreach attitude was adopted toward the team in initial discussions that involved learning about their ongoing work with Residential Schools and thinking about ways to intersect with the project objectives of engaging in cybercartographic mapping practices both online and on the ground. With respect to the online atlas development, the initial vision was to undertake six participatory mapping events at each of the six case study school sites over the course of the project, and later document the events in the atlas. However, over the first 2 years of this project, after many meetings with team members, including collaborators and RAs on issues ranging from the nature and scope of content, it became clear that extensive archival, historical and cartographic research would be required to develop the prototype atlas in a manner that would present archival records related to Residential Schools in innovative geospatial ways. In addition, we had begun with the strategy of focusing on two schools, which were the focus of collaborators: Shingwauk and Assiniboia Residential Schools.

2.5 Conclusion

The RSLMMP continues the work that has begun toward intercultural reconciliation via the various iterations of the LHTA since its inception as the Treaties Module of the GLSL Atlas in 2007. Part of this consistency has to do with iterative improvements in technology development; yet, at the same time, story drives technology in Cybercartography (Brauen et al., 2011). As such, oral knowledge transmission and collaborative research relationships are the true foundation upon which the RSLMMP is based, and from which it will grow. 'Cybercartography' has two important connotations: Whereas, the most immediate might be 'cyberspace', a more significant connotation is 'community' (Pyne, 2013, 2019; Taylor, 2005; Taylor and Lauriault, 2014). The iterative development of the conceptual and applied dimensions of Cybercartography, which began officially in 2003, resulted in the release of the Nunaliit software framework in 2006. The production of this software involved the collaborative participation of a diverse 'community' of researchers spanning many disciplines, all sharing the same purpose of designing a community-friendly online, interactive multimedia Atlas. Community interaction in the collaborative atlas-making context involves sharing knowledge and perspectives across disciplines and between cultures. Stemming from this, the iterative development of the LHTA project has given rise to the emergence of knowledge, relationships and—not only stories but also innovative ways to tell them. This particular approach to Cybercartography can be described as being emergent, experimental and involved with the creation of geonarratives with collaborators from different disciplines, including art, archival studies, critical museology, anthropology, sociology and of course, geography, and with other collaborators, including Residential Schools survivors and intergenerational survivors. Rather than being a project concerned with a complete approach to mapping content, the RSLMMP is more concerned with exploring different methods of mapping, which emerge through the transdisciplinary collaborations, and with producing what often amount to prototype sketch-type geonarratives (always in need of further development).

The role of emergence as an atlas design factor seems anathematic to the notion of planning, a feature that is necessary to a great degree in research projects. Due to constraints related to accountability for funding such as deliverables and ethics protocols, detailed plans are often required well in advance of the actual events. It can often take considerable effort to balance emergence with a carefully planned and budgeted project. I have found that in my ongoing work developing atlases in an iterative manner, the cybercartographic atlas framework is flexible enough to allow for the emergence of new concepts (mapping epiphany) (Pyne, 2013, 2014) and new approaches (geo-transcription) (Pyne, 2013); technology development has progressed in an iterative fashion as a function of narrative development; and, research relationships have grown as a function of intersecting interests and ongoing commitments to common goals such as reconciliation. In addition, incorporating emergence into the project and dealing with issues at the intersection of participatory and institutional ethics (see Chapter 10) has reshaped the project in terms of atlas module deliverables and process, including devoting the bulk of project funding to research and training.

This chapter has followed the reflexive approach of Bourdieu (1992; referred to in Chapter 1) by giving a brief tour of the iterative evolution of the prehistory of the RSLMMP in the interest of drawing attention to the importance of iterative processes and emergence, especially in research with intercultural reconciliation at heart. It has also appealed to David Turnbull's (2000) 'talk, template, tradition' approach (also referred to in Chapter 1), which provides a framework for documenting and interpreting the project's iterative development. At the same time, it constitutes an implicit approach to cartography in the form of an epi map, which interprets an ongoing cybercartographic project. As with explicit approaches to mapping (screenshots of which are included in the chapter), which include a limited selection of information, it is necessary to present only a portion of all possible events in iterative atlas development in a single chapter.

Since the inception of the RSLMMP, a variety of developments have emerged across several dimensions, including the narrative and conceptual dimensions. Many of these developments will be discussed in other chapters, for example the emergence of the first test map module in the Residential Schools Prototype Atlas involving participatory mapping: The Exhibit Map, which is discussed in Chapter 11; work with Jeff Thomas on the Where Are the Children Map discussed in Chapter 3; work with research assistants to develop the Shingwauk Schools and Assiniboia School Maps discussed in Chapter 8; and intersecting work with the Remembering Assiniboia Project, discussed in Chapter 9.

References

Assembly of First Nations, 2014. National Commemorative Marker Project. http://www.afn.ca/2014/02/20/national-commemorative-marker-project/.

Akerman, D., 1991. On the Shoulders of a Titan: Viewing the World of the Past in Atlas Structure. Unpublished PhD Dissertation. Pennsylvania State University.

Akerman, J., 1993. From books with maps to books as maps: the editor in the creation of the atlas idea. In: Editing Early and Historical Atlases: Papers Given at the Twenty-Ninth Annual Conference on Editorial Problems, University of Toronto, Toronto, November 5–6.

Aporta, C., Kritsch, I., Andre, A., Benson, K., Snowshoe, S., Firth, W., Carry, D., 2014. The Gwich'in Atlas: place names, maps, and narratives. In: Taylor, D.R.F., Lauriault, T. (Eds.), Developments in the Theory and Practice of Cybercartography: Applications and Indigenous Mapping. Elsevier, Amsterdam (Chapter 16).

Bourdieu, P., 1992. The practice of reflexive sociology. In: Bourdieu, P., Wacquant, L. (Eds.), An Invitation to Reflexive Sociology. University of Chicago Press, Chicago, pp. 217–253.

Brauen, G., Pyne, S., Hayes, A., Fiset, J.P., Taylor, D.R.F., 2011. Transdisciplinary participation using an open source cybercartographic toolkit: the atlas of the Lake Huron treaty relationship process. Geomatica 65 (1), 27–45.

Caquard, S., Pyne, S., Igloliorte, H., Mierins, K., Hayes, A., Taylor, D.R.F., 2009. A "living" atlas for geospatial storytelling: the cybercartographic atlas of indigenous perspectives and knowledge of the Great Lakes region. Cartographica 44 (2), 83–100.

Cartwright, W., 2013. 'Cartography as engagement'. The Cartographic Journal 50 (2), 149–151.

Corner, J., 1999. The agency of mapping: speculation, critique and invention. In: Cosgrove, D. (Ed.), Mappings. Reaktion Books, London, pp. 213–252.

Craib, R.B., Burnett, D.G., 1998. Insular visions: cartographic imagery and the Spanish–American war. Historian 61 (1), 100–118.

Crampton, J., 2001. Maps as social constructions: power, communication and visualization. Progress in Human Geography 25 (2), 235–252.

Crocker, D., 1999. Reckoning with past wrongs: a normative framework. Ethics and International Affairs 13, 43–64.

Crocker, D.A., 2008. Ethics of Global Development: Agency, Capability, and Deliberative Democracy. Cambridge University Press, Cambridge, New York.

Crocker, D., 2015. Obstacles to reconciliation in Peru: an ethical analysis. Unpublished English translation of Obtsaculos para la reconciliacion en el Peru. In: Giusti, M., Gutiérrez, G., Salmón, E. (Eds.), La Verdad nos Hace Libres. Sobre las Relaciones entre Filosofía, derechos Humanos, Religión y Universidad, Fondo Editorial de la Pontificia Universidad Católica del Perú, Lima, Peru.

Del Casino, V.J., Hanna, S.P., 2006. 'Beyond the 'binaries': a methodological intervention for interrogating maps as representational practices. ACME: An International E-Journal for Critical Geographies 4 (1), 34–56.

Dennis, J.S., 1851. Report, diary and field notes, survey of the Indian reserves on Lake Huron. In: Ontario Crown Survey Records, Ministry of Natural Resources (Toronto), Field Note Book (FNB), vol. 1, 832.

Dennis, J.S., 1853. Report, diary and field notes, survey of the Indian reserves on Lake Huron. In: Ministry of Natural Resources, Ontario Crown Survey Records, (Toronto), Field Note Book (FNB), vol. 1, 832.

Edney, M.H., 1997. Mapping an Empire: The Geographical Construction of British India. The University of Chicago Press, Chicago, London, pp. 1765–1843.

Edney, M.H., 2005. The origins and development of J.B. Harley's cartographic theories. Cartographica 40 (1), 1–143. Monograph 54.

Fox, J., Suryanata, K., Hershock, P., 2005. Mapping Communities: Ethics, Values, Practice. East-West Center, Honolulu, Hawaii.

Given, M., 2002. Maps, fields and boundary cairns: demarcation and resistance in colonial Cyprus. International Journal of Historical Archaeology 6 (1), 1–22.

Godlewska, A., 1995. Map, text and image – the mentality of enlightened conquerors – a new look at the Description de L'Egypte. Transactions of the Institute of British Geographers 20, 5–28.

Harley, J.B., 1988. Secrecy and silences: the hidden agenda of state cartography in early modern Europe. Imago Mundi 40, 57–76.

Harley, J.B., 1989. Deconstructing the map. Cartographica 26, 1–20.

Harley, B., 1990. Cartography, ethics and social theory. Cartographica 27, 1–23.

Harley, J.B., 2001. The New Nature of Maps: Essays in the History of Cartography. Johns Hopkins University Press, Baltimore, MD.

Harvey, F., 2000. The social construction of geographical information systems. International Journal of Geographical Information Science 14 (8), 711–713.

Hayes, A., Pulsifer, P., Fiset, J.P., 2014. The Nunaliit cybercartographic atlas framework. In: Taylor, D.R.F., Lauriault, T. (Eds.), Developments in the Theory and Practice of Cybercartography: Applications and Indigenous Mapping. Elsevier, Amsterdam (Chapter 9).

IRSSA, 2007. Indian Residential School Settlement Agreement see. https://www.aadnc-aandc.gc.ca/eng/1100100015576/1100100015577#sect1.

Johnson, J., Louis, R., Pramono, H., 2006. Facing the future: encouraging critical cartographic literacies in Indigenous communities. ACME: An International E-Journal for Critical Geographies 4 (1), 80–98.

Keith, D.1, Crockatt, K., Hayes, A., 2014. The Kitikmeot place name atlas. In: Taylor, D.R.F., Lauriault, T. (Eds.), Developments in the Theory and Practice of Cybercartography: Applications and Indigenous Mapping. Elsevier, Amsterdam (Chapter 15).

Kelty, C.M., 2008. Two bits: the cultural significance of free software. Experimental Futures: Technological Lives, Scientific Arts, Anthropological Voices Series. Duke University Press, Durham, NC; London.

Kitchin, R., 2008. The practices of mapping. Cartographica 43 (3), 211–216.

Kitchin, R., Dodge, M., 2007. Rethinking maps. Progress in Human Geography 31 (3), 331–344.

Lauriault, T., Taylor, D.R.F., 2005. Cybercartography and the new economy: collaborative research in action. In: Taylor, D.R.F. (Ed.), Cybercartography – Theory and Practice, Volume 4 in Modern Cartography Series. Elsevier, Amsterdam, pp. 181–210 (Chapter 8).

Lewis, G.M., 1998. Cartographic encounters: perspectives on Native American mapmaking and map use. In: Nebenzahl Jr., K. (Ed.), Lectures in the History of Cartography. University of Chicago Press, Chicago.

Ljubicic, G., Pulsifer, P., Hayes, A., Taylor, D.R.F., 2014. The creation of the Inuit siku (sea ice) atlas. In: Taylor, D.R.F., Lauriault, T. (Eds.), Developments in the Theory and Practice of Cybercartography: Applications and Indigenous Mapping. Elsevier, Amsterdam (Chapter 14).

MacEachren, A.M., 2013. Cartography as an academic field: a lost opportunity or a new beginning? The Cartographic Journal 50, 1–6.

Marlatt, M., 2004. The calamity of the initial surveys under the Robinson treaties. In: Wolfart, H.C. (Ed.), Papers of the Thirty-Fifth Algonquian Conference. University of Manitoba Press, Winnipeg, pp. 281–335.

Payne, C., Hayes, A., Ellison, S., 2014. Mapping views from the north: cybercartographic technology and Inuit photographic encounters. In: Taylor, D.R.F., Lauriault, T. (Eds.), Developments in the Theory and Practice of Cybercartography: Applications and Indigenous Mapping. Elsevier, Amsterdam (Chapter 13).

Pearce, M.W., 2008. Framing the days: place and narrative in cartography. Cartography and Geographic Information Science 35, 17–32.

Pearce, M.W., Louis, R., 2007. Mapping Indigenous depth of place. American Indian Culture & Research Journal 32 (3), 107–126.

Perkins, C., 2003. Cartography: mapping theory. Progress in Human Geography 27, 341–351.

Pulsifer, P.L., Hayes, A., Fiset, J.P., Taylor, D.R.F., 2008. An education and outreach atlas based on geographic infrastructure: lessons learned from the development of an on-line polar atlas. Geomatica 62 (2), 169–188.

Pulsifer, P., Laidler, G., Taylor, D.R.F., Hayes, A., 2011. Towards an indigenist data management program: reflections on experiences developing an atlas of sea ice knowledge and use. Canadian Geographer 55 (1), 108–124.

Pyne, S., 2006. Full Responsible Reason and Good Development. Master's Thesis, unpublished. Carleton University, Ottawa. https://curve.carleton.ca/904fd230-bd1a-4164-b21d-d613ac2dd66c.

Pyne, S., 2008. Linking Cybercartography and Indigenous tourism: the reconciliation trail. In: Paper Presented at the Canadian Association of Geographers Annual Meeting, Quebec City, May.

Pyne, S., 2013. Sound of the Drum, Energy of the Dance: Making the Lake Huron Treaty Atlas the Anishinaabe Way. Unpublished PhD Dissertation. Carleton University, Ottawa.

Pyne, S., 2014. The role of experience in the iterative development of the Lake Huron Treaty Atlas. In: Taylor, D.R.F., Lauriault, T. (Eds.), Developments in the Theory and Practice of Cybercartography: Applications and Indigenous Mapping. Elsevier, Amsterdam (Chapter 17).

Pyne, S., 2015. A reflexive approach to learning about Cybercartography. In: Presentation to Fourth Year Undergraduate Class in Digital Methods at the University of Manitoba, November 20.

Pyne, S., 2016a. The residential schools land memory mapping project. In: Paper Presented to the History Department of the University of Manitoba on February 10.

Pyne, S., 2016b. The evolution of the Lake Huron treaty atlas: from treaties to residential schools and back again. In: Paper Presented at the St. John's College Soup and Bread Lecture Series, March 2.

Pyne, S., 2019. Cybercartography and the critical cartography clan. In: Taylor, F., Anonby, E., Murasugi, K. (Eds.), Further Developments in the Theory and Practice of Cybercartography: International Dimensions and Language Mapping. Elsevier, London, Forthcoming.

Pyne, S., Taylor, D.R.F., 2012. Mapping indigenous perspectives in the making of the cybercartographic atlas of the Lake Huron treaty relationship process. Cartographica, Special Issue on Indigenous Cartography and Counter Mapping 47 (2), 92–104.

Pyne, S., Taylor, D.R.F., 2015. Cybercartography, transitional justice and the residential schools legacy. Geomatica 69 (1), 173–187.

Ruddick, S., 2009. Maternal Thinking: Philosophy, Politics, Practice. Demeter Press, Toronto (originally published 1989).

Sparke, M., 1998. A map that roared and an original atlas: Canada, cartography, and the narration of nation. Annals of the Association of American Geographers 88 (3), 463–495.

Sparke, M., 2005. In the Space of Theory: Postfoundational Geographies of the Nationstate. University of Minnesota Press, Minneapolis.

Sprake, J., 2012. Learning Through Touring: Mobilising Learners and Touring Technologies to Creatively Explore the Built Environment. Sense Publishers, Rotterdam, Boston.

Taylor, D.R.F., 1991. A conceptual basis for cartography/new directions for the information era. Cartographica 28 (4), 1–8.

Taylor, D.R.F., 1997. Maps and mapping in the information era. Keynote address to the 18th ICA Conference, Stockholm. In: Ottoson, L. (Ed.), Proceedings, vol. 1, pp. 1–10. In: https://icaci.org/files/documents/ICC_proceedings/ICC1997/icc1997_volume1_part1.pdf.

Taylor, D.R.F., 2003. The concept of Cybercartography. In: Peterson, M.P. (Ed.), Maps and the Internet. Elsevier, Amsterdam, pp. 405–420.

Taylor, D.R.F. (Ed.), 2005. Cybercartography: Theory and Practice, Volume 4 in Modern Cartography Series. Elsevier, Amsterdam.

Taylor, D.R.F., 2014. Some recent developments in the theory and practice of Cybercartography: applications in Indigenous mapping – an introduction. In: Taylor, D.R.F., Lauriault, T. (Eds.), Developments in the Theory and Practice of Cybercartography: Applications and Indigenous Mapping. Elsevier, Amsterdam (Chapter 1).

Taylor D.R.F. and Caquard S. (Eds.), Cartographica 41 (1), Special issue on Cybercartography, 2006, 1–5.

Taylor, D.R.F., Pyne, S., 2010. The history and development of the theory and practice of Cybercartography. International Journal of Digital Earth 3 (1), 1–14.

Taylor, D.R.F., Lauriault, T. (Eds.), 2014. Developments in the Theory and Practice of Cybercartography: Applications and Indigenous Mapping. Elsevier, Amsterdam.

Taylor, D.R.F., Cowan, C., Ljubicic, G., Sullivan, C., 2014. Cybercartography for education: the application of Cybercartography to teaching and learning in Nunavut, Canada. In: Taylor, D.R.F., Lauriault, T. (Eds.), Developments in the Theory and Practice of Cybercartography: Applications and Indigenous Mapping. Elsevier, Amsterdam (Chapter 20).

Turnbull, D., 2000. Masons, Tricksters, and Cartographers: Comparative Studies in the Sociology of Scientific and Indigenous Knowledge. Harwood Academic, Amsterdam.

Turnbull, D., 2007. Maps, narratives, and trails: performativity, hodology, and distributed knowledges in complex adaptive systems: an approach to emergent mapping. Geographical Research 45, 140–149.

Van Staveren, I., 2001. The Values of Economics: An Aristotelian Perspective. Routledge, London.

Vidal, A., Anderson, T.G., 1849. RG 1. Report of Commissioners A. Vidal and T.G. Anderson on Visit to the Indians on the North Shores of Lakes Huron and Superior for Purposes of Investigating Their Claims to Territory Bordering on Those Lakes, vol. 266. Library and Archives, Canada (Ottawa). Reel C-12652, 4.

Wood, D., Fels, J., 2008. The Natures of Maps: Cartographic Constructions of the Natural World. University of Chicago Press, Chicago.

Young, I., 2000. Inclusion and Democracy. Oxford University Press, Oxford, New York.

CHAPTER

3

Mapping Jeff Thomas mapping: Exploring the reflexive relationship between art, written narrative and Cybercartography in commemorating residential schools

Stephanie Pyne*,[1], Jeff Thomas[2]

[1]Postdoctoral Research Fellow, Geomatics and Cartographic Research Centre (GCRC), Carleton University, Ottawa, ON, Canada; [2]Independent Curator and Photographer, Ottawa, ON, Canada
*Corresponding author

OUTLINE

3.1 Introduction 58

3.2 Jeff Thomas: Mapping and art/art and mapping 60

3.3 Jeff Thomas and Cybercartography: A high-speed tour 64
   3.3.1 The Cybercartographic Atlas of Indigenous Perspectives and Knowledge (Great Lakes Region-St. Lawrence Region, GLSL Atlas) 64
   3.3.2 The Jeff Thomas Atlas: Journey with the Champlain monument 65
   3.3.3 Travels Map of Residential Schools Land Memory Atlas 68

3.3.4 The Jeff Thomas 2015 Road Trip overlay 69
3.3.5 Mapping the family camera and visitor information exhibits: Geo-locating composite photograph components 71
3.3.6 Residential schools volunteered geographic information: Battleford Indian Residential School 73
3.3.7 Transposing the Jeff Thomas Cahokia Mounds Journey overlay onto the Residential Schools Map 74
3.3.8 Jeff Thomas Mapping Thomas Moore overlay in the Nenboozhoo

Cybercartography in a Reconciliation Community, Volume 8
https://doi.org/10.1016/B978-0-12-815343-7.00003-8

57

© 2019 Elsevier Inc.
All rights reserved.

3.1 Introduction

It has been said that cartography is in the midst of an ontological crisis (Kitchin and Dodge, 2007), meaning that it is in the midst of a radical transformation away from a focus on representation, communication and objectivity (Crampton, 2001) and toward a focus on performance, reflexivity and narrative — in short, toward a relational approach to understanding (Crampton, 2001; Kitchin and Dodge, 2007; Pearce, 2008; Turnbull, 2007). The new moral consciousness that attends this transformation is characterized by an overarching preoccupation with promoting justice, which includes the enhancement of agency and empowerment, especially in those peoples who have historically been subjugated to a colonial authority reinforced through the use of maps (Brealey, 1995; Peluso, 1995; Sparke, 1998, 2005). However, cartography is only one of many areas on inquiry and expression that find themselves in this state of ontological transformation. Just as 'contemporary theories of mapmaking argue that it is a creative activity that focuses on the process instead of the object of maps' (Irwin et al., 2009), similar changes are occurring in the world of art:

> For several decades many artists have been interested in site-specific work and how the creation, installation and reception of an artwork are situated in the contextual conditions of a particular location. Furthermore, as Miwon Kwon (2002) argues, the term 'site' needs to be re/imagined beyond a particular location to 'sites' that are not geographically bound, but rather, are informed by context. This relational understanding is constituted through social, economic, cultural and political processes in what Nicholas Bourriaud (2002, 2004) calls relational aesthetics. For both Kwon and Bourriaud, 'sites' and 'situations' become social engagements that change conventional relationships between artists and their artworks and audiences. Rather than simply receiving and interpreting art, audience members become analysers or interlocutors, even active participants in the artworks. Art is no longer just about visual style but social purpose. Education is no longer just about individual achievement but social understanding and contribution (64).

Irwin et al. (2009) use art and cartography together through a method they refer to as 'a/r/tography' to begin to map, explore, reveal and understand identities and the rhizomatic nature of relationships that comprise the particular city of Richmond, British Columbia.

This use is in line with the positive potential for the relationship between cartography and art outlined by Sébastien Caquard and D.R. Fraser Taylor who provide some 'examples of maps designed by contemporary artists [...] in order to highlight the ways these artists challenge the objectivity of maps' (2005, p. 286).

A primary concern in this transformational period — for cartography at least — is with whether or not cartography is capable of meaningfully conveying such things as experience, indigenous perspectives and knowledge, and critical academic approaches to the status quo (Johnson et al., 2006; Turnbull, 2007). The multimedia, multisensory, multimodal, interactive and interdisciplinary nature of Cybercartography positions it well to be able to address critical concerns such as inclusivity and performativity (Pyne, 2013, 2019; Pyne and Taylor, 2012, 2015; Taylor and Pyne, 2010; Taylor and Lauriault, 2014; Taylor et al., 2019).

A holistic relationship exists between art, narrative and maps such that art can both include narrative and be constituted by it, and maps can be seen both in form and function in terms of art and narrative (Turnbull, 2007). The focus on storytelling and geonarratives in the cybercartographic approach to atlas-making acknowledges this relationship and is an important factor in overcoming incommensurability and enhancing agency:

> [Through] storytellings, authenticities and meanings of landscapes are (re)defined, relations of power are negotiated, and ultimately, structures and processes of neocolonial control are made visible. The stories informing and deriving from mapmaking thus provide frameworks for social and political assessment and agency (Taylor, 2003b); that is, if we view the performances associated with participatory mapping as art, then participatory mapping is implicitly and unavoidably action "intended to change the world rather than encode symbolic propositions about it". *Gell (1998, p. 6); Sletto (2009, p. 444)*

According to Turnbull (2007), the problem of incommensurability, 'of multiple, incompatible ontologies and perspectives' is a 'mapping problem' that is not only 'part of the broader problem of the relationship between the incommensurable knowledge traditions of science and Indigenous knowledge, but that it is also a problem right across the broad spectrum of the ways in which we have to deal with knowledge in this "transmodern" era (Dussel, 1993)' (140). Overcoming incommensurability is part of what reflexivity, or critical awareness, accomplishes.

Conventional cartographic representations often fail to adequately reflect and communicate experience, sense of place and diversity in world views in a nondominatory manner (Pickles, 2004; Pearce, 2008). Moving beyond the concept of traditional representational cartography, the cybercartographic atlas framework allows for the development of novel approaches to the mapping process with the potential to create geospatial modes of expression that are capable of presenting traditional and contemporary indigenous knowledge and understandings and critical post colonial perspectives in a nondominatory manner. Cybercartography combines art with science, allows for the exploration of multiple perspectives through geonarrative and has provided a vehicle for artists such as distinguished artist Jeff Thomas to develop innovative 'a/r/tographic' ways of unsettling conventional representations, thus contributing further to the ontological revisioning program of critical cartography and art (Irwin et al., 2009).

Consistent with a central purpose of this book — to make intersecting links apparent — this chapter follows suit by reporting on some of the intersecting research and experimental mapping that Jeff and I have collaborated on under the Cybercartography umbrella. Working in an intersectional manner is consistent with Jeff's aims as an independent curator and photographer, which reflect a concern with 'his own history and identity, [and with] issues of aboriginality that have arisen at the intersections of Native and non-Native cultures in what is now Ontario and northern New York state' (Jessup, 2019).

The chapter participates in reflexive practice by experimenting with a mosaic narrative approach. In an analogous manner to zooming in from a small-scale to a large-scale perspective on a digital map, the chapter includes this section as the most zoomed out level. Next, it zooms in a bit to introduce Jeff Thomas and his broader interests in maps and mapping. Zooming in one level further, Section 7.3 chronicles some details of Jeff's journey with his artwork and Cybercartography, including his work relating to residential schools and a map overlay I created to geo-transcribe Jeff's previously unpublished essay, 'I Have a Right to be Heard', which is a narrative mapping in itself in the way it retraces the path of the Where Are the Children Exhibition.

In true reflexive fashion, the chapter zooms in one level further to present Jeff's essay, which is a narrative mapping of Jeff's critical reflections as exhibition curator for the Where Are the Children Project. Initiated by the Aboriginal Healing Foundation (http://www.ahf.ca/), this project included an analog component in the form of an exhibition that began travelling across Canada in 2002, and a digital component, which continues to exist and is available at http://wherearethechildren.ca/en/. Jeff's curatorial approach to the archived residential schools-related photographic records involved interpretative strategies intended to 'connect the dots' between (1) often-staged photographic evidence, which tended to reflect little of the students' actual experiences, while telling more about the aims and objectives of the residential school system's administration and (2) the actual memories and experiences of former residential schools students, now termed 'Survivors'. His essay is a reflective tour through his memory of salient stops along his journey to create and curate the Where Are the Children Exhibition and includes some new interpretations of the Exhibition, not previously shared.

Finally, the chapter zooms back out to engage in some discussion about the nature of Jeff's approach as a hybrid implicit—explicit approach to cartography involving art, curatorial arrangement and text.

3.2 Jeff Thomas: Mapping and art/art and mapping

Jeff Thomas has been working for some time at the intersection of art and cartography as an independent curator and photographer concerned with asserting, exploring and expressing his identity as an 'Urban Iroquois' in the context of broader and related issues such as treaty-based relationships and the residential schools legacy. As Jeff recounts, his hybrid approach was inspired by a notable map encounter in his teen years. During a visit with his elder at Six Nations, Ontario, Jeff was shown a book on the Six Nations Haldimand Tract that included two pull-out maps. The first map showed the original boundaries, and the second showed where Jeff's ancestors had settled along the Grand River. Jeff remembered this early encounter and years later, he featured these maps in a show he curated at McMaster University in Hamilton, Ontario, Canada, that revolved around the book and its maps. Jeff's reflection on the nature of his mapping activities related to this exhibition reveals not only his concern with the historical geography of the land but also his readiness to incorporate mapping technologies in his artistic work:

> I even went as far as to find the disputed head of the Grand River. The British said the tract did not begin there, while Brant hired a surveyor to determine where the river began and the tract began. I probably came as close as I could to the head of the river. It was in a cow pasture. I also used my GPS receiver to mark my journey. *Personal communication, 16 February 2018*

Jeff incorporated an image of the Haldimand Tract map once more in a composite photo work entitled The Imposition of Order, which was commissioned for the Art Gallery of Ontario (AGO) exhibition entitled 'Every. Now. Then: Reframing Nationhood' (curated by Andrew Hunter, Fredrik S. Eaton Curator, Canadian Art, AGO, June 2017–February 2018; see Fig. 3.2.1). Jeff describes the exhibition in the following manner on his website:

> The work combines four significant images: Samuel de Champlain's map of New France (1612), the Champlain monument in Ottawa, William Berczy's Portrait of Thayendanegea or Joseph Brant (the Mohawk war chief who led the Haudenosaunee Confederacy to fight with the British during the American Revolution); The Haldimand Tract map defining the territory along the Grand River granted to the Haudenosaunee Confederacy in recognition of their service. The Haldimand Tract defined a twelve-mile wide area along the entire length of the Grand River, equaling about 950,000 acres. Today, only 48,000 acres remains of this original grant with much of the tract now occupied by settlers. Mounted to the front of the St. Lawrence Centre, Imposition of Order faces Berczy Park, named for the artist who produced several compelling portraits of Thayendanegea (Joseph Brant). This portrait is from The Thomson Collection at AGO. *Thomas (2019a).*

In the early stages of work on this chapter, Jeff sent me a working draft entitled 'I Refuse to Be Invisible'. The following excerpts from this draft describe his evolving practice with

The Imposition Of Order

FIGURE 3.2.1   Imposition of Order (2017, pigment print on archival paper). *Courtesy: Jeff Thomas.*

photography, archives and maps, including the way these aspects came together in his work curating the Where Are the Children Exhibition:

> I began working with a camera in the late seventies, and almost from the start my mantra was: I refuse to be invisible. This was in response to what I had not seen in the photo archive: indigenous representatives as image-makers. Also missing were images showing indigenous people like me, growing up and living in a city. It was as though all the Indians had truly vanished. But that was only in the public's mind, not in the mind of indigenous people. My career began with a challenge to initiate a conversation where none had yet taken place. But why did this omission exist in the first place? It begins with the reality of living in a society built for non-indigenous people. There was no reason to accommodate indigenous people in urban centres like the one I was raised in. To have such a conversation would require the blending of existing documentary information – from images to text, and more recently, the use of maps. Over the years I pursued a dual course of action – producing images with my camera that reflected who I am, and researching archival collections in search of 'the Holy Grail': an indigenous photographer from the 19th century. But I had to settle for engaging a 19th century photographer named Edward S. Curtis.
>
> Curtis began photographing indigenous people in his Seattle, Washington studio in the late 19th century. They were often people he met on the streets of Seattle, selling craftwork to white people. But Curtis was not content with only studio work and set out with a grand plan to produce a permanent record of the vanishing indigenous world. By the time Curtis began his project in 1900, the coveted tribal world had been gone for decades. So Curtis turned to the elders who had lived that lifestyle and his team began collecting tribal information while also producing photographs.
>
> Curtis' obsession would eventually lead to the publication of 20 volume series titled "The North American Indian". The series was composed of two parts. The first was the actual books that combined history, ethnography, musicology, and photographs. The second part was a folio of large format photogravure prints, one for each volume. The first photogravure print is titled "Vanishing Race," and shows a group of Navajo people riding in single file. ***Personal communication, 12 June 2018***

Jeff continues in a deconstructive manner by noting how Curtis participated in the social construction of knowledge about 'the North American Indian' through his publications, which contained history, ethnography, musicology and photographs, with a folio of large format photogravure prints in each volume. Jeff's comments on the first photogravure print, 'Vanishing Race' (Curtis, 1904) in Fig. 3.2.2, which shows a group of Navajo people riding in single file, provide further insight:

> Curtis embellished the image for total impact by creating a dark and lonely feel as the procession approached a canyon. The impact was one of moving towards the abyss or unknown land or future. The North American Indian begins here because what Curtis and his field work team produced was a record of what that procession was leaving behind.
>
> What was left behind? In Curtis's own words he talks about the fate of the survivors and their survival. He was not a total pessimist as the last print in the portfolio shows a group of riders emerging from a canyon and into the light of day. I feel my own story emerging here. As many indigenous families were forced to leave their homelands in search of work in the white man's world. My grandparents left the reserve and eventually settled in Buffalo, New York, where I was born and raised.

FIGURE 3.2.2   Digital copy of 'Vanishing Race-Navaho' obtained from the Library of Congress (Curtis, 1904).

I was a first generation urban Iroquois and my early photographic work used a documentary format to record my world, what I saw as an indigenous person. Around this time, Curtis played a pivotal role in the direction I would pursue. My interest in finding indigenous people working with a camera was to really get a sense of how they saw the world, not necessarily how they saw and photographed other Indians. So I set up a binary relationship between me and Edward Curtis; I would play off his Indian sitters and add with my work, what I perceived as "seen". Location was a vital part of my record keeping. Each image I made was documented, including any interesting stories that took place while I was working. I worked for many years, building an archive of new work and research with the hope of one day producing a conversation with Curtis.

Then the role of Curtis took on new currency in 2001 when I was commissioned by The Aboriginal Healing Foundation to curate the first exhibition on residential schools in Canada. When viewing the history of residential schools Curtis's image "Vanishing Race" resonated on a new level, becoming a waypoint that set up the thesis for my exhibition — what happened to the children at these schools? In fact the children were on horseback and leaving all that they had known, behind them as they entered Canada's social engineering experiment, converting Indian children to facsimile white children, but without the same rights as white children. *Personal communication, 12 June 2018*

Jeff takes a reflexive and narrative approach to maps and art that involves making links where they may not otherwise be apparent, a characteristic he discusses in more detail in his essay in Section 3.4.6. In my cartographic work with Jeff, I have found that he is a geographically minded artist who provides good map-related information, including travelogues with dates, coordinates, comments and photographic media. When it comes to disciplinary intersections, 'curating with a focus on place' is a strong intersectional node between Jeff's artistic work and Cybercartography — a situation that reflects a common link between art and cartography more broadly speaking. Further, as will become apparent through the discussion below, Jeff's work involves both implicit and explicit approaches to cartography.

3.3 Jeff Thomas and Cybercartography: A high-speed tour

This section follows the approach of Pierre Bourdieu in his Paris Workshop, which features a 'simple, unpretentious, and candid exposition of the work done' (1992, 218), and a 'high-speed tour' (220) of his work. This is consistent with the precedent established in Chapter 9 of my PhD thesis (Pyne, 2013) wherein I reflected on some of the editing decisions that were made to create the various map modules in the cybercartographic Lake Huron Treaty Atlas and discussed the iterative processes that went into their making, their significance and their relationship to some of the other maps in the Atlas — a method that also followed Bourdieu's reflexive, high-speed tour approach. It is hoped that the process of providing a brief review of Jeff Thomas' forays into Cybercartography — including a variety of our collaborations that involved experimental efforts to give cybercartographic expression to Jeff's work and visions — will contribute to a better understanding of this ongoing work.

Jeff Thomas has been working with the Geomatics and Cartographic Research Centre (GCRC) since 2007 on a variety of projects: from collaborations on the Cybercartographic Atlas of Indigenous Perspectives and Knowledge (Great Lakes Region-St. Lawrence Region; GLSL Atlas; see Chapter 2) to collaborations on this chapter and related work on a new map module that has emerged through the writing of this chapter in the development of the Residential Schools Land Memory Mapping Atlas: The Where Are the Children Exhibition Map Module. These collaborations provide a good example of transdisciplinary research at the intersection of art, including curatorial studies, and cartography that is at the same time reflexive in several respects, especially with respect to not only the interpretative potential of text as epi map (Pyne, 2013; Wood and Fels, 2008), but also its generative potential.

3.3.1 The Cybercartographic Atlas of Indigenous Perspectives and Knowledge (Great Lakes Region-St. Lawrence Region, GLSL Atlas)

The incorporation of art into the Cybercartographic Atlas of Indigenous Perspectives and Knowledge (Great Lakes-St. Lawrence region, GLSL Atlas) provides an example of a cybercartographic approach that conveys a variety of perspectives that fall outside of the conventional range of cartographically representable subjects. This pilot Atlas was developed 'to enhance the capability to recover the systemic nature of traditional Indigenous knowledge by electronically interrelating different forms of expressive culture, including various forms of art' (Caquard et al., 2009, p. 83). Although both the Treaties and Cultures modules of this Atlas incorporate art, the Culture Module is devoted specifically to this theme.

Jeff collaborated on the Culture Module with the GCRC team and Inuk scholar Heather Igloliorte (then, a graduate research assistant and now Associate Professor of Indigenous Art History and Indigenous Art History and Community Engagement Research Chair at Concordia University in Montreal, Quebec). Working with Heather, Jeff selected relevant photographs from a series of early-20th-century photographs related to the cultural heritage of the Six Nations of the Grand River reserve and recorded his interpretations of the photographs to accompany their presentation in the map module: The aim being to create 'links between the past and present by connecting historical photographs to living traditions and communities today' (Caquard et al., 2009, p. 91). In one case, Jeff's voice can be heard

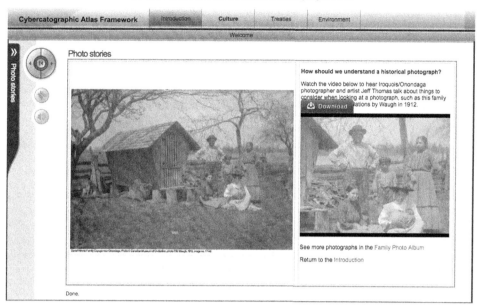

FIGURE 3.3.1.1   Screenshot from the photo stories section of the culture module (GLSL Atlas) showing the video playing in the right hand panel and the photograph being videoed in the larger frame to the left.

interpreting a video with the camera moving around and zooming in and out on a photograph in an analogous fashion to 'reading a map' (see Fig. 3.3.1.1). This work was reported on in Caquard et al. (2009), in which it was noted that improvements to the Nunaliit I software framework were needed to enhance the ability of those without specialized technical knowledge to add information.

Jeff's participation in the development of the GLSL Atlas led him to his next project involving Cybercartography: The pilot Jeff Thomas Atlas, which originally included many instances of geo-tagged photographs from his ever-growing collection, and eventually came to focus on the relationship between 'the Scout' and Samuel de Champlain: The Scout being the main photographic subject in Jeff's ongoing Scouting for Indians project (Thomas, 2019b); with Samuel de Champlain being the statue of said man standing behind the National Art Gallery in Ottawa, Canada, and formerly including a smaller partner statue of an 'Indian Scout' placed at its base. The Indian Scout statue was moved to the nearby Major's Hill Park in 1999.

3.3.2 The Jeff Thomas Atlas: Journey with the Champlain monument

In the section on Cybercartography on Jeff's website (https://jeff-thomas.ca/cybercartography/), he describes his artistic journey re-inscribing the relationship between the Scout and the Samuel de Champlain Monument — a journey that eventually involved Jeff and I collaborating via coffee shop meetings and email communications on experimental work to link Jeff's photographic journeys and curatorial endeavours through cybercartographic mapping:

> My journey with the Samuel de Champlain monument began in the fall of 1992, and came full circle in 2013 when National Gallery of Canada curator Greg Hill invited me to be in the Sakahàn exhibition. I contributed an outdoor installation of several images of my Champlain series Seize the Space and I also renewed the portrait sessions of Seize the Space during the run of Sakahàn by inviting the public to pose on the monument platform where the Indian Scout had once resided.

> The original plan for the monument did not include an Indian figure, but one was added in 1924 and designed to have the Indian figure placed in a canoe. The project however, ran out of money and the canoe was never added. The year 2013 was also marked by my participation as a panelist and artist-in-residence at the Carleton University Champlain colloquium: Champlain on the Anishinabe Aki: Histories and Memories of an Encounter. This coincided with the fulfillment of a long-time dream of working with Carleton's Geomatics and Cartographic Research Centre and incorporating their living atlas component into my website.

> These projects were realized with a grant from Carleton's History Department. For my artist-in-residence project, I brought a large photograph of the Indian Scout to the colloquium on the first day and offered the participants an opportunity to pose with the image. The idea was to produce a conversation about the role of monuments and indigenous history. And, as with my portrait project at the monument, I offered each sitter a print and emailed a jpeg version as well. My presentation at the Colloquium was a conversation between me and the Indian Scout, with Greg Hill reading the part of the Scout. The conversation was based on the phrase Lest We Forget and decades of service the Scout put in before his removal and transfer to Major's Hill Park in 1999. *Thomas (2019b)*

After this description, Jeff includes a link to his pilot atlas (http://atlas.jeff-thomas.ca/index.html), which was initially created with the help of Leah Snyder, whose Master's studies focus on the construction of national narratives in a digital era. I later extended this work in conjunction with Jeff in an effort to cybercartographically map his ongoing photographic journey of the relationship between the Scout and Samuel de Champlain.

In August 2016, Jeff and I began work to geographically structure — or geo-transcribe (Caquard et al., 2009; Pyne, 2013) — a series of photographs to correspond with an essay entitled 'Lest We Forget', which he had written for his presentation at the Carleton University Colloquium referred to above. We met on several occasions to discuss the mapping project, and Jeff followed up with the relevant media, text, geographic locations and other relevant resources, including instructive comments such as the following concerning his vision for the map:

> What I have in mind is presenting the text as part of the map site and then making links from the text illustrations to the map. I have three sets of images; first are the images identified in the text; a sample of the portraits I made at the colloquium, which have some text written by the sitters. Each sitter is holding an image of the Scout at Major's Hill Park, and the third set are the portraits I made at Major's Hill Park. The text may be too long for this map application, but I'm really not sure. I have started compiling the images that will be used. The images used in the text are from my Scouting for Indians series and will consist of sites I have photographed around Ottawa and Gatineau. *Personal communication, 29 August 2016*

Following up on Jeff's instructions, I created a point corresponding to the location of the symposium where the colloquium portraits were taken (in the Dunton Tower of Carleton

University), and attached the photographs associated with Jeff's Colloquium portrait project to the point (Fig. 3.3.2.1).

At the point for the Champlain Monument, I added media associated with portraits Jeff had taken over a number of years, this time with people posing at the base of the Champlain statue in the former position of the Scout Statue. Included in the portrait volunteers was Jeff's son, Bear, who posed by the monument before the Scout was removed (Fig. 3.3.2.2).

In the process of geo-transcribing Jeff's essay, I used coordinates provided by Jeff and researched coordinates for photographs of statues and other relevant Scout sitings, which were referred to in his essay, in addition to 'mapping' the essay itself. In an attempt to provide a curated experience of navigation, I used hyperlinks to guide map users through the mapped version of Jeff's essay, and — in consultation with Jeff — titled the map module 'Journey With the Champlain Monument'. Several issues surfaced, which remain to be addressed in a future project. For example, although it is possible to navigate the map of the essay entirely in the right hand panel of the map module, coordination with the actual 'map' in the central display remains strained, with the only option to find the point on being considered in the right side panel being to zoom and pan and/or by clicking on 'find on map', a right side panel option. This map module was created several years ago, and the GCRC now has the capacity to address design issues such as these. A future project could pick up where this pilot project left off and continue to build in an iterative fashion on the progress made to date.

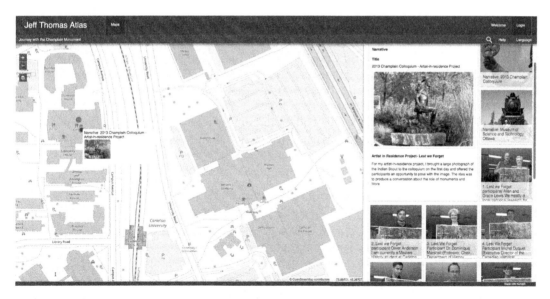

FIGURE 3.3.2.1 Screenshot from the Jeff Thomas Atlas showing a sample of the portraits from the Carleton University Champlain colloquium in the right hand panel and the location of the portrait sittings near the white text box in the central map panel.

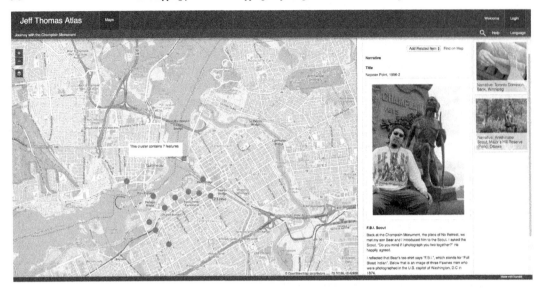

FIGURE 3.3.2.2 Screenshot from the Jeff Thomas Atlas showing Jeff's son Bear with the Champlain monument and the Scout in the right hand panel and a map display, and various 'Indian Scout' locations as sited by Jeff in the larger frame to the left. The photograph in the right side panel is indicated by the red-outlined square at the top of the map. In addition, a thumbnail of the Scout statue in its new location is also visible in the side panel.

Concurrent with the work to create the Journey with the Champlain Monument Map, Jeff and I discussed the Residential Schools Land Memory Project, including ideas for Jeff's participation as a project collaborator. Consistent with the reflexive approach of the project, a Travels Map module was created in the developing Residential Schools Land Memory Atlas with overlays for many of the project collaborators (see Chapter 2). This map would enable us to geo-document their work and travels over time; a central aim of this map module being to identify intersections between various collaborators, especially with respect to their work relating to residential schools. Jeff's CV provided an excellent resource to inform the development of the map overlay corresponding to his life-work path.

3.3.3 Travels Map of Residential Schools Land Memory Atlas

The Jeff Thomas overlay is currently the most developed overlay in the Travels Map module, which tracks the work paths of project collaborators as part of a geonarrative approach to identifying collaborator intersections, for example, attending the same residential schools event. There is still much work to do to complete this module, and it is possible that it will remain unpublished and experimental. The Jeff Thomas overlay maps Jeff's solo exhibitions and some of his group exhibition over time and includes relevant media and links. My work to create this overlay involved researching the exhibitions, and including links and more precise date information than was provided in his CV. In addition, MA research assistant, Kevin Palendat (currently an archivist with Library and Archives Canada; see Chapter

FIGURE 3.3.3.1 Screenshot from the Travels Map showing various exhibition locations, and profiling the location for 'Imposition of Order', referred to in Section 3.1, on the front of the St. Lawrence Centre (Toronto, Ontario, Canada).

8), mapped Jeff's group exhibitions, following my method. This provided Kevin with the opportunity to learn about Cybercartography (and Jeff's work) through practice. Although this map overlay is by no means complete, even in its initial sketched-in state, it is useful to get a geo-visual impression of a portion of Jeff's CV (Fig. 3.3.3.1).

Work on the Jeff Thomas overlay in the Travels Map was related to another cybercartographic map-making project: The Jeff Thomas 2015 Road Trip overlay in the Bagone-Giizhig Module of the cybercartographic Lake Huron Treaty Atlas (referred to in Chapter 2), featuring Jeff's visit to the Cahokia Mounds, Woodhenge, and other mound sites in St. Louis, in addition to a variety of other sites relating to sacred Native American material artefacts.

3.3.4 The Jeff Thomas 2015 Road Trip overlay

As referred to above, the Jeff Thomas 2015 Road Trip overlay lives in the Bagone-giizhig (Hole in the Day) Map module of the Lake Huron Treaty Atlas (Pyne, 2013). The Bagone-giizhig Map is a pilot module that was originally created to experiment with geo-transcribing *The Assassination of Hole in the Day* by Treuer (2011); it begins to interactively map some of the events in the story of Chief Hole in the Day's life, including his life in Minnesota, and extending to his travels throughout the United States. Drawing on the works of modern scholars, early writers and the stories of Anishinaabe Native American elders, Treuer's book traces the life and relations of Bagone-giizhig, and his father Bagone-giizhig the Elder from the 1820s through to the late 1860s. Interactively mapping out the story enables people to track Bagone-giizhig's travels over the water and across the land.

In addition to being important in itself, learning about Bagone-giizhig's life and his times provides insights into the 19th-century relationships between Indigenous people on both sides of the America—Canadian border, in addition to a variety of other relationships relevant to the pre- and the emerging history of residential and boarding schools in North America. Although much further work remains to be done in order to present a fuller geo-transcription of this book, more work was done in 2016 when the Beltrami Journey to America overlay was added to this Map. Giacomo Costantino Beltrami was an Italian man who travelled through Ojibwe (Chippewa) and Dakota (Sioux) Territories in 1823, around the time of Bagone-giizhig Jr. and Sr. Not only is Beltrami's journey to America spatially and temporally related to the Bagone-giizhig's life journey, but Beltrami is also noteworthy as an atypical European character for his times (see Chapter 11 for more on Beltrami and his travels through North America).

In 2017, the Jeff Thomas 2015 Road Trip overlay was added to this Map Module, since he traversed some of the same ground as the historical figures already sketched in on the map. I was personally interested in this case in transposing Jeff's contemporary 'Urban Iroquois' perspective, including his photographic interest in ancient history, onto the historical perspectives I had been tracking. The Road Trip overlay maps Jeff's travels southwest from his home in Ottawa, Canada to St. Louis, Missouri — a place that Beltrami had also travelled through almost 200 years prior. This overlay includes photographs taken by Jeff of the Cahokia Mounds and other mounds in St. Louis, in addition to photographs of monuments and material artefacts, which reflect intersections between age-old indigenous history and contemporary times, and form the basis for Jeff's ongoing work (Fig. 3.3.4.1).

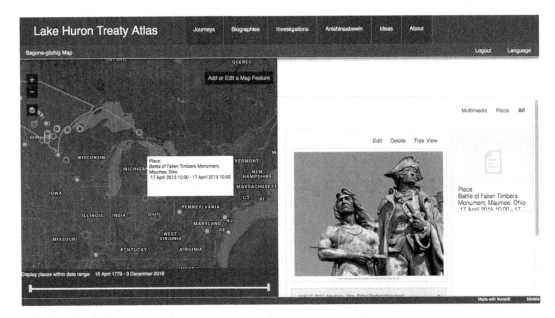

FIGURE 3.3.4.1   Screenshot showing three layers of the Bagone-giizhig Map Module: the Jeff Thomas 2015 Road Trip overlay in dark purple dots, the Bagone-giizhig overlay in blue dots and the Beltrami Journey overlay in light purple dots, with the point for Jeff's stop at the Battle of Fallen Timbers Monument, Maumee, Ohio, selected (in the central map display). While the side panel displays a photograph taken by Jeff of the monument.

Reflecting the cybercartographic principal of using place as an organizing factor (Pyne, 2013, 2019; Taylor, 1997, 2003a,b, 2005; Taylor and Pyne, 2010), this map module brings together journey maps of three individuals who travelled across the same territories at similar and different times. In addition to bringing Jeff Thomas into 'cartographic conversation' with both Beltrami and Bagone-giizhig (and all their relations), this emergent mapping practice is also a good example of the process of re-inscribing maps with previously excluded information (Pyne, 2013) and is consistent with Jeff's ongoing aim of initiating conversations where none had previously taken place referred to in Section 3.2. The benefits of combining Jeff's road trip with the mapped journeys of two compelling historical figures have allowed Jeff's perspective to be included in my application of a 'spatializing history' (Pyne, 2013; Pyne and Taylor, 2015) approach to teaching. This includes a Master's course in intercultural geography for an International Tourism program at the University of Bergamo where Beltrami was born and where a collection of artefacts he acquired in North America still lives (see Chapter 11 for more on the Beltrami overlay). The points for the various layers can be selected by following the logic of the timeline function, or by following spatial logic. This timeline functionality was developed early in the development of the Lake Huron Treaty Atlas as a result of transdisciplinary discussions between myself and Glenn Brauen (Brauen et al., 2011; Pyne, 2013).

3.3.5 Mapping the family camera and visitor information exhibits: Geo-locating composite photograph components

During our meetings in the spring of 2017, Jeff asked me to do some cybercartographic mapping associated with his contributions to the Family Camera Exhibit at the Royal Ontario Museum (Toronto, Ontario; 5 May to 17 November 2017) and the Visitor Information Exhibit at the Robert McLaughlin Gallery (Oshawa, Ontario; 5 May 5 to 19 September 2017). Both of Jeff's exhibit contributions featured composite portraits with three images — or triptychs — which had the potential to be geo-located, in addition to mapping the exhibition venues. After a meeting to discuss the map project, Jeff sent me the relevant media, location information and relevant comments. I proceeded to process the digital composites by breaking them apart into individual image files, which I, the geo-tagged, included in the appropriate map overlay, and linked to the point for the relevant exhibit location with the composite image. The process I was engaging in can be described as an attempt to 'geo-provenance' Jeff's composite images both on the map space in the central panel and in the associated information and media in the right panel. In contrast with conventional notions of provenance, geo-provenance involves privileging place over other descriptive categories and is consistent with Cybercartography's emphasis on place as an organizing factor for information (Pyne, 2013, 2019; Taylor, 2003a,b; 2005; Taylor and Pyne, 2010) (Fig. 3.3.5.1).

As I have done with many other emergent mapping exercises engaged in since the official launch of the Lake Huron Treaty Atlas in 2012, I created a new overlay for each exhibition in an existing atlas module: The Community Stories Map Module. Although the Lake Huron Treaty Atlas is no longer a funded project, it remains active and updatable and has served as an excellent space for experimental and pilot work. Points for these exhibition contributions will also be added to Jeff's overlay in the Travels Map of the Residential Schools Residential Schools Land Memory Atlas referred to in Section 3.3.3 (Fig. 3.3.5.2).

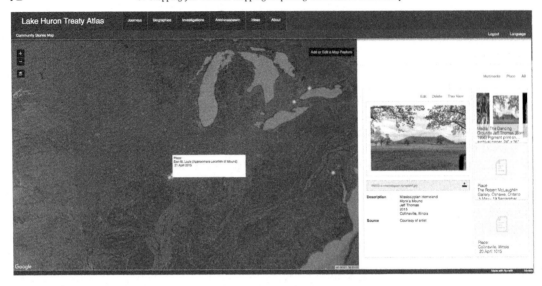

FIGURE 3.3.5.1   Screenshot of the Jeff Thomas Visitors Information Exhibit overlay in the Community Stories Map Module of the Lake Huron Treaty Atlas showing the image of Monk's Mound and an image of its source triptych, in the right side panel; with the location of Monk's Mound profiled in the central map window.

FIGURE 3.3.5.2   Screenshot of the Jeff Thomas Family Camera Exhibit overlay in the Community Stories Map Module of the Lake Huron Treaty Atlas showing a portrait of Jeff's son, Bear, with a thumbnail of the composite image the portrait is contained in, in the right side panel; with the location of the portrait highlighted in the central map window among other locations related to the images in the exhibit.

3.3.6 Residential schools volunteered geographic information: Battleford Indian Residential School

Throughout our meetings since the proposal stage of the Residential Schools Land Memory Mapping Project, Jeff and I discussed possible links between his work and the project, including ideas for his participation. In the spring of 2017, Jeff mentioned a photographic road trip that included a visit to the Battleford, Saskatchewan Indian Residential School site, and later shared three images that he had taken of the site (together with coordinates, which enabled me to fine-tune the school's location). I proceeded to add to the media to a relevant point in the pilot Residential Schools Map, which was first made available online as a pilot map module in the Lake Huron Treaty Atlas and is currently in further development under the Residential Schools Land Memory Mapping Project. Although this map intervention was not as involved as some of Jeff's other collaborations in Cybercartography, it is nevertheless valuable as an example of the ease and spontaneity with which Jeff and I collaborate in a cybercartographic context. Jeff's contribution enriches the information in the Residential Schools Map and documents another example of his participation in the ongoing Residential Schools reconciliation project. In our coffee shop conversations, Jeff reflected on his thoughts and feelings while at the site. In the future, we may make use of the audio and/or video recording functions available with the current Nunaliit software technology that underpins cybercartographic atlases (Fig. 3.3.6.1).

FIGURE 3.3.6.1 Screenshot from the Residential Schools Map showing photographic contributions (in colour) by Jeff Thomas alongside other contributions reflecting Internet research in the right side panel, with a blue dot indicating the location of the Battleford Indian Residential School in the central map window (Fig. 3.3.6.2).

FIGURE 3.3.6.2 Screenshot from the Residential Schools Map profiling one of Jeff's photographic contributions with a view of the partial remains of the now demolished Battleford Indian Residential School.

3.3.7 Transposing the Jeff Thomas Cahokia Mounds Journey overlay onto the Residential Schools Map

In the winter of 2015, I created the Concept Journeys overlay in the Travels Map module of the Lake Huron Treaty Atlas (LHTA) to accompany a paper I was writing on Cybercartography and narrative, which included discussion of the overlay. This was another experiment of the utility of cybercartographic mapping in the writing process, which I had initiated during my PhD thesis-writing process. Following suit in the writing of this chapter, I experimented with this strategy once more. Yet again, I was pleasantly surprised with the renewed energy this process yielded, which was combined with an interesting observation made possible only through the mapping process. I began by manually transposing the data from Jeff's 2015 Road Trip overlay (referred to in Section 3.3.4) onto the 'Communities near Residential Schools' overlay in the Residential Schools Map. My basic rationale was to use the cybercartographic mapping process, including a focus on place as an organizing factor for information, to explore the proximity between the points in Jeff's 2015 Road Trip and the American boarding schools identified on the Residential Schools Map.

Mapping the points of Jeff's journey through the centre of the United States to visit historical monuments and photograph ancient artefacts gave rise to an interesting insight when transposed onto the Residential Schools Map: Jeff's path cleanly transected American Indian Boarding Schools Territory. Making its way southwest through Ohio, Indiana, and settling in Illinois, his road trip went through some of the only territory that did not include any boarding schools. The process of 'mapping' the road trip onto the Residential Schools Map gave rise to new insights, helped to focus attention on the territories not having boarding schools and invoked new questions concerning the territoriality of the boarding school legacy in the

United States. In this regard, it is notable that Jeff's path concurrently tracked a portion of Indian removal territory, which is paradoxically also territory with a significant amount of indigenous material culture (as evidenced by the photographs included in the Jeff Thomas Road Trip overlay referred to in Section 3.3.4). Engaging in the mapping exercise of transposing Jeff's Road Trip overlay onto the Residential Schools Map proved to be an interesting way of bringing geographical techniques to historical questions concerning the broader historical geographical context of Residential and Boarding Schools in North America. Moreover, it provided yet another example of the reflexive relationship between text and map (this time in a creation context), in which text can be considered as an extension of 'the map' Pyne (2013); Wood and Fels (2008) and vice versa (Fig. 3.3.7.1).

In an email to Jeff, in which I shared screenshots that illustrated my observations, I suggested this mapping could somehow be a geospatial manifestation of his artistic approach: meaning that it might help him to better understand his own motivations for taking the road trip in the first place. Jeff responded that he liked where I was going with this interpretation, included some cartographic details related to his 2015 Road Trip and reflected on the relationship between that journey and residential schools:

> I like where [you're] going [with this]. I also made a number of new images during my last trip to St. Louis in March. There is only one outdoor mound site visit in Newark, Ohio. But I photographed pottery faces made by Mississippian artists, in several archives. I have to add the GPS coordinates for the museums.

> I have been thinking about the role of my St Louis work in relationship to my Residential Schools work. You are right, I didn't pass through any former Residential Schools sites in the Ohio or St. Louis area. Those areas had been pretty well gutted of Indigenous people by the 19th century. *Jeff Thomas, email communication, 8 May 2018*

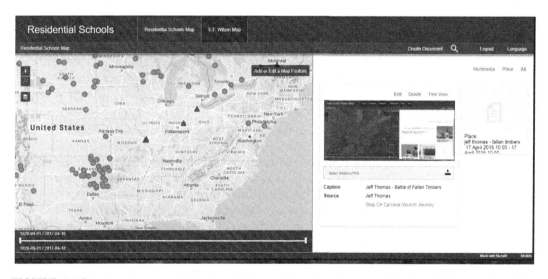

FIGURE 3.3.7.1 Screenshot showing American boarding schools (purple dots), the Jeff Thomas Road Trip overlay (blue triangles) in the central map window, with the point for the Fallen Timbers chosen and associated media from Jeff's journey displayed in the right side panel.

The process of reflecting on my 'transpositional mapping' of Jeff's 2015 Road Trip onto the Residential Schools Map included further ruminations by Jeff concerning his writing on previous work with Edward Curtis and his iconic image 'Vanishing Race':

> What it reminds me of is the divide, when traditional elements of indigenous tribal culture were quickly dying out with the passing of each knowledge keeper. And the children were being transported to Indian Schools. The image shows a line of Navajo people on horseback and they are riding single file, towards a very bleak and ominous backdrop (canyon). I refer to it as a divide because it meant that new skills would have to be developed, and a new landscape would have to be mapped and studied. I feel that is where we are, now.
> **Jeff Thomas, email communication, 8 May 2018**

Although transposing Jeff's Cahokia Mounds Journey onto the Residential Schools Map provided an interesting forum to begin discussions, it was not until Jeff provided me with his unpublished essay 'I Have a Right to be Heard' that another geonarrative possibility presented itself. Jeff's essay included images that could be mapped and inspired me to attempt a second experimental cybercartographic map module overlay in relation to the writing of this chapter: the Jeff Thomas Mapping Thomas Moore overlay in the Nenboozhoo Mindemoye Map of the Lake Huron Treaty Atlas (Pyne, 2013). I chose to create the overlay in this map for a number of reasons, including the fact that it employs an ordinal timeline, rather than being driven by particular dates; in addition to the fact that I have the knowledge of how to create new overlays in the LHTA, which has become a useful space for experimental and emergent mapping since its launch in 2012.

3.3.8 Jeff Thomas Mapping Thomas Moore overlay in the Nenboozhoo Mindemoye Map of the Lake Huron Treaty Atlas

Following the method I developed during the writing of my PhD thesis, where I incorporated mapping into the thesis-writing process (also referred to above), I created a map layer for Jeff's essay, 'I Have a Right to be Heard', which involved geo-locating the images (and associated text) in his essay and making decisions regarding how to locate the images.

'I Have a Right to be Heard' is a critical chronicle of Jeff's experiences and processes curating the Where Are the Children Exhibit with the Legacy of Hope Foundation (http://wherearethechildren.ca/), which he begins with an introduction to Thomas Moore, a student of the Regina Industrial Home in the late 19th century. The emergence of the idea to map Jeff's essay is a good example of the importance of iterative processes in cybercartographic atlas development, which involve both theory and practice (Taylor, 2003a,b; 2005; Taylor and Pyne, 2010) and can be understood in terms of David Turnbull's 'talk, templates and tradition' framework (Pyne, 2013, 2014, 2019; Turnbull, 2000). This became increasingly clear to me as I progressed through the process of geo-transcribing Jeff's chapter and found that mapping strategies I had developed in previous collaborations with Jeff (such as those described in Sections 3.3.2 to 3.3.7) were also useful in this mapping context. In addition, I became keenly aware of the reflexive holistic interplay between written text and map, as I continued to make progress through the writing of this chapter.

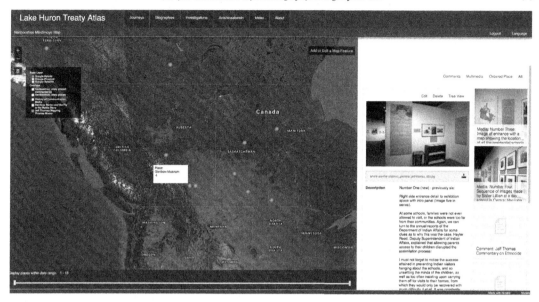

FIGURE 3.3.8.1 Screenshot showing the geographical extent of the geonarrative overlay created in the Nenboozhoo Mindemoye Map of the Lake Huron Treaty Atlas to correspond with Jeff's essay in the central map window, with media from the Jeff Thomas website for several exhibit panels in the Where are the Children Exhibition at the Regina Art Gallery in the right side panel.

With respect to intersections between Jeff's artistic work and the Residential Schools Land Memory Mapping Project, I began to think about how Jeff's work curating the Where Are the Children Exhibit and his reflections on this process were linked with the making of the map overlay and about the potential for a cybercartographic map of the exhibit itself (Fig. 3.3.8.1).

The map overlay I created to geo-transcribe Jeff's essay provided a way to interact with his story in a manner that involved more explicit knowledge of the locations referred to, in addition to being freed from the linear form of textual presentation, and opening up possibilities for the addition of new information such as links and screenshots from the Where Are the Children online exhibition (http://wherearethechildren.ca/). With the intention of including the text from Jeff's essay in this chapter, as I was creating the overlay to correspond with the essay, I made screenshots of each point on the 'Jeff Thomas Mapping Thomas Moore' overlay to be inserted into his essay at the relevant locations. However, as I was engaging in this process, I noticed an issue linked to the inability to map the same location at two separate (ordered) times. At the same time, once this initial map work was completed, I immediately became inspired to suggest to Jeff, Fraser Taylor (the Residential Schools Land Memory Mapping Project's principal investigator) and the GCRC technical team, that we create a map module modelled on my pilot work on the Jeff Thomas Mapping Thomas Moore overlay in the LHTA in our developing Residential Schools Land Memory Atlas. In this way, the chapter writing process became inextricably linked not only to experimental mapping in the LHTA, but also to ongoing design and development research in the RSLMMP.

3.3.9 The 'I Have a Right to be Heard' overlay in the Where Are the Children Exhibition Map Module

The mapping processes and decisions made in creating the Jeff Thomas Mapping Thomas Moore overlay in the Nenboozhoo Mindemoye Map of the Lake Huron Treaty Atlas involved ongoing consultation with Jeff and Fraser via email, and drawing on material from the Where Are the Children website; which led to the decision to create the Where Are the Children Exhibition Map Module and to remap the essay based on my first map efforts (Fig. 3.3.9.1).

The new map narrative comprised a slightly different cybercartographic interpretation of Jeff's essay chapter that aligned better with its chronology and included improved styling with a homepage in which to include the introduction, background and navigational instructions. At the same time, a broader vision for the new map module was developing that included reaching out to the Legacy of Hope Foundation (the host of the Where Are the Children project) to inform them of the chapter and the map work, and to welcome them to join us in ongoing future collaborations on the Where Are the Children Exhibition Map Module; which was beginning with initial work on the 'I Have a Right to be Heard' overlay, but which also had the potential to include a mapped version of the online Where Are the Children exhibition. This vision has begun to be realized, and we look forward to a productive collaboration in the near future.

In the meantime, we zoom in with the next section as Jeff shares his knowledge, experience and insights in relation to his role as curator of the Where Are the Children Exhibition in conjunction with screenshots from the I Have a Right to be Heard overlay to illustrate and provide geographical context.

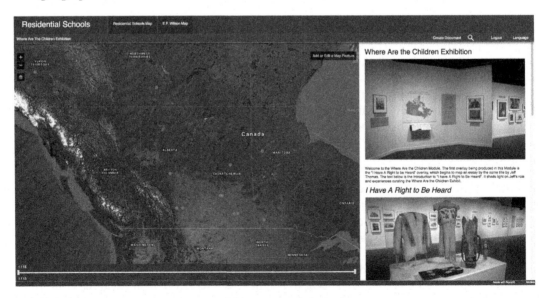

FIGURE 3.3.9.1   Screenshot showing the homepage of Where Are the Children Module of the emerging Residential Schools Land Memory Atlas, with the geographical extent of the geonarrative overlay created in the to correspond with Jeff's essay in the central map window, and styled media from the Jeff Thomas Website for several exhibit panels in the Where are the Children Exhibition at the Regina Art Gallery, and text corresponding to the essay 'I Have a Right to be Heard' in the right side panel.

3.4 Zooming in on Jeff Thomas' narrative mapping of the Where Are the Children Exhibit: 'I Have a Right to be Heard' (by Jeff Thomas)

I see it all about mapping and making the connection between past and present, beginning with the map we use in the exhibition. After all, the title is Where are the Children. So, in my mind, I see the journey as curator to artist (now) and tying into mapping in my personal work. *Personal communication, 25 February 2018*

3.4.1 Personal Bill of Rights (Fig. 3.4.1.1)

I have the right to freedom of speech.
I have the right to be heard.
I have the right to be respected.
I have the right to accept and own my own power.
I have the right to not disclose unless I am comfortable.
I have the right to feel my emotions.
I have the right to say no.
I have the right to challenge the status quo.
I have the right to ask questions.
I have the right to be me.

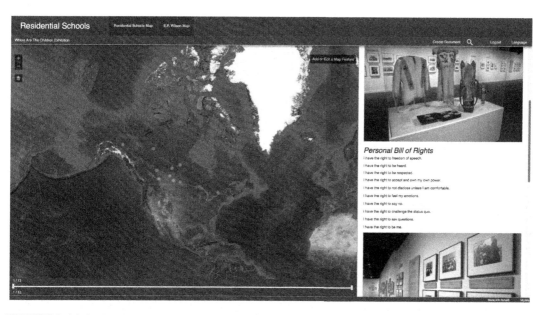

FIGURE 3.4.1.1 Screenshot showing the homepage of Where Are the Children Module of the emerging Residential Schools Land Memory Atlas, with the geographical extent of the geonarrative overlay created in the to correspond with Jeff's essay in the central map window, and styled media from the Jeff Thomas Website for several exhibit panels in the Where are the Children Exhibition at the Regina Art Gallery, and the Personal Bill of Rights text in the right side panel.

3.4.2 Meeting Thomas Moore

A young Indian boy stares out at me from the photograph. He is dressed in tribal clothing — leggings with beaded strips, a shirt decorated with trade items like metal tacks, bead necklaces, and a beadwork floral design on his breechcloth. He wears moccasins and his long hair is wrapped in fur. He holds a pistol in his right hand. He leans his left arm on what looks to be a table covered in a buffalo robe (Fig. 3.4.2.1).

Another photograph shows the same boy but appearing to be a bit older. The tribal clothing has been replaced with a military uniform, and his hair is cut short. He no longer holds a pistol. He stands in front of an ornate railing, probably a studio prop, which he leans his right arm on. His cap rests on the railing to his left. The body positioning is consistent in both images, as if he had been directed how to pose in both photographs. But what was most interesting to me was that the expression on his face had not changed: In both cases, it was one of a self-assured young boy (Fig. 3.4.2.2).

Who was this boy and what was his picture doing in the Department of Indian Affairs Annual Report for the year 1896? Was he being used as the poster child for the Indian residential school system? I found out that his name was Thomas Moore and that he had been a student at the Regina Indian Industrial School in Regina, Saskatchewan. A photograph of his school was included in the annual report. Perhaps Moore is one of the boys standing in front of the school. I try to blow it up and zoom in, but the photo is too grainy to reveal any details (Fig. 3.4.2.3).

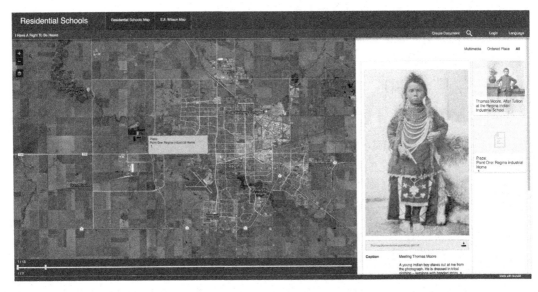

FIGURE 3.4.2.1   Screenshot showing the first point in the 'I Have a Right to be Heard overlay' of the Where Are the Children Exhibition Map Module in the central map window, with 'Thomas Moore, as he appeared when admitted to the Regina Indian Industrial School' (Department of Indian Affairs, 1896) and related media shown in the right side panel.

FIGURE 3.4.2.2   Screenshot of a darkened map display in the Where Are the Children Exhibition Map Module profiling an archival image of Thomas Moore after tuition at the Regina Indian Industrial School (Department of Indian Affairs, 1896).

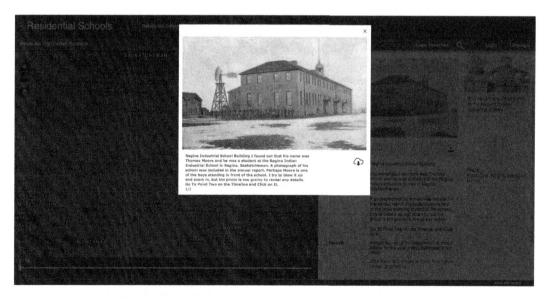

FIGURE 3.4.2.3   Screenshot of a darkened map display in the Where Are the Children Exhibition Map Module profiling an archival image of the Regina Industrial School Building (Department of Indian Affairs, 1896).

3.4.3 Silenced witnesses

When the Aboriginal Healing Foundation contacted me to curate a photo-based exhibition on residential schools in 2001, I knew that my challenge would be to take images that were originally intended to promote the so-called 'good' work being carried on by the schools and to re-purpose those images. My three decades of working with historical photographs of Indigenous people by white society has led me to conclude that we, as Indigenous people, must re-purpose these photographs to address indigenous self-determination. My challenge as a curator for the residential school exhibition was to make the photographs that did exist socially relevant and to use them to create an environment in which survivor stories could be told, the questions that children and grandchildren of survivors have could be answered, intergenerational impacts could be acknowledged, and, most importantly, resolution and healing could begin (Fig. 3.4.3.1).

During my research phase, I visited a psychologist who was working with survivors to ask for advice. She showed me a copy of the Personal Bill of Rights she gives to each client. I was profoundly moved by its simplicity. Like the photographs I was repurposing for the exhibition, it clearly visualized the substance of the pain. When each child passed through the school doors, they were systematically stripped of each of those rights. The second right, 'I Have the Right to be Heard', resonated with me. As I combed through church and other archives for thousands of photographs filled with unnamed children sitting in classrooms or lined up outside drab and cold school buildings, I noticed one common trait: a profound sense of silence. My role as curator was to weave a new story from a desperate and myopic

FIGURE 3.4.3.1 Screenshot showing the homepage of Where Are the Children Module of the emerging Residential Schools Land Memory Atlas, with the geographical extent of the geonarrative overlay created in the to correspond with Jeff's essay in the central map window, and styled media from the Jeff Thomas website for one wall of the Where Are the Children Exhibition at the Regina Art Gallery and the Silenced Witness text in the right side panel.

archive for a contemporary audience wanting to make sense of a system of education that was implemented to render tribal society invisible.

Why were children like Moore sent to residential schools? A clue may be found in the government annual reports themselves, which open a window onto the 19th century rhetoric that the government and churches used to justify their social engineering project to 'civilize' the Indian race. In 1883, as industrial schools were being established, Edgar Dewdney (Indian Commissioner for the Department of Indian Affairs) wrote about the desire to bring indigenous children into the 'circle of civilization':

> Experience has taught that little can be done which will have a permanent effect with the adult Indian, consequently, to create a lasting impression and elevate him above his brethren, we must take charge of the youth and keep him constantly within the circle of civilization. I am confident that the Industrial School now about to be established will be a principal feature in the civilization of the Indian mind.

> By the children being separated from their parents and properly and regularly instructed not only in the rudiments of the English language, but also in trades and agriculture, so that what is taught may not be readily forgotten, I can but assure myself that a great end will be attained for the permanent and lasting benefit of the Indian. *Dewdney (1883, p. 104)*

3.4.4 Listening to survivors

The before and after photographs of Thomas Moore open the exhibition *Where Are the Children? Healing the Legacy of the Residential Schools*. Blown up into mural size prints and flanking the entrance, they invite the visitor to enter Moore's world, and to imagine themselves as a child walking into those schools. Shirley (Pheasant) Williams was such a child. She was 10 years old when she entered the St. Joseph's Residential School at Spanish River, Ontario, in 1949, having had a traditional Ojibway education in early life and speaking only her native language. She tells us her story:

> When I saw the building [St. Joseph's] it was grey — a brick building that, when it rains, is dark and grey. It was an ugly day and the feeling was of a kind of ugliness. Then they opened the gate and the bus went in. I think when the gate closed something happened to me, something locked. It was like my heart locked, because it could hear that clink of the gates […] when we got there the bus stopped and then the sister or the nun — she wasn't dressed in a habit, they didn't dress in habits, they just wore regular clothing — came in. She sounded very very cross, and I could just imagine what she was trying to say, because my sisters had told me what she would probably say. We had to get off the bus and go two by two, through this little door and up the stairs […] it was four stories and no elevator and we had to walk up the stairs with our suitcases […] when we got up on top of the stairs there were tables in there and there were girls in there and so when you went in you were asked your name, and they looked it up. This was another thing my mother prepared me for […] so I was very proud to say yes, my name was Shirley Pheasant. Then they gave you a number and you went and got another bundle with something they called a chemise, and bloomers and stockings. Then you went to the next person […] The last person you saw was the nun who looked into your hair for bugs. *Personal communication*

Children like Shirley Williams and Thomas Moore did not enter these schools uneducated. In our discussion, Shirley described her education:

> I received the traditional education at home before going to the residential school. My parents taught me what was needed and how things were done in the traditional ways of life. We learned about medicines and when to pick them and what they were used for. In this way I was learning about science. about the biology of plants and trees—that was the environmental studies about trees. By picking and cutting birch bark we learned how to only take the bark off the tree without harming the tree and how to dry and look after the bark when it was stored. I learned about the science of the weather. I learned about seasons and how these seasons helped human beings to survive according to the season. *Personal communication*

They were not uneducated; rather, they were re-educated to fit a European model. This program of social engineering by the Canadian government can be, I argue, characterized as ethnocide, defined in the *Dictionary of Anthropology* in the following manner:

> Ethnocide refers to the deliberate attempt to eradicate the culture or way of life of a people [...] Ethnocide depends on the use of political power to force relatively powerless people to give up their culture and is therefore characteristic of colonial or other situations where coercion can be applied [...] The term is sometimes used to refer to any process or policy that results in the disappearance of a people's culture. *Barfield (1997, p. 156)*

Now imagine yourself as a parent and having to leave your child at those school gates, not knowing if you would ever see him or her again. One of the tools the government used to separate children from their cultures was to sever their ties with their parents and communities (Fig. 3.4.4.1).

Jim Abikoki was likely one of those parents. In this photograph, he is seated on a chair in front of the gate leading to the Anglican mission compound on the Blackfoot reserve in Alberta. He is dressed in a three-piece suit, with leather boots. His hair is long and braided. He looks directly at the camera. Seated next to him is a woman, probably his wife, dressed in tribal-style clothing: a cloth dress decorated with ribbon and a blanket draped over her shoulders. Her gaze is not as direct as his; she appears to be reluctant to look at the photographer. Standing behind the couple are two young girls, wearing the type of clothing that children would have worn at residential schools: long cloth dresses with straw hats on their heads. What is happening in this photograph? The gate is closed. Have Jim and his wife come to bring their children to the school? Have they come to visit their children? Have they not been allowed into the compound, with their daughters forced to visit them from the other side of the gate? Were the two very young girls seen peering through the fence waiting for their friends to come back in? Were they waiting for their own families to visit?

The gate is a powerful symbol of colonial control; this indigenous family did not control the gate or who was allowed in or out. It resonates with Shirley Williams' story about hearing the 'clink' of the gate as her bus entered the grounds of the residential school and how her 'heart locked' upon hearing that sound. At some schools, families were not even allowed to visit, or the schools were too far from their communities. Again, we can turn to the annual reports of the Department of Indian Affairs for some clues as to why this was the case. Hayter

FIGURE 3.4.4.1 Screenshot of a darkened map display in the Where Are the Children Exhibition Map Module profiling an archival image of Jim Abikoki and family in front of the fence surrounding the Anglican Mission on the Blackfoot Reserve, Alberta, ca. 1900, Glenbow Archives, NC-5-8.

Reed, Deputy Superintendent of Indian Affairs, explained that allowing parents access to their children disrupted the assimilation process:

> I must not forget to notice the success attained in preventing Indian visitors hanging about the schools, and so unsettling the minds of the children, as well as too often insisting upon carrying them off for visits to their homes, from which they would only be recovered with much difficulty if at all. It was constantly represented to me by those in charge of the institutions that to prevent such visits and to refuse to let parents take away their children as the whim might seize them, would bring the schools into bad repute, and render it impossible to secure new pupils. I felt convinced, however, that the Qu'Appelle and Battleford Institutions have now been so firmly established that such risk might well be incurred in view of the advantages to be derived from putting a stop to the practices referred to, and I am glad to say that the measures taken to that end have been attended with considerable success, without the direful consequences anticipated in some quarters. *Reed (1891, p. 201)*

3.4.5 Repurposing history

When I began research for the Where are the Children Exhibition, I found a large number of residential school photographs, but none that illustrated the kinds of experiences that Shirley Williams describes or that directly imaged the abuses survivors were disclosing. This was a challenge, and was also discouraging, because it meant that the exhibition would not be able to respond directly to their experiences. Instead, I had to weave a story from photographs that had been originally intended to show the so-called 'good work' taking place at the schools and make that story meaningful to the indigenous community; not only the survivors, but their children and grandchildren, as well as the youth that are feeling the inter-generational impacts of that horrific history. Although the residential school system had largely closed down by the 1970s, its legacy of systemic abuse of indigenous children is still felt today by Indigenous youth. In fact, it was the questions that Indigenous youth were asking about residential schools that led to the Exhibition's creation.

The challenge of repurposing the archival photographs was compounded by a traditional indigenous suspicion of so-called 'documentary' evidence, since photographs, written reports and tape recordings had been used for centuries by anthropologists and government agents in ways that did not speak to indigenous experiences or benefit their communities. The views of the indigenous people who went to those schools were not part of the archive. This situation meant that I had to take a different route. I chose to use the storytelling tradition I learnt from my elders as the exhibition framework: The photographs would tell a story, but not always the one they were originally intended to tell. Take, for example, the photograph in Library and Archives Canada of a classroom in the Lac la Ronge Residential School in Saskatchewan. Bud Glunz, a photographer for the National Film Board, presumably took the photograph to document everyday life in this school as part of a broader project of documenting everyday life across Canada (Fig. 3.4.5.1).

It looks like a typical classroom scene. A teacher stands amid a group of Cree students seated at their desks. At first glance, you would never suspect the kind of undercurrents indigenous students were being subjected to in these schools. The students look directly at the photographer. While they are not smiling, they do not look particularly unhappy. It is

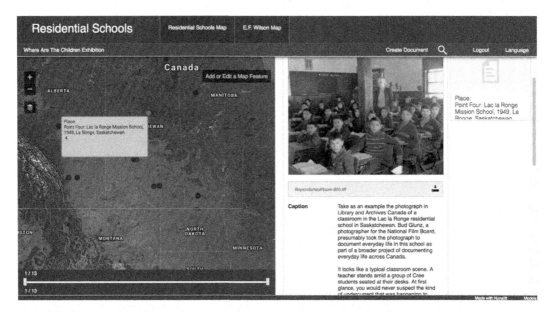

FIGURE 3.4.5.1 Screenshot highlighting the point for Lac La Ronge Mission School in the central map window of the Where Are the Children Exhibition Map Module, with following image (and related text) in the right side panel: 'Thou Shalt Not Tell Lies.' Cree students attending the Anglican-run Lac la Ronge Mission School in La Ronge, Saskatchewan, 1949. *Photographer: Bud Glunz, National Film Board of Canada, Library and Archives Canada, PA-134110.*

only when you take the time to study the photograph in detail that a new story emerges. Not only might you notice the stern face of the teacher, but you may also realize that the faces of the students could simply be reflecting the fact that a photographer was in the classroom that day. Everyone knows what it is like to pose for a photographer; no matter how you are really feeling, you are directed to look at a certain focal point, to smile or not smile, and not to move.

Then you notice something in the photograph that makes you stop. There, on the blackboard, are written the words 'Thou shalt not tell lies'. Obviously, the words were there to teach the students Christian dogma. But today, we can turn the question around. Who is really telling the lies? Isn't it the children that are being told lies about their own culture? They are being told that they are uncivilized savages, that their languages and cultural practices are inferior. And what lies are being told to Canadian society? Based on what we know today about what took place in many residential schools, the lies were many.

Journeying into the past is full of twists and turns. What should we believe and not believe about the photographs? Whose stories do they tell? Can we trust the vision of the white photographer? Indigenous people have an understandable suspicion of institutional archives. And, as my experience searching through these archives can attest, it is difficult to know what to trust and what to question. In the end, however, I believe that we can choose how much power we want to give these images. There are new stories waiting to come out of the photographs shown in the Where Are the Children Exhibition. Rather than dismissing the photographs simply as images of colonialism or racism, we can choose, as Indigenous people, to make them our own, to add them to our stories, and to give the children of residential schools a voice. I envisioned the exhibition space of Where Are the Children as one where the children's voices could finally be heard. One such voice is that of Virginia Bird (Cyr). During my research in Saskatoon, Saskatchewan, I asked a friend, Lori Blondeau, if she could put me in touch with a residential school survivor. She said her grandmother and mother had both gone to residential school but never talked about it. So, the three of us — Lori, her mother Leona Blondeau (Bird) and I — made a trip to the Gordon First Nation Reserve where Lori's grandmother lived. Sitting around the kitchen table, Virginia took out her photo albums and began talking about her family history. Surprisingly, this was the first time that her granddaughter Lori had heard the stories about the residential schools. I realized then how important family photo albums are and how the images can act as a catalyst for sharing family stories (Fig. 3.4.5.2).

Another voice is that of Mollie's grandmother, a former student/survivor of the St. Bernard Residential School in Joussard, Alberta. Her voice comes to us through a video made by her then 10-year-old granddaughter Molly Bellerose in 2007. The video was one component of an exhibition that had been developed by the local community to accompany the Where Are The Children Exhibition when it opened in Prince George, British Columbia. I remember being impressed and excited by her video because it was a perfect complement to the project concept: to stimulate a dialogue between survivors and their children and grandchildren. A list of questions, Molly asked her grandmother during the 22-minute interview, includes the following items:

- What was it like in residential school?
- What would happen if you accidently said a word in Cree?

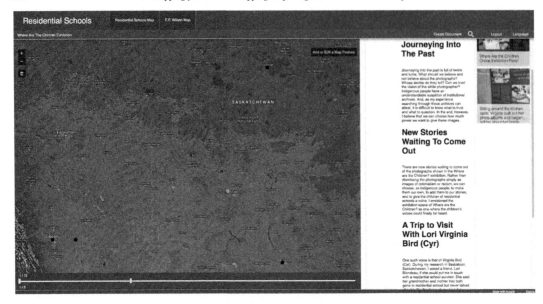

FIGURE 3.4.5.2   Screenshot showing the display for Point Five of Jeff's essay in the Where Are the Children Module of the emerging Residential Schools Land Memory Atlas, with the geographical extent of the first five points in the geonarrative overlay created to correspond with Jeff's essay in the central map window, and styled text with related media thumbnails in the right side panel.

- What did you get for holidays?
- Did you ever get to go home for Christmas?
- Question on being punished: Did you ever tell your parents?
- Did they treat you like slaves?
- Where there some good nuns?
- Where there any good times at residential school?
- Did you ever have dances there?
- Could you try out for a baseball team?
- Did you ever go to different places and play ball... at other mission schools?
- What kind of classes did you take?
- How come you didn't go back to school?
- Did other people from your family go to school there...mission at St. Bernard?
- Did you ever have a talent that they taught you at residential school?
- Did you have a talent of your own?
- What did your school look like?

In the video, Molly's grandmother responds to her last question by holding up a large framed photograph of her school and pointing out the different buildings. I listened to Molly's young voice and her grandmother's mature voice, which contrasted with the silence emanating from the residential school photographs I was studying. Listening to the audio, I imagined Molly's young voice becoming the voice of her grandmother as a child who finally has the opportunity to speak about her experiences at residential school. I wondered what

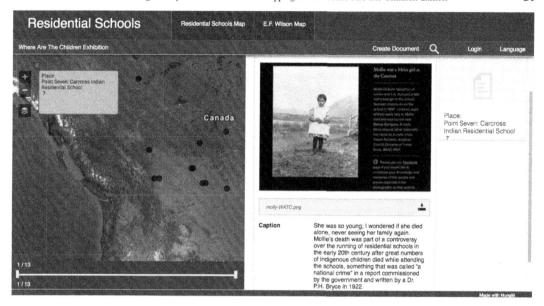

FIGURE 3.4.5.3   Screenshot highlighting the point for Carcross Indian Residential School in the central map window of the Where Are the Children Exhibition Map Module with a screenshot from the Where Are the Children website showing Mollie Dickson. *Original image from Yukon Archives, Anglican Church, Diocese of Yukon fonds, 86/61, #591.*

stories other children would tell if they afforded the chance. I began to see a link with all of the faces that had haunted me since I first began this project. I could hear the voice of Shirley Williams describing her entrance to the school; the voice of Virginia Bird describing how difficult it was seeing people from her community pass by her school and not being able to talk to them; and the voices and stories of the survivors I met during my tours of the exhibition, which together echoed the sound of the younger children crying during the night.

I see a link between the stories shared during my visit at Gordon First Nation and a photograph of another Mollie, a Métis girl at the Carcross Indian Residential School who became ill when cholera struck the school in 1907, died and was buried on the school grounds. She was so young; I wondered if she died alone, never seeing her family again. Mollie's death was part of a controversy over the running of residential schools in the early 20th century after great numbers of Indigenous children died while attending the schools, something that was called 'a national crime' in a report commissioned by the government and written by a Dr. P.H. Bryce in 1922 (Figs. 3.4.5.3 and 3.4.5.4).

3.4.6 Connecting the dots

Part of storytelling means being able to make connections, in this case among photographs that may not at first appear to be connected. Consider the example of three images that I framed together in the exhibition.

On the far left of the composite in Fig. 3.4.6.1 is a photograph taken on 16 October 1887 in northern Manitoba by J.B. Tyrrell (see also Fig. 3.4.6.2) as part of his survey for the Geological

FIGURE 3.4.5.4   Screenshot of a darkened map display in the Where Are the Children Exhibition Map Module profiling a screenshot from the Where Are the Children website showing Mollie Dickson. *Original image from Yukon Archives, Anglican Church, Diocese of Yukon fonds, 86/61, #591.*

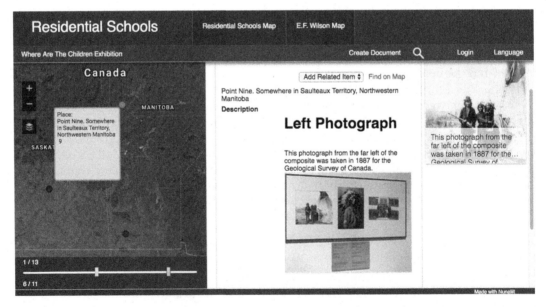

FIGURE 3.4.6.1   Screenshot showing approximate location for Saulteaux First Nation family (blue dot) in the central map window of the Where Are the Children Exhibition Map Module, with the composite image, individual image and text from 'I Have a Right to be Heard' in the right side panel.

FIGURE 3.4.6.2 Screenshot of a darkened map display in the Where Are the Children Exhibition Map Module profiling an archival image of Saulteaux First Nation family, Manitoba (Tyrell, 1887).

Survey of Canada. Entitled 'A Saulteaux First Nation Family' (Tyrell, 1887), it shows a Saulteaux First Nation family in an unidentified part of Manitoba. Treaties had just been signed earlier in the decade, and the transition from tribal society to reserve culture was just beginning. The Saulteaux family represents what I suggest is a form of acculturation. They have adopted Western-style clothing, yet they still live in a tipi. The male figure is holding a rifle that may well have been used in a traditional buffalo hunt, except for the fact that the great herds had by that time been slaughtered for sport, to feed the railway workers and to clear the land for settlement. The tipi, which would have customarily been made from buffalo hide, appears in the photograph to be made from canvas. The mother is seen holding a baby in the traditional cradleboard, used by Indigenous people across North America. They appear to be doing well, despite the government assertion that Indigenous people were not taking care of their children nor providing them a 'proper' lifestyle. This child could very well have gone to a residential school later in life.

The central photograph of the composite in Fig. 3.4.6.1 is entitled 'Wanduta' (Red Arrow) (Library and Archives Canada, 1913) (see also Fig. 3.4.6.3). As with the Saulteaux Family photograph, it was taken in Manitoba; this time around 1913, and farther south in the city of Brandon at the studio of H.W. Gould. The man is Wanduta, a Dakota from the Oak Lake area. He is photographed seated, from the waist up, against a neutral studio backdrop. He is turned slightly to his left to highlight the headdress and his distinctive profile. He has on a cloth shirt with a bit of ribbon around the cuff and what appear to be bands above his elbows. Wanduta also wears two important symbols that reveal the clash of cultures he witnessed. The first is an eagle feather headdress, a symbol of his status in Dakota society. The second is a treaty medal hanging around his neck, a symbol of the negotiations between

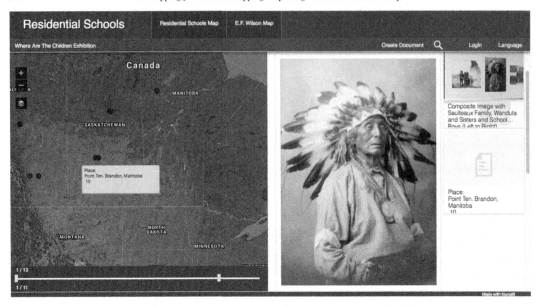

FIGURE 3.4.6.3   Screenshot showing approximate location for the Brandon Studio (where Wanduta photograph was taken) in the central map window of the Where Are the Children Exhibition Map Module, with the individual image and the composite image in the right side panel.

Indigenous tribes and the white government, negotiations that led to the loss of indigenous lands. Wanduta was considered a trouble maker by local Indian Affairs authorities and ended up in jail for organizing a public performance that the government interpreted as a give-away ceremony. The give-away ceremony was an integral part of many First Nations cultures but had been outlawed by the federal government as the concept of giving things away was anathema to the white Euro-Canadian economic system. Wanduta had to face a trial and was subsequently jailed for 4 months. An interesting link between this photograph and the other two in the triptych is that, in an ironic twist, Wanduta's son, a student from the Brandon Residential School, helped his father at his trial.

On the far right of the composite in Fig. 3.4.6.1 is a photograph that was made in the reserve community of Pukatawagan, Manitoba, entitled, 'Sisters outside the Pukatawagan day school with a group of boys wearing Plains Indian-style headdresses' (Library and Archives Canada, c. 1860), which is made from paper and attributed to Sister Liliane (see Fig. 3.4.6.4). The time is the 1860s, and 10 little Indian boys are posed in front of the Pukata-wagan day school. The clash of cultures is still evident here, but somewhat different from that in the Wanduta photograph. In this case, the treaty medal — the symbol of colonial power in the last photograph — has been replaced by the crucifixes around the necks of the Christian nuns; while Wanduta's feather headdress is replaced by the paper headdresses made by the boys under the supervision of the nuns. What story is being told here? I suggest that it is one of erasure: the substitution of the family by the Christian nuns and priests, the substitutions of land for treaty medals, the substitution of indigenous practices and cultures for paper symbols.

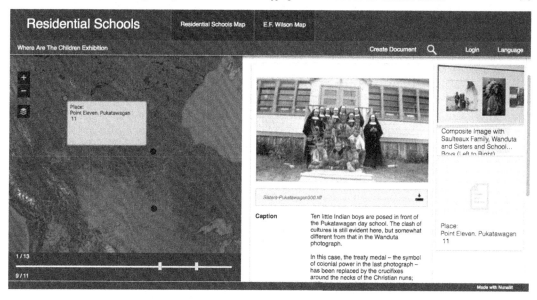

FIGURE 3.4.6.4 Screenshot showing location for Pukatawagan, Manitoba (where the photograph entitled 'Sisters outside the Pukatawagan day school with a group of boys wearing Plains Indian-style headdresses' (Library and Archives Canada, c. 1960) was taken), and the points for the Saulteaux Family and Wanduta photograph locations in the central map window of the Where Are the Children Exhibition Map Module, with the individual image and the composite image in the right side panel.

3.4.7 Is This Thomas Moore?

The goal of this exhibition is self-empowerment; to give Indigenous audiences the opportunity to begin to see the places Indigenous children were taken and to imagine their stories. In a sense, the images offer an opportunity to come full circle and move forward. By rethinking their role and purpose, perhaps these photographs can contribute to the healing process for those who attended residential schools as well as their children and grandchildren. I suggest these photographs can play a new role — one that speaks to the loss of imagination that occurred in those classrooms. How can we begin to imagine a different story — one where Jim Abikoki stands up and opens the gate, one where the white teacher is replaced by an elder who tells the children the history of their people, one where Wanduta is not forced to give up one culture for the other, one where Mollie receives proper care and returns to her community?

I want to end by returning to Thomas Moore. In 2011, new information about Moore was forwarded to me from the Saskatchewan Archives. I learnt what reserve he came from, his parents' names and when he entered residential school: Thomas Moore Kusick was admitted to the Regina Industrial School on 26 August 1891. His mother's name was Hanna Moore Kusick and his father, Paul Desjarlais, was listed in the Regina Industrial School register as 'St. Paul Desjarlais'. According to the archivist, the name Kusick may be of Aboriginal or Ukrainian origin. It is hard to tell, given the lack of uniformity in the spelling of the era. Moore's father died early. According to the Regina Industrial School register, Thomas Moore was protestant and had previously attended Lakes End School (formerly Muscowpetung Boarding School) in Saskatchewan. His state of education upon admission consisted in knowledge of the alphabet. He was 8 years old, 3'11" and weighed 54.5 lbs. A note in the register says to see page 20 of the discharge register. Thomas Moore was from the Muscow-petung Band of the Saulteaux Tribe.

With this new information about Moore, I went back to the 1896 Annual Report, and I saw something I had not noticed before: a portrait of the brass band from the Regina Industrial School, the same school Thomas Moore had attended. There, in the front row, seated on the ground third from the left, was a young boy who looked familiar. Could that young boy be Thomas Moore? If so, his face seemed to me to resemble his face in the first photo-graph presenting the boy in traditional clothing. Things came full circle, and he was back to being just a young boy, rather than a poster child for residential schools. A young boy with a story to tell (Fig. 3.4.7.1).

When you think about healing, it is about trying to give these children back the Indigenous names that their parents, grandparents or elders gave them when they were born; names that had meaning for them. It is about trying to piece together their stories. It is about trying to give them a way back home. Although we have not − yet − been able to give all these things back to Thomas Moore, the work continues. Looking back at the years the Where Are the Children Exhibition toured throughout Canada, I would like to believe that the exhibition has played a small part in helping those voices be heard. The final section upon exiting the exhibition is titled 'Role Models' and consists of five contemporary portraits of survivors

FIGURE 3.4.7.1   Screenshot of a darkened map display in the Where Are the Children Exhibition Map Module, profiling the archival image 'The Brass Band, Regina Indian Industrial School' (Department of Indian Affairs, 1896, p. 395).

FIGURE 3.4.7.2   Screenshot of a darkened map display in the Where Are the Children Exhibition Map Module, profiling the archival image 'Sisters outside the Pukatawagan day school with a group of boys wearing Plains Indian-style headdresses' (Library and Archives Canada, c. 1860).

I made in 2002. The large colour portraits stand out in sharp contrast to the black and white photographs seen to this point. They represent First Nation, Inuit and Metis people.

I wanted the visitor to connect back to one of the first photographs seen in the exhibition, 'Sisters outside the Pukatawagan day school with a group of boys wearing Plains Indian-style headdresses' (c. 1960s, paper, ref.) featured in Figs. 3.4.6.4 and 3.4.7.2. Imagine how different the residential school history could have been if one of these Indigenous role models were standing behind those children. In essence, this is the point I wanted to make with the exhibition: When we address the healing of this painful legacy, we must return to the indigenous principles that have given us the strength to survive. We must reaffirm the rights our ancestors lived by that guided how they treated their children and others.

3.5 Discussion and conclusion: Reflexive interplay between implicit and explicit Cartographic approaches

This chapter can be understood as an example of reflexive multi-genre crystallization, drawing on Ellingson (2009). The chapter is a story about itself, as much as it is a story of an emerging mapping project, and a story about the Where Are the Children Exhibition. The purpose of including Jeff's essay in this chapter is to share his curator's perspective; thus enriching the potential of both the digital and travelling exhibitions, and providing the reader with a deeper sense of the stories being mapped in the Residential Schools Land Memory Mapping Project. The reflexive relationships that abound in this chapter and its subject matter can best be understood in terms of layers or dimensions of reflexivity. For example, a reflexive relationship exists between the 'digital' exhibition and the 'analog' travelling exhibitions — including map displays reflecting each of these dimensions will aid in understanding this relationship. There are many more examples including the relationship between implicit and explicit approaches to cartography in terms of purpose and method.

Jeff employed an 'implicit cartographic approach' in (1) in curating the Where Are the Children Exhibition, (2) in writing his essay, 'I Have a Right to be Heard' and (3) in interpreting the images presented in the map of his essay in a decolonizing, culturally aware way that provides important insights into their meaning (see Section 8.4). Consider Jeff's description of the Wanduta image (see Fig. 8.4.1.5.3), which navigates the image by first setting the context, and then focussing attention on salient features located in different regions of the image:

> [Wanduta] is photographed seated, from the waist up, against a neutral studio backdrop. He is turned slightly to his left to highlight the headdress and his distinctive profile. He has on a cloth shirt with a bit of ribbon around the cuff and what appear to be bands above his elbows. Wanduta also wears two important symbols that reveal the clash of cultures he witnessed. The first is an eagle feather headdress, a symbol of his status in Dakota society. The second is a treaty medal hanging around his neck, a symbol of the negotiations between indigenous tribes and the white government, negotiations that led to the loss of indigenous lands.

Throughout his essay and in his practice generally, Jeff reads the photograph in an analogous fashion to reading a map, pointing out features of the photograph in a similar fashion to pointing out features of a map, and then asking questions about what the features reveal or about what may be missing from the photograph. This implicit cartographic approach runs through Jeff's photographic interpretation style, as the examples in this chapter demonstrate.

I used an 'explicit cartographic approach' (1) to begin exploring the relationship between the stops on Jeff's 2015 Road Trip and the locations of boarding schools in the United States (see Section 8.3.4); (2) to map Jeff's essay (see Sections 8.3.9 and 8.3.10) and to chart the geographical aspects of Jeff's practice in various ways, as is evidenced in the high-speed tour in Section 8.3. Making this distinction does not mean that Jeff does not participate in explicit cartography nor does it exempt me from practising implicit forms of cartography. For example, my high-speed tour of Jeff's collaborative cybercartographic mapping with me could be considered as a form of implicit cartography insofar as the tour is a written text, with map screenshots inserted to illustrate points made in the text: the map is read in the text and presented in a linear fashion; and the fact that Jeff participated in the creation of the cybercartographic map exercises mentioned in the tour is evidence of his involvement with explicit forms of cartography. Both approaches have worked together where the chapter writing and the essay mapping yielded new map module. Understanding written texts about maps as extensions, manifestations and performances of those maps has helped to focus both the mapping project and the chapter (Wood and Fels, 2008).

Providing a good example of iterative processes and emergence in cybercartographic mapping, the work to map the Jeff Thomas Mapping Thomas Moore overlay led to the idea to map the Where Are the Children Exhibition itself. Jeff's project of 'repurposing history' referred to in Section 3.4.5 where '[t]he photographs would tell a story, but not always the one they were originally intended to tell' is similar to the ongoing cartographic revisioning and deconstruction project in critical cartography (Harley, Kitchin, Perkins, Louis, etc.).

The approach to both implicit mapping in Jeff's essay and in his reading of photographs and the explicit approach to mapping in the development of the Where Are the Children Exhibition Map Module share the aim with participatory mapping identified by Bjorn Sletto where these 'performances' are 'intended to change the world rather than encode symbolic propositions about it' (Gell, 1998, p. 6); (Sletto, 2009, p. 444).

Several new labels for narrative cartographic methods were generated in the collaborative mapping practices I have engaged in with Jeff over the years. These can also be considered as templates. For example, I identified the method of mapping individual images in a triptych as 'geo-provenancing' (see Section 8.3.5), and later used this method in the mapping of the triptych referred to in Sections 3.4.6.2, 3.4.6.3 and 3.4.6.4. In contrast with conventional notions of provenance, geo-provenance involves privileging place over other descriptive categories and is consistent with Cybercartography's emphasis on place as an organizing factor for information. The initial work to geo-transcribe Jeff's essay 'Lest We Forget' referred to in Section 3.3.8 resulted from talk and functioned as a template for the mapping of Jeff's essay 'I Have a Right to Be Heard' in a new map module for the Residential Schools Land memory mapping project and for this chapter. Including Jeff's essay in this chapter provides a way to concretize the more abstract and methodological discussion in the chapter, literally giving the reader an idea of 'what' is being mapped, in addition to 'how'. Consistent with the understanding of reconciliation in a transitional justice framework (Crocker, 1999, 2015; Pyne and Taylor, 2015) introduced in Chapter 1, this reflexive approach contributes to a broader awareness of the colonial legacy, while presenting previously excluded perspectives and introducing important questions. In addition, the notion of 'cartographic conversation' referred to in Section 3.3.4 is consistent with Jeff's aim to initiate new conversations through his work and provides an interesting framework for considering 'multi-genre' 'conversations' involving art and cartography.

The chapter writing process involved a couple of apparent 'false' starts, which nevertheless served as templates in the ongoing development of the chapter and a newly created map module. This is consistent with the indigenous ethic of not wasting. There are important links between cartography and art and in the broader context of intercultural reconciliation and decolonization. These include reflexivity, emergence and iterative processes. The 'talk' component, which is essential to cybercartographic atlas development, has been reflected in this chapter in part, via quotations from email communications focused on various interrelated mapping and writing projects. Through this ongoing process, templates have emerged, including the practice of geo-provenancing.

As Jeff recounts in his draft essay, 'I Refuse to be Invisible' (referred to in Section 3.2), his work with photographs and maps in curating the Where Are the Children Exhibition brought him into relation with the Residential Schools Legacy:

Although I did not have any family who attended a Residential School, I was brought into the circle by the thousands of photographs I reviewed during my research. And in turn my own fears and pain of growing up indigenous in a non-indigenous world, set in. I realized that the Residential Schools system was symptomatic of a much larger and ongoing issue, systemic racism and rendering a people invisible.

So how does one define mapping in the context of a complicated history of oppression and forced assimilation? In Where Are the Children, the visitor is greeted by a map at the entrance. It shows all the known Residential Schools across Canada, with information about the years each school operated, and the religious domination that ran each school. The idea is to give the visitor a sense of the number of schools. But more important for survivor families, a map provides a departure point for experiencing the entire exhibition. *Personal communication, 12 June 2018*

The Where Are the Children Exhibition is still travelling and is currently installed in the MacOdrum Library at Carleton University (https://library.carleton.ca/library-news/category/exhibits). Through the writing of this chapter, experimental mapping has been engaged in, which has both enriched the chapter and yielded a new map module for the Residential Schools Land Memory Mapping Project. Further work includes revisiting many of the points in the various maps during a meeting with Jeff and having him video or audio record additional comments, which can be done directly in the online Where Are the Children map module. In addition, we are in the early development stages of cybercartographically mapping the Exhibition venues and the online Where Are the Children Exhibition Project in collaboration with the Legacy of Hope Foundation, and exploring sketch mapping of survivors' stories presented in the stories section of the Where Are the Children website. Stay tuned.

References

Barfield, T., 1997. The Dictionary of Anthropology. Blackwell, Oxford, Malden, Mass.

Brealey, K., 1995. Mapping them 'out': Euro-Canadian cartography and the appropriation of the Nuxalk and Ts'ilhqot'in first Nations' territories, 1793—1916. Canadian Geographer 39 (2), 140—156.

Bourriaud, N., 2002. Relational Aesthetics. Les presses du reel, Paris.

Bourriaud, N., 2004. Berlin letter about relational aesthetics. In: Doherty, C. (Ed.), Contemporary Art: From Studio to Situation. Black Dog Publishing, London, pp. 43—49.

Brauen, G., Pyne, S., Hayes, A., Fiset, J.P., Taylor, D.R.F., 2011. Transdisciplinary participation using an open source cybercartographic toolkit: the atlas of the Lake Huron treaty relationship process. Geomatica 65 (1), 27—45.

Caquard, S., Taylor, 2005. Art, maps and Cybercartography: stimulating reflexivity among map-users. In: Taylor, D.R.F. (Ed.), Cybercartography: Theory and Practice, Volume 4 in Modern Cartography Series. Elsevier, Amsterdam, pp. 285—307 (Chapter 12).

Caquard, S., Pyne, S., Igloliorte, H., Mierins, K., Hayes, A., Taylor, D.R.F., 2009. A "living" atlas for geospatial storytelling: the cybercartographic atlas of indigenous perspectives and knowledge of the great Lakes region. Cartographica 44 (2), 83—100.

Crampton, J., 2001. Maps as social constructions: power, communication and visualization. Progress in Human Geography 25 (2), 235—252.

Crocker, D., 1999. Reckoning with past wrongs: a normative framework. Ethics and International Affairs 13, 43—64.

Crocker, D., 2015. Obstacles to reconciliation in Peru: an ethical analysis. Unpublished English translation of Obtsaculos para la reconciliacion en el Peru. In: Giusti, M., Gutiérrez, G., Salmón, E. (Eds.), La Verdad nos Hace Libres. Sobre las Relaciones entre Filosofía, derechos Humanos. Religión y Universidad, Fondo Editorial de la Pontificia Universidad Católica del Perú, Lima, Peru.

Curtis, E.S., 1904. The Vanishing Race-Navaho. photographer. Photograph. ca. 1904. https://www.loc.gov/item/2004672871/.

Department of Indian Affairs, 1896. Annual Report of the Department of Indian Affairs for the Year Ended June 30th, 1896. Government of Canada, Ottawa.

Dewdney, E., 1883. Annual Report of the Department of Indian Affairs for the Year Ended December 31st, 1883. Department of Indian Affairs, Ottawa.

Dussel, E., 1993. Eurocentrism and modernity. Boundary 2, 65—76.

Ellingson, L., 2009. Engaging Crystallization in Qualitative Research. Sage, Thousand Oaks, CA.

Gell, A., 1998. Art and Agency: An Anthropological Theory. Clarendon, New York.

Irwin, R., Bickel, B., Triggs, V., Springgay, S., Beer, R., Grauer, K., Xiong, G., Sameshima, P., 2009. The City of Richgate: A/r/tographic cartography as public pedagogy. Jade, 28(1) 61–70.

Jessup, L., 2019. About Jeff Thomas. https://jeff-thomas.ca/bio/.

Johnson, J., Louis, R., Pramono, H., 2006. Facing the future: encouraging critical cartographic literacies in Indigenous communities. ACME: An International E-Journal for Critical Geographies 4 (1), 80–98.

Kitchin, R., Dodge, M., 2007. Rethinking maps. Progress in Human Geography 31 (3), 331–344.

Kwon, M., 2002. Feminist visualization: Re-envisioning GIS as a method in feminist geographic research. ? is this it – check: Kwan, M. (2002). Annals of the Association of American Geographers 92 (4), 645–661.

Library, Archives Canada, 1860. Sisters outside the Pukatawagan Day School With a Group of Boys Wearing Plains Indian-style Headdresses Made from Paper. ca. Attributed to Sister Liliane Library and Archives Canada. PA-195120.

Library, Archives Canada, 1913. "Wanduta" (Red Arrow), Dakota First Nation, Oak Lake Area, Manitoba, Ca. 1913. Photographer: H.W. Gould. Library and Archives Canada. PA-030027.

Pearce, M.W., 2008. Framing the days: place and narrative in cartography. Cartography and Geographic Information Science 35, 17–32.

Peluso, N.L., 1995. Whose woods are these? Counter-mapping forest territories in Kalimantan, Indonesia. Antipode 27 (4), 383–406.

Pickles, J., 2004. A History of Spaces: Cartographic Reason, Mapping and the Geocoded World. Routledge, London, New York.

Pyne, S., 2013. Sound of the Drum, Energy of the Dance: Making the Lake Huron Treaty Atlas the Anishinaabe Way (Unpublished PhD Dissertation, Carleton University, Ottawa).

Pyne, S., 2014. The role of experience in the iterative development of the Lake Huron Treaty Atlas. In: Taylor, D.R.F., Lauriault, T. (Eds.), Developments in the Theory and Practice of Cybercartography: Applications and Indigenous Mapping. Elsevier, Amsterdam (Chapter 17).

Pyne, S., Taylor, D.R.F., 2012. Mapping indigenous perspectives in the making of the cybercartographic atlas of the Lake Huron treaty relationship process. Cartographica, Special Issue on Indigenous Cartography and Counter Mapping 47 (2), 92–104.

Pyne, S., Taylor, D.R.F., 2015. Cybercartography, transitional justice and the residential schools legacy. Geomatica 69 (1), 173–187.

Pyne, S., 2019. Cybercartography and the critical cartography clan. In: Taylor, F., Anonby, E., Murasugi, K. (Eds.), Further Developments in the Theory and Practice of Cybercartography: International Dimensions and Language Mapping. Elsevier, London. Forthcoming.

Reed, H., 1891. Sessional Report. Sessional Papers. Parliament of Canada, Ottawa.

Sletto, B., 2009. We drew what we imagined: Participatory mapping, performance, and the arts of landscape making. Current Anthropology 50 (4), 443–476.

Sparke, M., 1998. A map that roared and an original atlas: Canada, cartography, and the narration of nation. Annals of the Association of American Geographers 88 (3), 463–495.

Sparke, M., 2005. In the Space of Theory: Post foundational Geographies of the Nation state. University of Minnesota Press, Minneapolis.

Taylor, D.R.F., 1997. Maps and mapping in the information era. In: Ottoson, L. (Ed.), Keynote Address to the 18th ICA Conference, Stockholm, vol. 1, pp. 1–10. Proceedings.

Taylor, D.R.F., 2003a. The concept of cybercartography. In: Peterson, M.P. (Ed.), Maps and the Internet. Elsevier, Amsterdam, pp. 405–420.

Taylor, D.R.F., 2005. Cybercartography: Theory and Practice. In: Modern Cartography Series, vol. 4. Elsevier, Amsterdam, p. 574.

Taylor, D.R.F., Pyne, S., 2010. The history and development of the theory and practice of Cybercartography. International Journal of Digital Earth 3 (1), 1–14.

Taylor, D.R.F., Lauriault, T. (Eds.), 2014. Developments in the Theory and Practice of Cybercartography: Applications and Indigenous Mapping. Elsevier, Amsterdam.

Taylor, F., Anonby, E., Murasugi, K. (Eds.), 2019. Further Developments in the Theory and Practice of Cybercartography: International Dimensions and Language Mapping. Elsevier, London. Forthcoming.

Taylor, J., 2003b. The story of Jimmy: the practice of history on North Pentecost. Vanuatu. Oceania 73, 243–259.

Thomas, J., 2019a. Jeff Thomas. Imposition of Order. https://jeff-thomas.ca/2018/02/imposition-of-order/.

Thomas, J., 2019b. Jeff Thomas. Scouting for Indians. https://jeff-thomas.ca/2014/04/scouting-for-indians/.

Treuer, A., 2011. The Assassination of Hole in the Day. Minnesota Historical Society Press, Saint Paul.

Turnbull, D., 2000. Masons, Tricksters, and Cartographers: Comparative Studies in the Sociology of Scientific and Indigenous Knowledge. Harwood Academic, Amsterdam.

Turnbull, D., 2007. Maps, narratives, and trails: performativity, hodology, and distributed knowledges in complex adaptive systems: an approach to emergent mapping. Geographical Research 45, 140–149.

Tyrell, J., 1887. Sauteaux Family. Library and Archives Canada, The Canadian West, PA-050799. Available at: https://www.collectionscanada.gc.ca/canadian-west/052920/05292025_e.html.

Wood, D., Fels, J., 2008. The Natures of Maps: Cartographic Constructions of the Natural World. University of Chicago Press, Chicago.

Reimagining archival practice and place-based history at the Shingwauk Residential Schools Centre

Krista McCracken[*,1], *Skylee-Storm Hogan*[2,3]

[1]Researcher/Curator, Shingwauk Residential Schools Centre, Algoma University, Sault Ste. Marie, ON, Canada; [2]MA Candidate, University of Western Ontario, London, ON, Canada; [3]Shingwauk Residential Schools Centre, Algoma University, Sault Ste. Marie, ON, Canada
*Corresponding author

4.1 Introduction

The final report of the Truth and Reconciliation Commission (TRC) of Canada asserted that healing, reconciliation and restoring the relationship between Indigenous and non-Indigenous Canadians was a critical priority (Truth and Reconciliation Commission, 2015). The Commission called directly on Canada's museums and archives to work with Indigenous communities to better present Indigenous cultures and histories, including histories of residential schools, assimilation, cultural loss and resilience. By presenting effective historical narratives, heritage sites have the potential to inspire critical discussion about reconciliation and to stimulate nuanced historical thinking. The Shingwauk Residential Schools Centre (SRSC) at Algoma University is taking up this call for community engaged practice by making accessible the story of Indian Residential Schools and the story of the Shingwauk School. This chapter accessibly explores the story of Indian Residential Schools and the story of the Shingwauk School.

This chapter explores the narrative building techniques used by the SRSC in its collaborative work with the team at the Geomatics and Cartographic Research Centre (GCRC, Carleton University) to develop cybercartographic map modules related to the Shingwauk Residential School under the Residential Schools Land Memory Mapping Project (RSLMMP). By framing this work through a decolonial lens and utilizing critical archival theory, SRSC staff have worked to personalize, contextualize and bring to life the history of the early years of the Shingwauk School. All historical sites have the potential to tell unique narratives that are tied to place and have the ability to share 'the imprint of the lives of a community that went before us that gives the place its significance in our lives today' (O'Neill, 2008, p. 146). The Shingwauk site has the potential to connect people to the history of residential schools, intergenerational trauma and colonialism in Canada through the process of developing the cybercartographic map modules related to the Shingwauk Residential School. Throughout the work of the RSLMMP, the SRSC has been working to reinterpret historical records, create nuanced place-based narratives and explore new ways of sharing the history of the Shingwauk Indian Residential School; with a view to creating a tool for Indigenous people to access their own histories with their own voices tied to the land.

4.2 Positionality of the SRSC

The physical location of Algoma University and the SRSC has greatly shaped the educational approach, archival mandate and cross-cultural learning strategies practised on the site. Algoma University is the only university in Canada located in a former residential school building. The uniqueness of this position does not come without responsibilities — Algoma is called upon to do better, to respect the heritage of the site it embodies and to reflect on what it means to inhabit a space that is directly connected to intergenerational trauma.

In the spring of 1970, the Shingwauk Residential School, located in Sault Ste. Marie, Ontario, Canada, closed its doors. The closure of Shingwauk was part of the Canadian government's wider efforts in the 1960s and 1970s to phase out Indian residential schools across the country. Following the School's closure and the vacancy of the Shingwauk Hall building, Algoma University College in partnership with the Keewaitnung Institute moved onto the Shingwauk site in 1971. Since its relocation to the Shingwauk campus, Algoma University has undertaken many initiatives in cross-cultural education and prides itself on working with local Indigenous communities to create an inclusive and multicultural education environment. The undertaking of cross-cultural programming is directly tied to the residential school legacy, which Algoma became part of upon its relocation. One of the most enduring and significant cross-cultural efforts by Algoma has been the establishment and development of the SRSC. The SRSC, previously known as the Shingwauk Project, is a cross-cultural research and educational development initiative of Algoma University and the Children of Shingwauk Alumni Association (CSAA).

The Shingwauk Project was founded in 1979 by Professor Don Jackson in collaboration with Dr. Lloyd Bannerman of Algoma University, Chief Ron Boissoneau of Garden River First Nation, and Shingwauk alumnus and elder Dr. Dan Pine Sr. of Garden River First Nation. The first major initiative undertaken by the Shingwauk Project was hosting a residential school survivor reunion for former students of the Shingwauk School. In 1981, over 400 students, family, staff and community members gathered on the Shingwauk site to begin to address their communal past. After attending the 1981 reunion, many students, families and former staff felt compelled to share photographs, scrapbooks and documents with each other. As a means of facilitating this sharing, Jackson established the Shingwauk archives to promote the sharing of residential school records and resources. The archives were established with joint governance between the CSAA and Algoma University. From 1981 to 2008, the archives were staffed by volunteers and coordinated by Jackson. Funding for the project was minimal, and it initially had no dedicated space. The archives had no governing policies, no real organizational system existed for the collections and no one with archival experience was associated with the initiative. Its main focus was to provide copies of materials to Indigenous communities and act as a community repository for materials relating to residential schools.

In the late 1990s and early 2000s, Jackson received a small grant to hire staff. He hired community members and allies of the First Nations communities. Under this initiative, a card catalogue was created in an attempt to organize the collection. The focus of the archives was still on preserving as much material as possible and providing research services to residential school survivors. The staff dedicated most of their time to community outreach and liaising with survivors. The professionalization of the SRSC began in 2008 when operational funding was secured from the Aboriginal Healing Foundation. From 2010 to 2012, Ken Hernden, university librarian, acted as codirector of the SRSC, shepherding it from an externally funded operation to a unique resource with its own staff and mandate. The period from 2010 to 2012 also marked the hiring of a trained archives technician and several student assistants. During this period, the collection was made compliant with the Canadian Rules for Archival Description, digitization of the archival holdings was started and file-level description of the material began being placed online. Today, the SRSC is jointly governed by the CSAA and Algoma University. A heritage committee comprising members from both organizations provides guidance on best practices, outreach programming and policy development.

The current mandate of the SRSC includes the maintenance of an archives, support of residential school survivor healing initiatives, the development of exhibit materials relating to residential schools and delivering educational programming on residential schools. The SRSC is an example of a community archives born out of a desire to see the history of residential schools told from the survivor perspective. Recognizing the importance of involving the people whom the historical documents represent has contributed to SRSC programming which actively engages residential school survivors, families and communities in collection development, description and education programming. As Caswell et al. (2018, p. 82) have noted, many individuals view 'community archives metaphorically as home [...] home is a space where their experiences and those of their ancestors are validated. For others, still it is a space where intergenerational dialog — sometimes difficult and unsettling — occurs'. The work engaged in by the SRSC is informed by intergenerational connections to the Shingwauk site and the need to provide space for truth telling in relation to residential schools. The community-based approach to archival management used by the SRSC has aided the Centre in becoming a trusted archival space among survivor communities in Canada.

4.3 Place-based learning

The archival collections of the SRSC, combined with the physical location of the Algoma University, allow for an unparalleled approach to archival practice, experiential learning and analysis of the past. The main building of Algoma University — Shingwauk Hall — was the primary structure of the Shingwauk Residential School from 1935 to 1970. Faculty offices, classrooms and administrative facilities are all located in spaces that were inhabited by students as part of their daily life at Shingwauk. Many of the physical spaces in Shingwauk Hall have been renovated extensively, erasing tangible pasts and shifting the building into a modern university space. However, through the memories of the CSAA and the work of the SRSC, it is possible to reimagine Shingwauk Hall and begin to understand the spaces occupied by Algoma University. Walking through the halls of Algoma University with a residential school survivor significantly alters how you perceive the space. Classroom spaces are recalled as dormitories, memories of Christmas and isolation are triggered in the auditorium and office space becomes spaces of physical labour and mistreatment. Oral history and archival images make it is possible to imagine how spaces were once used. In addition, physical reminders such as the main doors to Shingwauk Hall remain as a tangible reminder of the history of the building.

One of the challenges posed by the Shingwauk site is that much of the history extends beyond the bounds of the Shingwauk Hall building. The original Shingwauk site encompassed 90.5 acres and stretched from the St. Mary's River to what is currently known as Wellington Street. The early years of the Shingwauk site witnessed the development of the first Shingwauk Home 1874, the Shingwauk Chapel 1883, an extensive farming operation and the establishment of numerous outbuildings. Today, the only remaining structure from before the turn of the 20th century is the Chapel. Other spaces have left no physical footprint on the site. For example, what was once farm fields is now occupied by student residences and a sciences building, and the location of the original Shingwauk Home building (1874–1935) on Algoma University's front lawn is marked by two monuments, which contain minimal text to contextualize the history of the site.

How do you commemorate and teach about spaces that no longer physically exist? As a means of documenting the early history of the Shingwauk site, SRSC staff examined archival documents and photographs to reconstruct narratives, experiences of the spaces and the lives of students involved in the early years of the Shingwauk School. To facilitate experiential learning opportunities, SRSC staff have created a historical walking tour that exposes participants to the long history of the Shingwauk/Algoma site. The walking tour discusses the 90.5-acre site as a whole, with an emphasis on contextualizing the Shingwauk experience within the broader landscape of colonialism while privileging the survivor perspective on the residential school system. Walking tours remain one of the most frequently accessed programs of the SRSC, with over 2000 participants attending tours in 2017.

Despite the popularity of the tours, one of the ongoing challenges of SRSC staff is how to engage visitors with the early history of the Shingwauk site when little physical evidence remains. Staff have done everything from verbally describing former buildings to situating tour stops in spatially significant locales, to holding up historical photographs of the site. These approaches all have their merits and limitations. For example, standing where the first

Shingwauk Home building was located allows participants to see how much closer the School was to the river and spatially situate the building using present-day landmarks. However, Sault Ste. Marie is located in Northern Ontario and receives a substantial amount of snow on annual basis, resulting in some tour stops being physically inaccessible for four or more months of the year. Thus, as much as SRSC staff would like to talk about the original Shingwauk building while standing where it was once located, actually being in this physical space is not always feasible. The cybercartographic approach utilized in the RSLMMP provides an alternative option for engagement with the Shingwauk site. The digital participatory mapping efforts undertaken in this project have facilitated new ways of thinking about space and education on the Shingwauk site and have indeed offered a new mode of interpretation and interaction with the past.

4.4 Constructing narratives in archival-based Cybercartography

Archival and historical work is directly connected to what (and whose) stories are remembered, accessible and told. The Shingwauk Project and the SRSC were founded on the desire to preserve and tell stories from the perspective of residential school survivors. Today the SRSC continues to narrate the history of the Shingwauk site from a survivor perspective and aims to extend its interpretation beyond the colonial words and images captured by archival records. For their part, archival records need to be understood as partial representations of historical truths: 'Not all events are recorded; not all records are incorporated into archives; not all archives are used to tell stories; not all stories are used to write history', and, perhaps most significantly, 'power is implicated in each of these moments; which stories get told, which get forgotten, when and by whom, is inextricably linked to the power to tell and remain silent' (Caswell, 2014, p. 10). Archival material is representative of historical narratives, popular opinions and societal relationships; who and what is left undocumented matters. Staff at the SRSC have often considered what is not said in archival records, as archival silences often speak to larger power dynamics and values. Historical truths are variable, they are constructed and representatives of the cultural forces in which they are created, read and reinterpreted (Brown and Vibert, 2003).

The archival records relating to residential schools are artefacts from the day-to-day administration of the residential school system. The views of school staff are evident in monthly reports, correspondence and general operational records of the schools, while the administration perspective is evident in the photographs and newsletters created to promote and record the efficacy of residential schools (McCracken, 2017). These records were not created to document individual student experience at residential schools, yet, in many cases, they are the only archival evidence of a student's attendance and the daily routine of residential school. The lack of information from the student perspective has implications for our current ability to understand the residential school system. Missing context and sparse student perspectives impairs the work of those seeking to analyze the 'average' student experience at residential schools. Survivors who performed well or who were considered 'above average' as pupils often have different feelings than the rest of the community when reflecting on their time at residential school. These are the students who dominate archival records and that clergy and school staff wanted to show off as good examples.

The students themselves had no perspective included in records, aside from form letters. This case of missing student perspective can be seen in the records that document the early years of the Shingwauk School, many of which were authored by Rev. E.F. Wilson, the first Principal of Shingwauk. Although Wilson's photographs and writings discuss students, they are not truly representatives of student perspectives. Indeed, Wilson's writing in the *Canadian Indian* publication highlights his Eurocentric value system. Wilson believed that the disappearing 'Indian' was part of natural progress and that 'the races which for thousands of years trod this continent, will, in the not distant future, be known only in the same as the bison, which has as suddenly disappeared' (Wilson, 1890, p. 57). By likening Indigenous communities to the bison — which were exterminated by settlers — Wilson's words illustrate a perspective of assimilatory practices and colonial sentiments. Wilson's need to record languages and cultural practices of these groups was done for future 'civilized Indians'. He believed that once 'they' had come to their senses and abandoned their primitive ways, they would find an appreciation for their history. Likening these future Indians to the English reading about the ways of the ancient Greeks, Wilson said it was his duty to record this dying way of life (Wilson, 1890, p. 190). Studies of Indigenous groups and culture were also performed to understand best approaches to convert them within the schools. Given these motivations and societal conditions, records relating to Indigenous people and communities cannot be understood outside of the colonial context in which they were created. Findlay (2016, p. 155) has argued that, 'Archivists/record keepers know that every recordkeeping act [...] occurs in and is influenced by its layers of context, from the systems and people that are directly associated with the act, to the motivations of the organization that funded it, to the expectations and norms of the wider society in which it occurs'. Archival records cannot be viewed in isolation, both their creation and why they were preserved by archives needs to be examined. Material authored by Wilson and other school administrators must be viewed within the colonial systems of power and control which they were authored.

The SRSC's involvement in the RSLMMP represents a navigation of archival power and a desire to illuminate survivor experiences in an instance where scant archival records represent survivors directly. As a first step, research assistants navigated the SRSC's archival management system with the goal of identifying all the archival holdings relating to the early years of the Shingwauk School. This identification project, looked at within the context of decolonizing archives, can be seen as a representation of the fundamental disconnect between Western archival arrangement principles and the needs of Indigenous community members. Archival concepts of original order, provenance and access often create barriers for Indigenous peoples looking to utilize archival records. As Fraser and Todd (2016) have argued, '[t]o reclaim, reshape, and transform the archives to meet the needs of Indigenous peoples requires an honest and blunt engagement with the bureaucratic and arcane structures that govern and shape research today'. Archives are not organized thematically or by location, and looking at all records relating to the Shingwauk Home required examining over 80 distinct collections spread across the SRSC. By having research assistants engage in this extensive source finding project, SRSC staff hope to simplify access to Shingwauk Residential School records held in the SRSC. Indeed, the RSLMMP itself provides an alternative option to a strict application of archival theory and structures for accessing residential school archival material.

In addition to facilitating new methods of accessing archival material, the RSLMMP has provided SRSC staff with the opportunity to develop new narratives about the construction of the Shingwauk Home and to situate residential school student experiences within geographic spaces. Research assistants conducted an extensive survey of principal letter books, administration records, annual reports and photographs. By examining thousands of pages of archival documents, staff were to compile a timeline of construction and demolition dates for additions for numerous buildings associated with the early years of the Shingwauk Home. This survey also enabled staff to develop a clearer sense of the physical footprint of the Home, including room dimensions, changes in room usage over time and the evolution of the Shingwauk site as a whole. This research ultimately contributed to the creation of a digital representation of the first Shingwauk Home building in a cybercartographic map (see also Chapter 8).

Digitally reconstructing physical spaces from archival documents is one thing, but how do you make those spaces speak to the past? How do you tell the lived experiences of the students who lived, worked and sometimes died within those spaces? As an attempt to tie residential school survivor experiences to the digital representation of the Shingwauk Home, SRSC staff began looking at photographs, student registers and oral histories, for evidence of daily life within the residential school system. This three-pronged approach to archival research has allowed for a range of survivor perspectives to be found and assisted in creating a more complete understanding of the early years at the Shingwauk Home.

4.5 Photographic narratives

Archival photographs can provide insight into the residential school experience, staff and student relationships and the traumatic impacts created by the system's legacy. Photographs have the power to evoke memories and inspire discussion, and they make it possible for researchers to critically analyze facets of residential school life that were not documented textually. In an age of reconciliation where conversations about student life and treatment at residential schools are prominent in educational settings, the media and in the public sphere, these photographs can help reconstruct school life and make student experiences more comprehensible to present generations. However, photographs cannot always be taken at face value, and any analysis of archival photographs should consider why particular photographs were taken and not others, what messages are conveyed by the images and why the photographs were preserved.

Oftentimes, in residential school images, the photographer and the archival process silence the student point of view (see Chapter 3 for a related discussion). Individuals who took photos of residential school students, such as Indian agents, mission staff, teachers or visitors, frequently arranged the students in a specific way for photo 'opportunities' and often failed to record the students' names with the images. This silencing of the student experience has been carried into the archival description process, resulting in residential schools photographs in archives containing little contextual information about the daily events and the students themselves. For many of the photographs documenting the early years of the Shingwauk Home, the photographer decided how the students appeared and how they were to be presented to the world. The students' natural postures, mannerisms and personalities are clouded by the photographer staging the photograph (Tinkler, 2013, p. 37).

The images we see of Shingwauk students in the 1800s and early 1900s are mediated through the lens of professional photographers, staff and administrators. Though they depict students at Shingwauk, they do not fully expose the lived experiences of the students. As

Caswell argues, 'the camera, particularly when it's employed by the institutions of the state, performs the political work of turning individuals into silent objects to be measured and indexed' (Caswell, 2014, p. 50). Residential school photographs removed the student perspective and obscured student experiences, and, instead, depicted scenes that reflected well on the residential school system and its staff. For example, the student portraits contained in the E.F. Wilson photo album held by the SRSC are formal images, which are heavily mitigated by the lens of the photographer.

Fifty-four out of the seventy-two images that comprise the Wilson album — or 75% — were taken in a portrait studio and feature Shingwauk boys in dress uniforms. The fit of the uniforms speaks to the staged nature of the photographs, many of the formal coats are much too large or small for the boys pictured. The coats did not belong to a single boy, but rather were shared amongst the students for special occasions such as portraits. An examination of the complete set of photographs reveals standard staging, with studio photographs posing male bodies upright or sitting in a refined posture with legs crossed and hands clasped. Another form of staging can be found when examining the larger group photographs in the Wilson album. There are 11 photographs of students groups within the Wilson album; 9% or 81%, of these images were taken outdoors and use props such as furs and snowshoes to evoke feelings of ruggedness and the far north. The contrast of formal uniforms with the outdoor props can be viewed as highlighting the 'civilizing' impact of residential schools and an attempt to showcase the 'success' of education at Shingwauk. These images are 'an implied comparison to an uncontained undisciplined former state, and an assumption that the sight of children in Western dress was a novelty, an indication of "progress"' (Farrell Racette, 2009, p. 56). Photographs of students in uniform, framed as representing the success of the Shingwauk Home, were often used by Shingwauk Principal Wilson in reports, newsletters and as a means of soliciting financial and in-kind support for the Shingwauk School. For many years, these images helped to obscure the realities of the residential school system and promoted a contrived view of the students as acclimating well to their new surroundings. By combining photographs with contextual information, textual records and survivor testimonies, it is possible to examine these images in a new light.

4.6 Written archival narratives

Textual records written about residential school students can be used to supplement archival photographs; however, they too need to be viewed within the context in which they were created. Even meticulously detailed records, which seemingly document student experiences — such as the student ledger created by Shingwauk Principal Wilson — are mediated through the Eurocentric worldviews of those who created them. When marginalized communities are represented in archival records written by from the view of a missionary, there is often a wide disparity between the perception of lived experience and the actual lives of marginalized communities (Van Kirk, 1980). Traditional archival records represent only one avenue for discovering the past. Looking beyond written texts and constructing narratives around what has not been recorded can be an act of reclaiming the past and exposing marginalized narratives. Archival provenance and a clear understanding of the relationships of creation and use can provide an avenue for exploring more complete versions of the past

(Nesmith, 2002). Gaining a more complete representation of student life at the Shingwauk Home requires reading Wilson's work against the grain and combining his perspective with contextual information and other sources such as photographs. According to Bastian, '[r]eading the records of colonialism in ways that truly illuminate all facets and all participants in colonial society speaks not merely to political correctness but, more to point, to the ability of records to reflect, to mirror, to define and to uncover the communities that create them' (Bastian, 2006, p. 268). Examining the context of how and why historical records were created can allow for new interpretations of archival material. Taken together, a record's content, structure, social relationships and the context of its creation can illuminate a range of relationships and histories beyond what might be seen in a surface reading.

In using the Shingwauk Home register, maintained by Wilson, research assistants were faced with navigating the language and perspectives of Wilson. Phrases and educational judgements such as 'wholly untaught' and 'little to no progress' were repeatedly used by Wilson to describe student experience prior to and during their time at Shingwauk. These statements need to be examined within the context of their creation. Wilson's assessment of students represents his role in assimilation and dismisses the wealth of traditional knowledge, Indigenous languages and the ways of knowing students brought with them to the Shingwauk School. Reading the descriptors used by Wilson in today's context is challenging, perhaps doubly so for intergenerational survivors and Indigenous community members. Wilson's descriptors paint their relatives as pagans, ignorant failures of a dying culture. It takes an active effort to read beyond these Eurocentric view points and discover what the student register can tell us beyond Wilson's perception of the students.

Research assistants under the RSLMMP connected student register records and student experience to physical school spaces by examining the 'trade', 'progress made' and 'after record' notes included by Wilson. The trade notations often listed the type of manual labour trade training students received at Shingwauk. Trades listed for male students included carpenter, printer, tinsmith, bootmaker, school teacher, blacksmith, farmer, tailor, baker, mason, telegraph operator, waggon maker, weaver, minister, mechanic and organist. The trades listed for female students were limited to laundress, domestic service and dressmaker. The recorded trades were used to connect students to physical spaces. For example, a student learning carpentry would have spent time in the woodworking shop and those listed as tailors would have spent time in the sewing room. Female students would have spent substantially more time in kitchen, dining and laundry spaces. By connecting student names to spaces and researching individual experiences, staff have been able to create student narratives to bring physical spaces to life. Instead of talking generically about labour within a space, SRSC staff can now use personal narratives as an entrance point into the past. In some cases, the SRSC holds photographs of the students who have been researched, and it is possible to pair archival images with personalized historical narratives. The pairing of photographs with physical spaces, student names and student experiences allows for the development of rich narratives that are centred on student experience not on staff perceptions of the residential school experience. Historical interpretation grounded in individual histories facilitates connection between lived experiences and archival records or artefacts. The use of individualized accounts in SRSC programming and in the RSLMMP represents an attempt to shift narratives away from administrative histories of residential school and focus on lived experience from the survivor perspective.

4.7 Survivor voices

The third mode of illuminating student narratives focused on connecting survivor testimony to physical spaces on the Shingwauk Hall. This work involved examining transcripts of oral history interviews and audiovisual records from survivor reunions and looking for any discussion of daily life at Shingwauk that could be connected to interior or exterior spaces. The reliance upon oral testimony and photographs represents the continued acknowledgement that written records for residential schools are often written from the administrative perspective and not representative of student experience. Miller (2004, p. 86) has argued the need to supplement textual records with other evidence, such as photographs, to more fully expose and understand what actually happened in the schools:

Written sources and collections cannot do the job as fully and as effectively as one would like. Other forms of evidence and techniques of inquiry are needed to recapture native experiences of residential schools as fully as possible. In particular, the contributions of visual and oral history are vital components in the multidisciplinary research strategy that is necessary to tell the story of residential schools.

The SRSC's use of oral history represents a desire to incorporate numerous ways of knowing. It also represents a desire to place Indigenous voices at the forefront of interpretation. This emphasis on community knowledge is based on the following assumption: 'interpretation of individual communities that is directly meaningful to the members of those communities [...] can only come from those communities themselves' (Hoerig, 2010, p. 66). Interpretation of residential school history has the potential to be the most impactful when directed by survivors and their descendants. Their experiences within the institution and after leaving it give a fuller view of the impact that continues in Indigenous daily life. Residential school oral histories also have the power to give intergenerational survivors, or any contemporary Indigenous person, the ability to access the voices of elders whenever they need to.

The importance of recorded oral histories from survivors who have passed on is invaluable for contemporary Indigenous peoples who want to know more about their own history. Tying these oral histories to a place gives meaning and purpose back to the land. The land has stories and history woven within it, and elders are the keepers of that history (Tuck and Yang, 2012). Through systems of colonization and genocide, access to these once readily available truths have become limited. As residential school survivors pass on, there is a threat of their lived history passing away with them. The RSLMMP offers Indigenous people the opportunity to experience an oral history connected to the land and to experience the words of survivors who may not be alive today. One of the barriers to Indigenous researchers accessing their own histories is how they are recorded. Linking the physical points of the map to survivor voices adds a new component for Indigenous users while resonating with the forever histories of that land; as histories were passed down since time immemorial through the words of elders. This adds a cultural familiarity to research that incorporates indigenous ways of knowing through oral history, while keeping the memories of that physical space alive.

The oral history material used in the RSLMMP included interview transcripts and audiovisual footage from the Shingwauk 1981 and 1991 Reunions. For many survivors, the Reunions represented the first or second time they returned to the Shingwauk site since they left the School as children or youth. The power of place in eliciting memories, emotions and reflections are evident throughout the recorded words of the survivors. For example, the impact of the strict regimented schedule of residential school comes across in survivor interviews, where one survivor noted 'part of our way of life is to be in a circle but I can't remember once being here that we ever went in a circle. Even to play games, we never had a circle. It was always rows and line up' (Fletcher Luther, 1991). Even in times of recreation and while outdoors, Shingwauk students experienced space through administrative structures and rules. The changed landscape of the Shingwauk site was also recalled by survivors who spoke about the work regime at Shingwauk: '[W]e used to have to go out and pick potatoes because we had a big field in the back here [...] now during the summer as the potatoes were ripening a lot of guys got hungry during the daytime. So we used to go out, dig these potatoes, build a fire in the bush and just throw the potatoes in there and cook them' (Rogers, 1991). These testimonies of farming, bush and life outside the classroom provide insight into the Shingwauk site beyond the walls of Shingwauk Hall and personalize the residential school experience.

Survivor Donald Sands spoke at the 1981 Shingwauk reunion opening ceremonies where survivors and former staff were asked to reminisce and share. Like many other survivors who spoke at that event, Donald's experience was emotionally charged and connected to lived experience as he shared a story about how boys would look for food outside of the school: '[W]hen we were here in the winter time we didn't always eat so good. We used to have to supplement our food by going to the dump, or the city incinerator' (Sands, 1981). These experiences of hunger, moments of hidden comradery amongst students and the ways in which students existed outside of the purview of school staff are not well documented in photographs or archival documents. Even staff themselves do not recall factual experiences had by students in their care. We can only learn about these experiences by listening to those who lived them. Listening and valuing these spoken words can be seen as a form of respecting Indigenous experience and honouring the survivors who have passed on (Thomas, 2005). The oral testimony and personal truths recorded at the Shingwauk 1981 and 1991 reunions provide unique insight into the daily realities of Shingwauk students. By pairing these oral records with other documents, it is possible to further understand life at Shingwauk and begin to tell Shingwauk history from the survivor perspective.

4.8 Conclusion

Embedded within the geographic space, physical walls, personal memories and archival records are a multitude of narratives relating to the lived experience of the Shingwauk Residential School site. In our collaborative work on the RSLMMP with the GCRC team, we utilized a three-pronged approach to exploring the history of the land occupied by Algoma University. By combining survivor testimonies, archival documents and historical photographs, it is possible to create additional layers of understanding of the residential school experience. These new dimensions of understanding provide pathways for SRSC to explore in its public programming, tours and exhibitions. Work undertaken as part of the RSLMMP has grown beyond points on a map to become integrated into community reclamations of space, the preservation of survivor experiences and increasing understanding of residential schools.

References

Bastian, J.A., 2006. Reading colonial records through an archival lens: the provenance of place, space and creation. Archival Science 6, 267—287. https://doi.org/10.1007/s10502-006-9019-1.

Brown, J., Vibert, E. (Eds.), 2003. Reading Beyond Words Contexts: For Native History, second ed. Broadview Press, Peterborough.

Caswell, M., 2014. Archiving the Unspeakable: Silence, Memory and the Photographic Record in Cambodia. University of Wisconsin Press, Madison.

Caswell, M., Gabiola, J., Zavala, J., et al., 2018. Imagining transformative spaces: the personal-political sites of community archives. Archival Science 18, 73—93. https://doi.org/10.1007/s10502-018-9286-7.

Farrell Racette, S., 2009. Haunted: first Nations children in residential school photography. In: Lerner, L. (Ed.), Depicting Canada's Children. Wilfrid Laurier University Press, Waterloo, pp. 49—84.

Findlay, C., 2016. Archival activism. Archives and Manuscripts 44 (3), 155—159.

Fletcher Luther, F., 1991. Reunion Transcript [document] Shingwauk 1991 Reunion fonds, (2017-002/001(001)). Shingwauk Residential Schools Centre, Algoma University, Sault Ste. Marie, Canada.

Fraser, C., Todd, Z., 2016. Decolonial Sensibilities: Indigenous Research and Engaging with Archives in Contemporary Colonial Canada. L'Internationale. http://www.internationaleonline.org/research/decolonising_practices/54_decolonial_sensibilities_indigenous_research_and_engaging_with_archives_in_contemporary_colonial_canada.

Hoerig, K.A., 2010. From third person to first: a call for reciprocity among non-Native and Native museums. Museum Anthropology 33, 62—74. https://doi.org/10.1111/j.1548-1379.2010.01076.x.

McCracken, K., 2017. Archival photographs in perspective: Indian Residential School images of health. British Journal of Canadian Studies 30, 175—177.

Miller, J.R., 2004. Reading photographs, reading voices: documenting the history of native residential schools. In: Miller, J.R. (Ed.), Reflections on Native—Newcomer Relations. University of Toronto Press, Toronto, pp. 82—104.

Nesmith, T., 2002. Seeing archives: postmodernism and the changing intellectual place of archives. American Archivist 65, 24—41.

O'Neill, J., Holland, A., Light, A., 2008. Environmental Values. Routledge, London.

Rogers, C., 1991. Reunion Transcript [document] Shingwauk 1991 Reunion fonds, (2017-002/001(002)). Shingwauk Residential Schools Centre, Algoma University, Sault Ste. Marie, Canada.

Sands, D., 1981. Reunion Video [audio-visual] Shingwauk 1981 Reunion fonds, (2016-037/002(007)). Shingwauk Residential Schools Centre, Algoma University, Sault Ste. Marie, Canada.

Tinkler, P., 2013. Using Photographs in Social and Historical Research. Sage Publications, London.

Thomas, R.A., 2005. Honouring the oral traditions of my ancestors through storytelling. In: Brown, L., Strega, S. (Eds.), Research as Resistance: Critical Indigenous & Anti-oppressive Approaches. Canadian Scholars' Press, Toronto, pp. 237—255.

Truth and Reconciliation Commission of Canada, 2015. TRC Final Report. http://www.trc.ca/websites/trcinstitution/index.php?p=890.

Tuck, E., Yang, W., 2012. Decolonization is not a metaphor. Decolonization Indigeneity Education & Society 1, 1—40.

Van Kirk, S., 1980. Many Tender Ties: Women in Fur-Trade Society, 1670-1870. University of Oklahoma Press, Norman.

Wilson, E.F., 1890. Indian folk-lore. The Canadian Indian 1, 57—58.

5

The Carlisle Indian Industrial School: Mapping resources to support an important conversation

Susan Rose*,1, James Gerencser2

1Charles A. Dana Professor and Chair of Sociology, Dickinson College, Carlisle, PA, United States;
2College Archivist, Dickinson College, Carlisle, PA, United States
*Corresponding author

5.1 Introduction

The Carlisle Indian Industrial School in Carlisle, Pennsylvania is a major site of memory for Native Nations across the country and for those interested in the history of American education and colonization. Lieutenant Richard Henry Pratt founded the school in 1879 as a means to implement his vision for solving the so-called 'Indian Problem'. During the school's 39 years of operation, over 8000 students were taken to Carlisle in an attempt to assimilate them to the dominant Euro-American culture. As the flagship school, Carlisle served as a model for other off-reservation boarding schools across the country and in Canada (including

Shingwauk Industrial Home referred to in Chapter 4 and elsewhere in the book); and though Carlisle closed in 1918, similar schools operated well into the latter half of the 20th century. The lasting and often traumatic impact of Carlisle and the Indian boarding school movement is an important part of American history that warrants continued exploration, interrogation, dissemination and discussion.

By the late 1800s, the federal government was entering the final stages of Native disposses-sion and American conquest. By the time Carlisle opened its doors, most of the military fight-ing was over. With Native Nations now sequestered on reservations, white Christian reformers, who called themselves 'Friends of the Indian', presented the policy of education and assimilation as a more enlightened and humane way to deal with Native populations. Yet the purpose of the education campaign matched previous policies: to dispossess Native peoples of their lands and extinguish their existence as distinct groups that threatened the nation-building project of the United States. These destructive objectives were effectively masked for the white public by a long-established American educational rhetoric that linked schooling to both democracy and individual advancement, and by a complementary and unquestioned commitment to the American republican experiment. For Pratt, the main task therefore was to convince white Americans that his mission to transform Native children from 'savagery' to 'civilization' was both desirable and possible.

Richard Henry Pratt implemented his vision for educating and 'civilizing' Native American students by removing them from their communities west of the Mississippi and bringing them east to Carlisle, Pennsylvania. Once at the school, Native students were forbidden to speak their own languages, wear their traditional clothing or practice their own customs and reli-gions. Pratt's infamous statement, 'Kill the Indian in him, and save the man', echoed through the chambers of Congress and historical texts for generations. Between 1879 and 1918, more than 8000 Native American students from nations and tribes all over the United States and its territories were enrolled at Carlisle.

The first contingent of students to arrive at the newly opened Carlisle Indian School was 84 Lakota children from the Rosebud and Pine Ridge Indian agencies in Dakota Territory. They were to become the subjects of an educational experiment that would soon be extended to include Native children from across the United States and Canada. Their train drew into the railroad station in Carlisle, Pennsylvania close to midnight on 5 October 1879, and the chil-dren were greeted by a large contingent of local townspeople, eager to catch a glimpse of these exotic young people from the western territories. The children had travelled over a 1000 miles by river and rail, and this great distance was fundamental to Carlisle's mission (Fig. 5.1.1).

Richard Henry Pratt, the school's first superintendent, was determined to remove Native children as far as possible from their families and communities, to strip them of all aspects of their traditional cultures, and instruct them in the language, religion, behaviour and skills of mainstream white society. Pratt's objective was to prepare Native youth for assimilation and American citizenship. He insisted that in schools like Carlisle this transformation could be achieved in a generation. An acting army officer, Pratt had secured government support to establish and run this first federally funded, off-reservation Indian boarding school. Carlisle provided the blueprint for the federal Indian school system that would be organized across the United States, with 24 analogous military-style, off-reservation schools, and similar boarding institutions on every reservation.

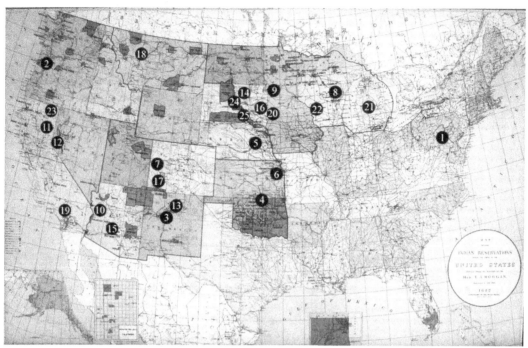

FIGURE 5.1.1 Map of Federal Off-Reservation Indian Boarding Schools, in order of founding (see Appendix for Names of Schools and Dates of Founding) http://carlisleindian.dickinson.edu/teach/locations-indian-boarding-schools-throughout-united-states.

For Native communities, Pratt's experiment at Carlisle initiated processes of diaspora, dislocation and rupture deeper and more profound than he had envisaged. These processes had many immediate impacts as well as long-term legacies. For all Native nations, physical and spiritual well-being was anchored not just within their communities but also within the environment and land that surrounded them. When Native children were transported hundreds, and sometimes even thousands, of miles to Carlisle, Pennsylvania, they were not only subjected to a strict 'civilizing' program to strip them of their cultures, they were also forced to live in an alien place devoid of familiar cultural, spiritual and geographical markers as well as the support and succour of kin and community.

Young people were brought from all over the country: California and the Carolinas, New Mexico and New York, Arizona and Alaska. The nations sending the highest number of children were the Sioux (Lakota, Nakota and Dakota) and the Chippewa (Ojibwe).

Carlisle students were enrolled initially for a period of 3 to 5 years. Most did not return home at all during that time, and many spent far longer at Carlisle. Pratt's goal was to immerse them in the dominant White Anglo-Saxon mainstream culture. Speaking to a convention of Baptist ministers in 1883, he used the image of baptism to explain his philosophy for transforming Native children so that they could be made to emulate white men and women:

> In Indian civilization I am a Baptist because I believe in immersing the Indians in our civilization and when we get them under holding them there until they are thoroughly soaked (Utley, 1964, p. xxiii).

The force and suggestion of drowning contained in Pratt's metaphor were not accidental; he believed every necessary measure should be taken to impose 'civilization' through total immersion. His slogan was: *To civilize the Indian, get him into civilization. To keep him civilized, let him stay* (Utley, 1964, p. 283). And this was the rationale for his 'outing' program. Almost all Carlisle students experienced multiple dislocations when, instead of returning home to their own families for the summer breaks, they were sent 'out' into local communities to work for white families, typically as farmhands or domestic servants. Some stayed 'out' much longer, and even attended local public schools alongside the children of their outing patrons. The outing program that Pratt instituted was a critical piece of the Carlisle Indian School experience, with many students being sent on multiple outings to different locations throughout their years of enrolment at Carlisle.

Thousands of Native children and youth followed that first group from Dakota Territory over the next 40 years, transported from Indian agencies across the continent on rail networks that built and connected the markets of the United States. The vast majority, however, did not assimilate into mainstream society as Pratt had envisioned but instead returned to their reservation homes, often feeling caught between two cultures and languages. Cut off from the nurture of tradition, family and community, they experienced a rupture in their affiliations, affections and identities. For many this began a legacy of trauma and disenfranchisement that was passed down through the generations.

Although the stories of Carlisle and its legacies are complex, the sources through which these can be tracked are typically very one-sided because the official record was created and preserved by white officials. Few students left any written records. Those who did leave behind textual evidence wrote mostly for school publications that were under the scrutiny of white editors. On their return home, many students did not speak about their experiences, and oral stories that were passed down the generations often remained closely guarded within the communities; for understandable reasons they are not widely accessible. Yet it is an indisputable fact that the Indian School initiated a large-scale diaspora of Native children, and that the geo-spatial-cultural dislocation they experienced as part of settler colonialism was grounded in a new and foreign place-name that would soon become infamous in all Native communities as a major site of cultural genocide: Carlisle.

For N. Scott Momaday, the Kiowa writer, artist and Pulitzer Prize winner, the name Carlisle carries an historical significance parallel to Gettysburg and Wounded Knee within America's national memory and history: 'Carlisle, in a more subtle and obscure story than that of Gettysburg, is a place name among place names on a chronological map that spans time and the continent' (Fear-Segal and Rose, 2016, p. 44—45).

5.2 Building accessible resources to support sharing and scholarship

The Carlisle Indian School (CIS) and the indigenous boarding school movement represent a very active area of research among Native and non-Native scholars, teachers, students, area residents and descendants from around the United States and internationally (especially Canada, Australia and New Zealand). Scholars are working hand in hand with descendants of the students, who are learning from and contributing to this research. In the last decade, not only have many scholarly and popular books, articles and documentaries related to the Carlisle Indian School been produced, but also a number of symposia and community events have taken place.

Using recent new frameworks for understanding American Indian education, including settler-colonialism and genocide studies, we are working to position Native American boarding school history within the wider history of the United States and American education. In exploring the ways in which this troubled and veiled history can be used to understand and illuminate current educational concerns, practices and community struggles, we plan to map out pathways for this history to be taught within the context of Native and non-Native schools and colleges, and within community organizations such as libraries, cultural centres and museums.

The Carlisle Indian School Digital Resource Center (CISDRC, found online here: http://carlisleindian.dickinson.edu/) represents an effort to aid the research of descendants and scholars alike by bringing together, in digital format, a variety of resources related to the Carlisle Indian School that are physically preserved in various locations around the country. Through these resources, we seek to increase knowledge and understanding of the school and its complex legacy, while also facilitating efforts to tell the stories of the many thousands of students who were enrolled there. With the CISDRC, the intention is not merely to share archival material, but to further build and develop the archival record by providing a platform for Native Americans to add both their voices and their personal documentary collections to the conversation. By doing so, the preserved and shared record is enriched, and the opportunity for increased learning, understanding and healing is supported.

The Carlisle Indian Industrial School operated for 39 years, but it closed in 1918 with little planning or advance notice due, in part, to the pressures exerted by the First World War. The school was situated on the site of a former military barracks. Needing space for convalescing soldiers returning from the fighting, the War Department requested in July that the school be closed and that the barracks be returned to military use as soon as possible. On 1 September 1918, the Department of the Interior formally relinquished control of the barracks.

In a rather hurried fashion that summer, arrangements were made to transfer students to other Indian boarding schools or send them home to their families. In many cases, students needed to be recalled from their various outings throughout the Mid-Atlantic region. Meanwhile, school administrators did their best to organize the ledgers, paper files and other records that documented the operations of the school and the experiences of its students. Most of the administrative files of the Carlisle Indian School were sent to the Bureau of Indian Affairs offices in Washington, DC. In time, those files were deposited at the US National Archives and Records Administration (NARA), where they continue to be preserved and made accessible for researchers.

While the extant files are incomplete, reflecting irregular gaps in the documentation, they do still provide a wealth of information about the manner in which the school operated. Enrolment ledgers detail, in chronological order, where in the United States students were travelling from as well as some information about their physical condition and their family background; financial ledgers reveal the routine expenses, including how much was spent on food, school uniforms, books and supplies, heating fuel, and transportation for the students and their chaperones to and from the school; outing ledgers show the towns where students were being sent to live and work with white families, part of the effort to more fully immerse them in the dominant American culture; and student files provide a variety of information about each individual student's experiences, including admissions applications, medical records and alumni surveys reflecting where former students settled in later years.

The effort to digitize these extant Carlisle school records began in 2013 through a Mellon Foundation grant to Dickinson College to support work in the digital humanities. Following a preliminary visit to NARA to assess the volume and formats of the material as well as to test workflows for scanning, the digitization process began in earnest. Research teams of four have typically made 2-week visits to NARA twice a year, once in the winter and once in the summer. With each visit, roughly 20,000 pages of original files have been digitized, and in the subsequent months, those pages have been processed, catalogued and then uploaded to the project website so that interested researchers have free, easy access to these documents.

The first records to be scanned were the student files, numbering 6129 folders, as well as oversized ledgers showing student enrolment and outing activity. Several boxes of information cards were also scanned early in the process. These cards supplement the student files, providing details about the students' experiences that have been copied from ledgers. As the student files were being processed, the handwritten entries in the ledgers were being transcribed to enable easier browsing and searching by researchers. These core records, representing some 140,000 pages of material, were fully digitized in about three and a half years. Most of that content was uploaded to the website within 6 months of being scanned, although some of the more than 16,000 individual information cards are still in the process of being uploaded.

Other materials that reflect the student experience were scanned next. Three volumes of meeting minutes of one of the young men's debating societies, the Invincibles, have been preserved, offering insight to issues of concern to the students at that time. Also preserved are financial ledgers that document student earnings and spending. Carlisle students were generally paid a modest wage for work done at the school or while on an outing, and these funds were available to the students for personal use, with the approval of the school administrators.

General school administrative records and other miscellaneous materials rounded out the documentation of the Carlisle Indian School maintained at NARA. Personnel files for just a handful of employees can be found, two of whom were themselves former Carlisle students. Several ledgers detailing the finances of the school remain, one of those dating from the very beginning of the school. Day books, showing the day to day enrolment changes and movements of students, provide details for just half a dozen years. And finally, copies of school calendars, magazines, catalogues and event programs document the later years of Carlisle's operations.

With all of the original administrative records of the school itself digitized, we next turned our attention to broader files of the Bureau of Indian Affairs, and particularly the files of the Commissioner, with whom the Carlisle school superintendents maintained a constant correspondence. These files, which are not particularly well indexed and thus have not previously been thoroughly explored by researchers, are proving to be a treasure trove of additional information about the school's operations and about individual student experiences. Questions about how to handle parents' requests for their children to be returned home and comments about student discipline are filed alongside blueprints for new school buildings and deeds for adjoining properties that the superintendent intends to purchase. Although we still have a great deal to digitize from this cache of records, this material opens the door to new areas of inquiry and promises to support robust research in the coming years.

While the CISDRC is an ongoing work in process, there are already more than a quarter million pages of digitized material freely available to website visitors, and new content is being added regularly. The student files and cards described earlier are searchable by name, and can also be browsed by nation or tribe, date of enrolment, and document type. Photographs and other images can be searched in much the same manner. Transcribed admissions, outing and student finance ledgers can be browsed page by page, or they can be downloaded as Excel files to allow for more detailed searching and sorting. Newspapers, magazines and newsletters that were produced by Carlisle students on the school's own printing presses are full-text searchable, can be browsed online and can be downloaded and viewed as PDF files.

This wealth of documentation is being accessed daily by people researching their own family history and genealogy. Students at various grade levels and in college are using the site for their own exploration and to complete school assignments examining the boarding school movement. Professional researchers across different disciplines are mining the website for information that will support their scholarly work, and they and others are requesting copies of material for use in publications. Through its first 5 years, the CISDRC has been visited more than 175,000 times and users have accessed and downloaded over one million individual pages and documents.

5.3 Mapping to support teaching

With a sizeable amount of original documentation available for access and exploration, which was our first objective, we moved on to providing teaching modules and lesson plans that enable teachers and students to more easily navigate and make effective use of the resource center while exploring and interrogating the school's history (http://carlisleindian. dickinson.edu/teaching). A National Historical Publications and Records Commission (NHPRC) grant 'Public Engagement, Discovery, and Contribution: Teaching and Sharing via the Carlisle Indian School Digital Resource Center', also supported a week-long Teachers' Institute for twenty educators from across the country to come to Carlisle to explore the site, and they also contributed to the lesson plans online.

Among the first teaching resources added to the website was a map of the grounds of the Carlisle Indian School, which was first prepared by Jacqueline Fear-Segal in 2000. The map she created shows changes in the grounds over time, contrasting the opening of the school in 1879 with the closing in 1918. A map that shows the grounds as of 2000 also includes a key indicating which of the various buildings date back to the time of the school's operations, and when each of them was constructed. Though static in its presentation, this map provided an opportunity to visualize the changes and growth to the Carlisle Indian School over time.

In order to demonstrate the early impact of Carlisle as a model for additional off-reservation boarding schools across the country, we also developed a simple map that pinpointed each of those subsequent schools and numbered them in the order in which they were founded. (See Fig. 5.1.1, pictured earlier.) The map shows two dozen schools being started in less than 20 years. Unlike Carlisle, however, all of these schools are located closer to the Native populations that they were designed to serve, which raises interesting questions about Pratt's notion that greater distance from students' homes was an important part of the assimilation-through-education process.

As has been mentioned, the outing program was another integral part of Pratt's method of assimilating Native American boys and girls. Once students learnt enough of the English language to be able to communicate reasonably well, and once they had learnt some of the important foundational skills and trades at the school, they would be sent away from Carlisle to live and work with white families, particularly during the summer months. Boys were typically sent to work on farms, while girls were sent to work as domestics in households.

School documents reveal where students were sent, their files indicating the names of the families with whom they stayed and where they lived. Files will also sometimes include reports about individual students' time while on outing, forms having been filled out by the outing family patrons themselves, or by Carlisle employees who made site visits to check on both the students and the patrons. Besides the outing information in students' personal files, ledgers were also maintained to record each of the outings. These ledgers generally listed a student's name, the date the student departed from Carlisle, the patron of the outing, the town (or in some cases, railroad station) where the patron lived and finally the date when the student returned to Carlisle.

In order to help visualize some interesting details of the outing system, we set about to map some of the information from these outing ledgers. The extant ledgers do not show all outings throughout the school's history, but they represent well over half. They also reflect dates from across the school's history. Using this information, we hoped to represent the places where students were sent most frequently, to compare differences between outing locations for boys and girls, and to compare the change over time in where students were sent.

The ledgers themselves had been transcribed by undergraduate students shortly after they were digitized. This initial transcription effort required several hundred hours of labour, as the ledgers included more than 6000 lines of data. As the ledgers were created in tabular form, the transcription was done in a spreadsheet. Initial transcription was done as true to the original as possible, with best guesses being made in cases of handwriting that was difficult to decipher. In order to facilitate searching, handwritten abbreviations were replaced with the full spellings for names and places, and ditto marks were also replaced with the text they represented.

In order to prepare these ledgers for mapping, we needed to make some additional adjustments to the available information. First, we added a field to note the sex of the student, so that we might be able to see any differences between the outing locations of boys and girls. Second, we worked through all of the place names to ensure that we had a correct county identified for every outing. (In some cases, towns or railroad crossings would not include the county, so we needed to consult contemporary sources to identify the proper place listed) Third, we made certain that the place names were all represented in the same manner — consistent spellings for county names and modern two-digit abbreviations for state names.

With the primary data points normalized, we sought the assistance of Kayla Kahan, an undergraduate intern with experience utilizing GIS tools. She set about the task of beginning to map out the information, giving consideration for how best to represent numbers of outings, how to frame the time parameters and how to differentiate the outing for boys and girls. After a few meetings to discuss various details, the data were presented (Fig. 5.3.1).

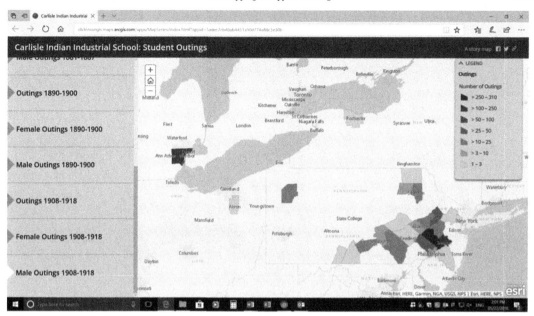

FIGURE 5.3.1 Map showing the number of outings by county for male students, 1908–18.

The resulting maps reveal a number of interesting features about the outing program and its changes over time. Throughout the school's history, the outings took place most often in central and southeastern Pennsylvania, as well as parts of New Jersey and northern Maryland. The heaviest concentrations of outings were in Philadelphia and its surrounding counties. There were occasional outings in more distant locations, including New York, parts of New England, Ohio, Indiana, and Michigan, and even Missouri and South Dakota.

The outings in Michigan are exclusively for male students, and they all take place during the later years of Carlisle's operation. All but one of these outings take place around Detroit. This is evidence of the growth of factories in that area, and the movement toward having young men spend outings in manufacturing instead of farming. The only two counties that register more outings during that time period are Bucks County in Pennsylvania (which hosted more outings for young men than any other, by far, throughout the school's history) and Mercer County in New Jersey.

In the other counties in and around Philadelphia, there are interesting distinctions by gender. While Chester County hosted the most outings for female students, only a handful of outings for male students ever took place there. The situation is similar in Delaware County, and also in Burlington County in New Jersey, with very high numbers of young women having outings, but very few young men. Montgomery County sees a little more of a mix, but still shows outings for female students at six times the number for men (Fig. 5.3.2).

While it will take other types of research to determine precisely why we see these disparities between the sexes in outing locations, the mapping itself does help us to formulate the questions. Future researchers may want to drill into the data more deeply, breaking things down

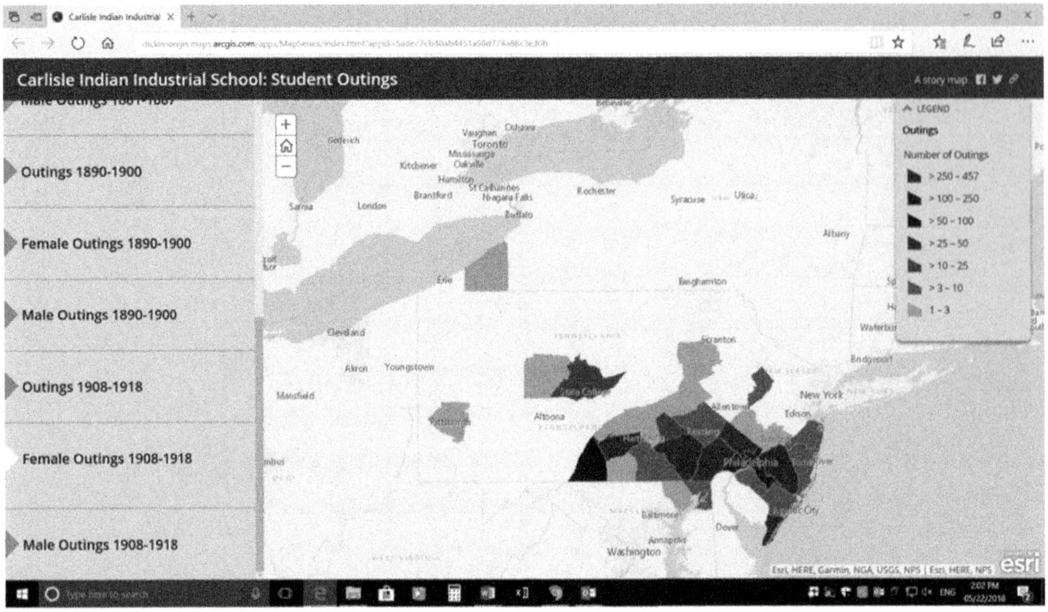

FIGURE 5.3.2   Map showing the number of outings by county for female students, 1908–18.

into smaller time increments, or looking at individual towns instead of counties, to see if there are even higher concentrations of activity than appear to us already.

5.4 Mapping to tell a story

When the granddaughter of a US Cavalry officer sent a copy of a photograph to the Cumberland County Historical Society in 2002, she was looking for information about the two children pictured. The boy and girl had been provided a kind of foster home by her grandparents, but she knew nothing further about them. That question sparked the research of Jacqueline Fear-Segal, who was able to piece together the stories of Jack and Kesetta with the help of the documentary sources and the oral histories of members of the Lipan Apache Band of Texas (Fear-Segal and Rose, 2016, p. 201–233).

When the US Army attacked Jack and Kesetta's village in Mexico, most of the men were away on a hunting expedition. The two children had been hidden by their mother not far from the village, and as a result, they survived the massacre. The soldiers took Jack and Kesetta as prisoners, and a family with that Army unit began to care for the children. A few years later, in March 1880, these two Lipan Apache children were sent to the Carlisle Indian School.

While most Carlisle students had a home to return to eventually, Jack and Kesetta had none, all ties to their tribe having been severed. Jack studied at Carlisle for a few years before being adopted by Sarah Mather, an older woman who had long been a friend and supporter of the school. He was sent to her home in Florida for a short while, but he fell ill and was returned to Carlisle. Jack died there and was buried in the school cemetery in 1884.

Kesetta was a few years older than her brother, and she was sent on an outing to work for a family in Schuylkill Haven, Pennsylvania in April 1883. Over the next 20 years, Kesetta would be sent on numerous other outings, and she would eventually earn the distinction of being the longest enrolled student at Carlisle. She became pregnant during her last outing, and she gave birth to her son Richard (Dick) in Philadelphia in 1903. When her son was only 3 years old, Kesetta died from tuberculosis at the age of 39. Dick was subsequently sent to the Carlisle Indian School, becoming the youngest child enrolled there.

The stories of Jack, Kesetta and Dick are recounted by Jacqueline Fear-Segal as a chapter in her 2007 monograph *White Man's Club* (Fear-Segal, 2007, p. 255–282; Fear-Segal, 2016a). Daniel Castro Romero, Jr., General Council Chairman of the Lipan Apache Band of Texas, collaborated with Fear-Segal to further tell this story in the 2009 documentary 'The Lost Ones: The Long Journey Home', produced by Susan Rose. Both the documentary and the published monograph utilize oral history and extant records to share what is known about the experiences of these three individuals.

In order to help illustrate this story, Kate Tanabe, another undergraduate with an interest in GIS tools, tracked the movements of Jack, Kesetta and Dick. Using photographs, the records from the school and the information in the documentary, Kate mapped out a timeline depicting the key moments in Jack and Kesetta's lives. From their capture in Mexico, to their enrolment at Carlisle, and Kesetta's various outings, the story map allows audiences to more easily visualize the great distances that were travelled, as well as the continuous movement that characterized Kesetta's outings over a period of 20 years (Fig. 5.4.1).

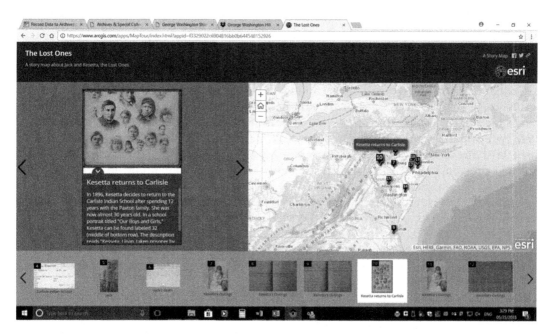

FIGURE 5.4.1 Story map showing the movements of Jack and Kesetta.

5.5 Mapping to answer research questions

The Carlisle Indian Industrial School saw its first students arrive in October 1879. Within weeks Amos LaFramboise, a boy of 13 from the Sisseton Sioux Agency, fell ill and died, 'the first and we sincerely hope it may prove the only death in the school'. (*Carlisle Herald*, 4 December, 1879). A second student, Abe Lincoln from the Cheyenne Agency, died less than 2 months later. Since the local cemetery did not admit the burial of non-Whites, the school established its own cemetery early in 1880. By the time of the school's closing in 1918, this cemetery held more than 190 occupants.

A few years after the US Army took control of the Carlisle Barracks, it sought to have that cemetery moved to another part of the grounds. By that time, many of the headstones and other markers left by the graves had deteriorated so that the names and dates were no longer legible. After the cemetery was moved during the summer of 1927, there were 14 fresh new headstones that read, simply, 'Unknown'.

Over the years, staff at the Cumberland County Historical Society, project partners with the CISDRC, have sought to identify those 14 unnamed individuals. Using documentary clues collected over the years, Barbara Landis had identified roughly half of these people (Landis, 2016, p. 185−197). From digitized correspondence between the superintendent and the commissioner, a few additional names were found in 2016, but some 'Unknowns' remained.

During the summer of 2017, CISDRC team members Frank Vitale and George Gilbert began an effort to map the burial locations of the individuals in both the original cemetery and the new location. Through this work, we wanted to make information about the deaths and burials of Carlisle's students more easily discoverable, but we also hoped that the mapping itself might provide new clues about the unnamed burials.

A diagram found at NARA dating from February 1927 showed 186 burial plots in the old cemetery. By using the existing data about individuals and their dates of death, the details of how the cemetery grew over time became more evident. Using that information, we surmised that the dates of death on existing headstones were incorrect in a few cases, and we were able to find documentation to corroborate those assumptions.

On that diagram of 186 burial plots, names were recorded for only 178 of the spaces, leaving 8 'unknowns'. By confirming all of the known individuals' dates of death, we were able to determine who was likely to have been buried in a few of those plots. In one case, there were two burials less than a month apart, but there was also an 'unknown' plot between them. With the date narrowed down, our research team was able to find a source that identified that previously unknown student to have died at the school, a Pueblo girl of 16 years named Ella Soisewitza.

Taking all of the information they had been able to gather to that point, Frank and George set about mapping both the old and new cemeteries. Using the diagram of the old cemetery, they plotted each burial in chronological order so that visitors to the site could visualize the incremental growth of the cemetery over its 39 years of use. They created one view of the cemetery colour-coded by the nation or tribe of each of the students, and created others showing burials in 5-year and 10-year groupings.

For the relocated cemetery (Fear-Segal, 2016b), Frank and George created similar views reflecting nation or tribe and groupings by decade, but they also took advantage of basic GIS

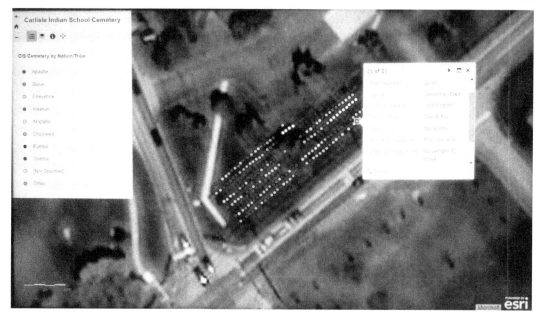

FIGURE 5.5.1 Interactive plot map of the cemetery.

mapping to pinpoint each headstone as it currently stands. Each point on this interactive map shows the information about who is buried there; filters allow the user to again visualize the burials by the individuals' nation or tribe, or by decade groupings for dates of death (Fig. 5.5.1).

Accompanying these various maps and diagrams are the different sources that were used to confirm each person's death and burial (The cemetery resources are available here: http:// carlisleindian.dickinson.edu/cemetery-information). Each individual has a webpage on which is mounted a compilation of the available documentation found thus far, including items from the student files as well as notices in newspapers, mentions in a day book, or notes in enrolment ledgers. Names, tribal affiliations, dates of decease and burial plot locations are also available in tabular form as a downloadable Excel spreadsheet. Finally, other sources that have documented the cemetery over time, each of which provides additional clues to the burial sites of the students, are included. This work has been useful as various nations have discussed the possible disinterment of their family members.

5.6 Conclusion

In exploring the ways in which this troubled and veiled history can be used to understand and illuminate current educational concerns, practices and community struggles, we plan to map out pathways for this history to be taught within the context of Native and non-Native schools and colleges, and within libraries, museums and other community-based cultural organizations. The NHPRC-funded Teachers' Institute mentioned earlier focused not only on how and why Indian boarding schools developed and educated young people but also on how this information can be effectively taught to enable a better understanding of how this intertwined history continues to affect the present. Many Native communities continue to suffer from the negative effects of the boarding school movement, and helping younger generations understand more about this history is an important part of the healing and reconciliation process.

In developing digital and curricular resources that better enable this history to be widely disseminated, the following questions guide our discussions:

- How do we confront, discuss and teach the history of Indian schooling effectively — not as a static historical exercise, but as a living history that affects all of us today?
- How should the experience of Carlisle and the other boarding schools be shared and understood, not only by those directly affected by them but also by those who know little about them?
- How can Native young people contribute their own work and voices (through original research, oral histories, photographs and creative works) to inform this history and its impact on their families and communities? How can their engagement with the CISDRC contribute to their own and others' understandings of our shared histories?
- How can non-Native young people learn more about Native American history and education in ways that also help them to interrogate their own histories and educational experiences?
- How can the scholarly and creative contributions of both Native and non-Native young people enrich the CISDRC for other students, teachers, scholars and communities?
- How may this multicultural history be taught so that students, teachers and community members can better understand shared pasts and shared futures?
- How may what we learn from these conversations influence current teaching content and practices, especially as the US population is becoming increasingly diverse?
- What additional resources and content would benefit users of the CISDRC, and how might these materials best be created and presented? How can we make the CISDRC more usable for students, teachers and scholars?

During the academic year 2017–18, we conducted additional workshops across the country at more than a dozen different Native communities and schools using the materials described in this paper. We are encouraging students, their families and community members to become citizen archivists, contributing their own family stories and collections to the site. In so doing, we will promote the use of primary sources by teachers, students and the public and encourage further citizen contributions to the historical record.

We hope that some of the modest mapping projects we have undertaken for the CISDRC will serve as useful and interesting examples of how to share and view the available information in ways that help to answer some questions while sparking new ones. We look to such efforts to inspire students and scholars to consider how the data made accessible through the CISDRC, as well as the information they may collect and hold themselves, can be viewed in different ways and tell different stories, which in turn may spark further enquires and deepen the conversation.

The Indian schools legacy is an important part of American educational history and the history of settler-colonialism. It needs to be taught because it reveals the ways in which education can be used to enforce conformity of subordinated groups as well as promote democratic opportunity. The research team, composed of dedicated archivists, teachers and scholars who have been at the forefront of research on this subject, are committed to disseminating this information to broader audiences. The purpose is not only to inform, but also to raise awareness that may inspire people to engage their own histories and contribute to a more holistic understanding of this entangled history. The project team knows well how this history has been obscured by the dominant and benign narrative surrounding American schooling that promises equal opportunity and upward mobility. Through this collaborative work with citizen archivists, we are eager to open up a dialogue concerning the challenges and possibilities of contemporary, multicultural education to the benefit of all.

Appendix (see Fig. 5.1.1)

Locations of Federal Off-Reservation Indian Boarding Schools throughout the United States:

1. Carlisle Indian Industrial Boarding School, Carlisle, Pennsylvania — Founded 1879
2. Chemawa Indian School, Salem, Oregon — Founded 1880
3. Albuquerque Indian School, Albuquerque, New Mexico — Founded 1882
4. Chilocco Indian Agricultural School, Chilocco, Oklahoma — Founded 1884
5. Genoa Indian Industrial School, Genoa, Nebraska — Founded 1884
6. Haskell Indian Industrial School, Lawrence, Kansas — Founded 1884
7. Grand Junction Indian School, Grand Junction, Colorado — Founded 1886
8. Wittenberg Indian School, Wittenberg, Wisconsin — Founded 1886
9. Morris Indian Boarding School, Morris, Minnesota — Founded 1887
10. Fort Mojave Indian School, Fort Mojave, Arizona — Founded 1890
11. Greenville School & Agency, Greenville, California — Founded 1890
12. Stewart Indian School, Carson, Nevada — Founded 1890
13. Santa Fe Indian School, Santa Fe, New Mexico — Founded 1890
14. Pierre Indian School, Pierre, South Dakota — Founded 1891
15. Phoenix Indian School, Phoenix, Arizona — Founded 1891
16. Flandreau School, Flandreau, South Dakota — Founded 1892
17. Fort Lewis Indian School, Hesperus, Colorado — Founded 1892
18. Fort Shaw Industrial Indian Boarding School, Fort Shaw, Montana — Founded 1892
19. Perris Indian School (became the Sherman Indian School), Perris, California — Founded 1892
20. Pipestone Indian School, Pipestone, Minnesota — Founded 1892
21. Mt. Pleasant Indian Industrial Boarding School, Mt. Pleasant, Michigan — Founded 1893
22. Tomah Indian School, Tomah, Wisconsin — Founded 1893
23. Fort Bidwell Indian School, Fort Bidwell, California — Founded 1898
24. Rapid City Indian School, Rapid City, South Dakota — Founded 1898
25. St. Joseph's Indian School, Chamberlain, South Dakota — Founded 1927

References

Fear-Segal, J., 2007. White Man's Club: Schools, Race, and the Struggle of Indian Acculturation. University of Nebraska Press, Lincoln, pp. 255–282.

Fear-Segal, J., 2016a. The lost ones: piecing together the story. In: Fear-Segal, J., Rose, S.D. (Eds.), The Carlisle Indian Industrial School: Indigenous Histories, Memories, and Reclamations. University of Nebraska Press, Lincoln, pp. 201–233.

Fear-Segal, J., 2016b. The history and reclamation of a sacred space: the Indian school cemetery. In: Fear-Segal and Rose (Eds.), pp. 201–232.

Fear-Segal, J., Rose, S.D. (Eds.), 2016. The Carlisle Indian Industrial School: Indigenous Histories, Memories, and Reclamations. University of Nebraska Press, Lincoln.

Landis, B., 2016. Death at Carlisle and naming the unknowns in the cemetery. In: Fear-Segal, Rose (Eds.), pp. 185–197.

Utley, R.M. (Ed.), 1964. Battlefield and Classroom: Four Decades with the American Indian, 1867–1904. University of Nebraska Press, Lincoln.

Charting the intimate terrain of Indigenous Boarding Schools in Canada and the United States

Andrew Woolford*

Professor, Department of Sociology & Criminology, University of Manitoba, Winnipeg, MB, Canada

*Corresponding author

6.1 Introduction

The space of destructive assimilative education is uneven and fractured. Though developed around a vision of Indigenous peoples as a 'problem' to be resolved, and with a general acceptance this would be best dealt with through forced assimilative education directed toward severing Indigenous young people from their families and communities, the institutions themselves adapted to local geographies, ecologies, cultures and politics. For this reason, much like a cybercartographic map, any analytical framework developed to compare and contrast these schools—a name which is, itself, something of a misnomer, or, at minimum, only a partial admission (see Woolford and Gacek, 2016)—must be modular, fluid and dynamic. Any attempt to set rigid boundaries around assimilative education is likely to misrepresent the varied experiences within Indigenous boarding schools, the relations between Indigenous communities and the school and the specific interactions of staff and students, among other factors. Thus, while for many a term like genocide captures the destructive ends sought through these schools, a processual and nuanced application of the term must be insisted upon, so that the historical particularity of Indigenous Boarding Schools is not unduly circumscribed and reduced for the satisfaction of legal accusation. In this chapter, I present the metaphor of the 'settler colonial mesh' as means for capturing this complexity. In addition, I show how what is sometimes referred to as 'intimate colonialism' can be captured through this metaphor. Finally, I discuss how Cybercartography opens map-making to the intimacy and unevenness of settler colonial genocide.

The space of Indigenous boarding schools is, of course, broader than the schools themselves. To disconnect the boarding schools from other assimilative and carceral institutions, from day schools to forced child removals and prisons, not to mention from the broader field of settler colonial control over Indigenous lives, would be to overlook how boarding schools, as a specific response to an imagined Indian problem, emerge from this distinctive history. Though this chapter focuses on assimilative boarding schools, it does so with an eye to broader processes of settler colonial domination.

6.2 The settler colonial mesh

In *This Benevolent Experiment: Indigenous Boarding Schools, Genocide, and Redress in Canada and the United States* (2015), I chart the uneven development of Indigenous boarding schools in North America through a multilayered approach that is attuned to temporal and regional variations in assimilative schooling, as well as their location in a broader settler colonial project. Settler colonial practices of assimilative education are represented as a series of nets that operate at the macro-, meso- (or middle), and micro-societal levels. These nets tighten or slacken as they stretch across time and space, and when combined, one on top of the other, form a *settler colonial mesh*, which entraps Indigenous peoples within a settler colonial eliminationist project (Wolfe, 2006; for other approaches to the notion of a mesh, see Bush, 2014; Deleuze and Guatari, 2004; Wacquant, 2009; Woolford, 2014). This mesh, however, is prone to snags and tears, such as when relations between the actors and institutions engaged in processes of settlement and assimilation create a gap through which resilience and subversion might be asserted (Woolford, 2015).

In the book, I use these various layers of colonial netting to represent the different levels of analysis from which I approach the topic of Indigenous boarding schools. At the *macro-societal level*, my focus is on the broader social terrain, which is comprised of fields such as economics, law, governance, culture and science (Bourdieu, 1990; Bourdieu and Wacquant, 1992; Wacquant, 2009). At this level, the Indian as a 'problem' was formulated. Leaders within these social fields portrayed Indigenous peoples as obstacles to progress, throwbacks to a primitive era or too stubborn to engage in progress. In all cases, they were understood as an issue for which a policy solution was required.

These policy visions begin to take actual form at the *upper meso- or institutional level* of the settler colonial mesh. Here, the Indian bureaucracies in each country, as well as their education systems, were viewed to have an obvious role to play. But their full force could only be put into effect with assistance from other relevant governmental institutions. The law, for example, was eventually called upon to compel Indigenous attendance at boarding schools, when Indigenous parents proved less than eager to send their children. Welfare systems, as well, were deployed to deny social supports to parents who did not ensure their children's enrollment in boarding schools. The health-care system also had an interesting role to play, as the promise of better health for children, if they attended boarding schools, was used to sway parents to bring their children to the schools (even if these promises eventually proved false). Finally, police forces were called upon to apprehend truants, nonattenders and unenrolled students. Assimilative schooling was thus a multi-institutional undertaking, a fact made even more complicated in Canada because of the role of missionary societies in operating the schools.

The *lower meso-level* consists of the various boarding schools, as well as their competitor and feeder organizations and is where the broader visions and policy coordination of the higher levels of the mesh start to assume a more specific shape. In particular, this network is where multiple schooling organizations—boarding schools (both on- and off-reservations), mission schools, public schools and day schools—worked in competition and cooperation to provide assimilative education to Indigenous children. The intersection of these various schooling units created a mesh that sought to capture all Indigenous children within some facet of assimilative education. It is by channeling children through these networks that the everyday practice of assimilative schooling was formed.

This everyday or intimate practice of assimilative education occurred at the micro-level. This is where the individual assimilative school, which is itself comprised of a network of interactions, forms a unique assemblage of practices. At these localities, school officials (i.e. principals, teachers and staff) innovated specific techniques to interact with students, their parents and their communities. As well, all participants formed relationships and alliances with other agents, as well as nonhuman actors such as territory, food and disease (Callon, 1986; Latour, 1987), as a means to try to either advance or subvert assimilation. For example, Indigenous territory was mapped, policed and roaded to ensure more efficient capture of Indigenous children, but Indigenous parents also used their knowledge of territory to hide and protect their children from the schools.

These multiple levels of analysis are treated in the book as a series of nets coordinated to forcibly transform Indigenous peoples and thereby destroy these groups. The suggestion is made that this coordinated effort is consistent with sociological understandings of the concept of genocide (see Powell, 2007; Powell, 2011; Short, 2010; Woolford, 2009). But these nets did not simply tighten and enact their destructive potential in an undifferentiated and unilinear manner. Instead, resistances happened in different times and places, and at different levels, forcing openings in the mesh that allowed for the continued survival of Indigenous groups within North America. Therefore, it should be stressed up front that genocide is conceived here as a process and not an absolute or inevitable outcome. In most cases, Indigenous groups were not wholly destroyed, although many experienced destruction in part and all, to some extent, experienced (and continue to experience) the settler colonial mesh.

6.3 Indigenous Boarding Schools in Canada and the United States

As the fluid and variable mesh metaphor suggests, it is a challenge to provide a general history of assimilative education through Indigenous boarding schools in Canada and the United States. This is in part because one cannot speak of a distinct or single experience of assimilative schooling in either country. Although each country developed assimilative schooling policy and created the institutional and organizational infrastructure to carry out assimilation, in practice, the schools took particular shape depending on a variety of local factors, including the power of local Indigenous communities and the attitudes of principals, superintendents and other staff. Assimilative education was an uneven experience with the brunt of suffering felt more strongly in some regions and periods than others. It is also the case that Indigenous boarding schools were situated in complex settler colonial networks and were deployed alongside other techniques of forced transformation, for example laws against Indigenous spiritual practices, efforts to impose individual property ownership and interventions to destroy traditional forms of Indigenous governance. In order to avoid oversimplifying the impact of settler colonialism on Indigenous peoples by focusing solely on boarding schools, it is necessary to understand this issue in the broader historical geographical context in which it is embedded.

With these cautions in mind, it is possible to make a few generalizations about Indigenous boarding schools in the United States and Canada. In both countries, the late 19th century witnessed concerted use of assimilative schooling, which had existed in isolated and experimental form from the 17th century onward. At this time, discussion of the so-called 'Indian problem' turned toward ideas of 'assimilation' and 'civilization' as solutions. The American Indian Wars and forced removals convinced leaders in both countries that these more outwardly forceful approaches were untenable. Critics pointed to the costs of such violent encounters, arguing that educating Indians was cheaper than engaging them in war (e.g. Walker, 1874). In addition, Indian rights advocates raised concerns about reports from the Indian Wars that horrific atrocities were being committed by US troops and pressed the government to end the violence, though they rarely challenged the notion that Indians required assimilation (Hagan, 1985).

In this context, education—a technique long used by missionaries in efforts to convert Indigenous people—came to be viewed as the primary solution to the Indian problem. Lieutenant Richard Henry Pratt's experiment with industrial or vocational education was particularly influential. Pratt (1840–1924) implemented his vision of education as a tool of cultural transformation first at the Fort Marion prison, where he oversaw the custody of 72 captured Indigenous warriors. He moved some of his prisoners to the Hampton Normal Agricultural Institute, a school for former slaves in Virginia, before establishing his own school in Carlisle, Pennsylvania, in 1879 (Adams, 1995).

Pratt's model involved placing the Indigenous child close to 'civilization'. Such proximity to European settlers was seen as a necessary means to sever Indigenous children from their families and to attach them to American life. Pratt disciplined the children through a combination of military regimentation, work training and basic education. Students spent each day of the school year rising early to assemble on the grounds for marching and inspection before dividing the remainder of their day between work and school.

Pratt's ideas travelled across the border to influence Canada's ongoing efforts to assimilate Indigenous peoples. In the aftermath of the 1871 Red River Resistance, and with concern about Indigenous discontent in the West disrupting ambitions of national consolidation, Prime Minister John A. MacDonald's Conservative government dispatched Nicholas Flood Davin in 1879 to study Indian education in the United States. Davin advised that Canada should embrace American-style industrial education with one exception—he felt that a fully state-run system was impractical and that Canada should administer its system through the missions that were already established near Indigenous communities. Moreover, though Canada initially followed the US model by building large schools, the costs of this system eventually led Canada to opt for a residential school system featuring smaller schools located closer to Indigenous communities (Carney, 1971).

The roles of Catholic and Protestant churches in running the Canadian system mark an important difference between Indigenous boarding schools in the United States and Canada. In the United States, religious 'contract schools' for Indigenous children became much less numerous in the late 19th century when the government removed their funding (Szasz, 1999). In contrast, the Canadian government relied on Christian missionaries to operate schools. This resulted in a dependent relationship that had negative consequences for the children attending these schools. First, the churches maintained a great deal of influence over educational policy. This resulted in less change in policy over the history of residential schools in Canada than was true in the United States (Nichols, 1998). For example, the US Bureau of Indian Affairs (BIA) shifted its policies alongside changing governmental trends, moving from a militaristic schooling model to one based on management standards that encouraged students to govern themselves. Meanwhile, Canada's schools were consistently defined by monastic discipline oriented first and foremost toward shaping the souls of the students (Woolford, 2015).

Second, the relationship between government and churches in Canada was an unhealthy one. The government provided the minimum funding required to operate schools and the churches overcrowded schools to access more funds (Miller, 1996; Milloy, 1999). US schools were generally better funded than Canadian schools, though one would not mistake them for being well funded. In both countries, children suffered overcrowding, poor nutrition, inadequate clothing and deficient health care.

US boarding schools operated more systematically than those in Canada. According to the American policy, young children were to begin their education at reserve-based day schools before graduating to reserve-based boarding schools for middle school and then enrolling at nonreservation boarding schools for high school (Reyhner and Eder, 2004). Therefore, most students were in their teens at the latter institutions. In Canada, the policy that determined who would and would not be sent to a day or boarding school was more haphazard. After 1920, when the *Indian Act* clarified that Indigenous children belonged to the Christian denomination assigned to their home community, children were sent to a school run by that denomination whether it was a day or boarding institution. Under this model, children were frequently sent to boarding schools at ages as young as five or six years (Woolford, 2015).

Finally, the temporal shifts in Indigenous boarding school policy were less dramatic in Canada than the United States. In Canada, there was little turnover among bureaucrats running the system. For example, Deputy Superintendent of Indian Affairs Duncan Campbell

Scott, in particular, oversaw residential schools for almost 20 years between 1913 and 1932, and did so in a manner quite consistent with those who came before and after him (Titley, 1986). In contrast, the US BIA experienced multiple and sometimes radical leadership changes, conspicuously when John Collier, a former Indian rights advocate, became Commissioner from 1933 to 1945. Under Collier, the BIA sought to address concerns raised in the 1928 Meriam Report, also known as *The Problem of Indian Administration*, which lambasted the many failings of boarding schools (Philp, 1977).

6.4 Assimilative education and genocide

Several scholars, activists and survivors have described the damage done to Indigenous groups through boarding schools as genocide (e.g. Chrisjohn and Young, 1997; Churchill, 2004; Tinker, 1993). It is worthwhile to take a moment to reflect on this assertion. Raphael Lemkin (1900–59) coined the term genocide to address his concern that no law existed to protect the life of groups. For Lemkin (1944: 79), genocide represents a purposeful effort to destroy the 'essential foundations of the group'. This effort might occur through the physical murder of group members but could also take place through the destruction of cultural institutions (e.g. family, politics, art, religion). Lemkin was a tireless advocate for his concept, and after escaping the Holocaust he fled to the United States where he helped influence the 1948 *United Nations Convention on the Prevention and Punishment of the Crime of Genocide* (UNGC). It was during the United Nations General Assembly debates on later drafts of the UNGC that Lemkin saw his genocide concept receive legal recognition, but also watched as his notion of cultural genocide was largely excised from the Convention. Lemkin felt this to be an unfortunate but necessary compromise. For him, culture was essential to the group, binding members to one another through sharing traditions, practices, languages and other resources of group life. He hoped later versions of the UNGC would include cultural destruction in the crime of genocide (Lemkin, 2013).

Article III of the draft UNGC, on cultural genocide, was opposed by several General Assembly members on the grounds that it was too vague to be practical and that its protections could be covered by other minority protection laws and the Universal Deceleration of Human Rights that was also under discussion in the UN General Assembly. Some members further noted that it would be ridiculous to protect the cultures of those groups perceived to be backward and primitive. Canada and the United States were among the members opposed to Article III. Ottawa advised the Canadian delegation: 'You should support or initiate any move for the deletion of Article III on "Cultural" Genocide. If this move is not successful, you should vote against Article III and if necessary, against the Convention' (quoted in Brean, 2015: np). They were also instructed to look toward the United States for support on this issue. In the end, 26 nations voted in favour of excluding cultural genocide from UNGC versus 16 against and 4 abstentions (Abtahi and Webb, 2008).

The political self-interest that was transparent in the creation of the UNGC leaves social scientists, historians, lawyers, philosophers and others engaged in the study of genocide in conflict over what exactly the term seeks to protect. Does it merely protect individuals in the aggregate? Or is it the group? And if the latter, what is a group? How does one protect something that is dynamic and subject to change? In brief, a sociological approach to genocide is not beholden to a politicized genocide definition, though it is quite reasonable to argue that assimilative education is genocidal even within the limitations of the UNGC (see MacDonald, 2014). But for the purposes of this chapter, what is of concern is how groups persist as groups. A group is not a fixed or closed entity. It is a collective project whereby group identity and membership are matters of ongoing co-constitution. Culture is a key element in the continuing reproduction of the group. For this reason, some assaults upon cultural aspects of group life can be as detrimental to group survival as assaults on individual group members. The key cultural foundations of group life will differ group by group, but often include elements of language, spirituality and knowledge.

Returning to our example of Indigenous boarding schools, two ways that the schools sought to destroy Indigenous groups in a manner that might be termed genocide will be briefly discussed. First, boarding schools in both the United States and Canada sought to sever ties between the student and his or her community and family. Soon after arrival at boarding schools, children were stripped of the Indigenous names that connected them to community and territory. The clothes they wore were taken away and destroyed. Siblings of the opposite sex were housed in separate wings of schools, and there was little contact between brothers and sisters. Survivors recount that they were told not to refer to their parents as 'mom' and 'dad', while teachers, matrons and other supervising staff were directed to act as a surrogate family. Additionally, the physical distance between the school and the community, as well as the school's fence, kept children in and visiting relatives out. Interventions such as these destroyed Indigenous social networks. Children were deprived of a sense of collective identity, but also a sense of being grounded within a family, community and territory. Former students experienced difficulty returning to their communities in summers and at the end of their schooling.

Similarly, one can examine how culture was attacked through boarding schools. A primary assault was linked to the prohibition against speaking Indigenous languages. Noun-based languages that sought to fix and delimit the world replaced these verb-based languages. This effort at linguistic destruction overlapped with efforts to alter Indigenous cultural practices related to land. The assimilative education students were taught that land was not something alive and related to the student; instead it was to be surveyed, marked, farmed, landscaped and individually owned. Students learned that land was a resource rather than a relation. Likewise, animal life was treated as a resource through instruction in husbandry, altering the traditional sense that animals were to be lived with in harmony. Children lost out on much of their cultural learning, as entire story cycles were missed while students boarded at school over the winter. As well, songs, practices, traditions, games, jokes, ceremonies and other aspects of Indigenous cultural learning were not transmitted. Upon leaving school, students often lacked the cultural resources required to maintain a feel for Indigenous group life.

These experiences were in addition to the physical and sexual violence prevalent at boarding schools, and the combined suffering from these experiences severely upset group life. Experiences such as these made relationships between group members, in a manner that preserved their group existence, near to impossible. Assimilative schools in both the United States and Canada thus targeted the very ability of Indigenous groups to persist as distinct groups. This targeting took different forms in different times and places. In some circumstances, Indigenous groups managed to subvert educational processes designed to dilute their cultures into the dominant society. But, in general, these destructive impulses were present at most schools.

6.5 Modular and mobile maps: Representing intimate colonialisms in Indigenous Boarding Schools

Remaining at the micro-level of the settler colonial mesh, another example of the differential experience of settler colonialism can be found in how commentators at times grapple with the spectrum of staff behaviour that ranges from harsh to seemingly benevolent. In practice, cultural destruction can occur through both the 'carrot' and the 'stick'. In other words, assimilative strategies can vary between efforts to tempt children to take on new cultural patterns through bribes and compliments and brutalizing interventions to force them to accept an assumed inevitable transformation. Various techniques along this range were on display in assimilative boarding schools in the United States and Canada, though the intensity with which the carrot or stick were applied depended on the spatial and temporal conditions noted above.

The ambivalence of assimilative education can cause some confusion among those who expect that patterns of destruction should feature only the stick and not the carrot. Many authors have, for example, registered the challenge of dealing with seeming good intentions among staff in the IRS system. Some, such as Bays (2009) and Glenn (2011), treat these as redemptive moments. Likewise, Canadian Senator Lynn Beyak (quoted in Kirkup, 2017: np) stated, 'I speak partly for the record, but mostly in memory of the kindly and well-intentioned men and women and their descendants—perhaps some of us here in this chamber—whose remarkable works, good deeds and historical tales in the residential schools go unacknowledged for the most part and are overshadowed by negative reports'. In this framing, testimony that features the carrot is used to construct an image of good people working in a bad system. This view has led to further debates about whether or not the Canadian Truth and Reconciliation Commission (TRC) final report is too negative about the schools, ignoring those teachers and staff who legitimately sought to help Indigenous children. It has provoked further criticism of the TRC for referring to residential schools as 'cultural genocide', based on the assumption that the genocide concept lacks the capacity to contend with diverse experiences and interactions (TRC, 2015; for criticisms see Clifton and Rubenstein, 2015; Giesbrecht, 2017). Of course, genocide scholars have long grappled with the coexistence of violence and seeming exceptions to this violence within overarching destructive settings. Browning (1998), for example, compares the men of Police Battalion 101, who ravaged the Polish countryside and committed gruesome massacres of its Jewish occupants, to the participants in Philip Zimbardo's Stanford Prison Experiment in that not all participants were consistent in the levels of violence they displayed. Zimbardo (Haney et al., 1973) describes his guards as falling into one of three groups: kind guards who sought to ease the suffering of inmates to the extent possible in the existing structural conditions; guards who followed orders but were not excessive in their cruelty toward prisoners; and guards who were devious and creative in the cruelties they imposed on prisoners. In other studies, the fact that some participants in genocides are both perpetrators and rescuers has been brought to light, such as in instances where a Hutu identified individual in Rwanda, after spending his or her day participating in deadly violence, might return to the house where he or she hid a Tutsi family (see Straus, 2006).

Aside from the inherent reductiveness of redemptive analyses of the IRS system, and their refusal to admit the nuance contained within the genocide concept, critics who suggest that the TRC ignores kinder staff members are simply factually incorrect. The TRC final summary report does acknowledge that people of good intentions existed within the IRS system:

For the most part, the school staff members were not responsible for the policies that separated children from their parents and lodged them in inadequate and underfunded facilities. In fact, many staff members spent much of their time and energy attempting to humanize a harsh and often destructive system. Along with the children's own resilience, such staff members share credit for any positive results of the schools (TRC, 2015, 129–130).

Rather than paint all staff members as predators and perpetrators, the report offers them as a potentially humanizing force in this bleak system. The TRC cannot be criticized for ignoring such good intentions; however, it can be taken to task for not adequately contextualizing them and showing how, within such a system, kindness can serve to reinforce and intensify assimilation.

In colonial historiography, it is broadly accepted that colonialism operates not solely through brute oppression but also through rewards, pleasures and seductions. Colonialism carves racial cleavages through force of arms, but also through intimate relations, such marriage, child-rearing, sex and other areas of everyday life (Stoler, 2001). The latter are examples of 'intimate colonialism' or 'colonial intimacies'. These concepts arise from a historical perspective that takes the politics of knowledge as its starting point as a means to disrupt the taken-for-granted taxonomies and Eurocentric gaze of the colonial archive (Stoler, 2002a).

As Laura Ann Stoler's work shows, the colonial state makes investments in the carnal, the sensual, intimate pleasures and other dimensions of everyday life that are part of what she calls, paraphrasing Foucault, the 'microphysics of colonial rule' (Stoler, 2002b, 7). It is in this microphysics that one can best understand the quotidian techniques of discipline and domination enacted within Canadian and US assimilative boarding schools. Scholars such as Jacobs (2009, 2014) and Cahill (2011) have shown how in this context 'Federal mothers and fathers' were charged with enacting a paternal intimacy to lure Indigenous children away from Indigenous patterns of life.

The role of sports at US and Canadian assimilative boarding schools provide but one example of intimate colonialism in action. Many Canadian survivors recall sports as one of the occasions where they could escape the drudgery of their education (TRC, 2015). But the rationale for allowing sports in US and Canadian schools was more disciplinary than benevolent. Sports were seen as instilling rule-following, group cooperation beyond the family or tribal unit, and an embrace of the school as family/community through instillation of school spirit (Woolford, 2015). For the boys and young men, baseball, football and, in Canada, hockey were regularly played. While for girls and young women, domestic activities were typically emphasized, though some did compete in baseball, curling and other activities. Coaches of these sports could be figures of violence and abuse, or role models depending on the school and time period. Likewise, the teams themselves, because of the pleasure they allowed, could amount to a loosening of the settler colonial mesh, as students reframed their sport teams as a source of Indigenous pride, such as when defeating a local non-Indigenous team, or as providing limited freedoms beyond the dull compulsion of boarding school life. In short, the intimacy of sports, initially harnessed to pursue assimilative goals, could intensify or lessen the settler colonial mesh, depending on the local conditions at a particular school.

The settler colonial mesh thus can help map some of these intimate complexities in assimilative schooling. Since it does not describe to a simplistic intentionalist paradigm, whereby perpetrators of such evil must be wholly or in every situation evil themselves, it allows for both human and individual behaviour to vary across the time and space of destructive systems. It also makes it very possible for otherwise well-intentioned people to participate in acts of group destruction, even if their daily practices may have, at an individual level, been intended to mitigate harm. Of course, in the context of severing children from their families, even care and kindness can operate to ease the path of destruction. Resistances can be overcome with a gift or kind gesture, just as they can with a slap.

Based on the variable and intimate nature of settler colonial destruction, the question arises as to how one possibly maps such a differentiated history. Mapping, as a practice, typically provides a synchronic snapshot of the spatiality of a region or phenomenon. Residential schools can be plotted, as they existed in a particular year. Other maps can be juxtaposed to show their increase or decrease in number at different times. But the model remains static. In contrast, the fluidity introduced through Cybercartography better reflects the dynamism of the settler colonial mesh. Here, the pulsing shifts in assimilative schooling can be shown in their 'chronotopic' or spatiotemporal dimensions (Valverde, 2015), as movement through an interactive map that reveals different layers of information at specific temporally located sites. As Pyne (2014, 246) notes of the cybercartographic maps in the Lake Huron Treaty Atlas, such maps reflect a 'nonlinear, nonteleological [and] spatialized approach' to collaborative, fluid and multileveled map design that is sensitive to both change and singularity within the broader scope of the map's space.

Applied to assimilative boarding schools, the networks of meaning that contour experiences at particular schools might be better mapped through such techniques, showing how the mesh tightens or loosens at particular schools at particular times. Such movements can even be mapped as they occur simultaneously, as the mesh may tighten for certain students (those from specific communities targeted as more resistant to civilization, such as the Diné, or those perceived to be 'darker' than the favoured students) and loosen for others (such as the Pueblo, whose communities possessed greater negotiating power with the US government, in part because they were perceived by some to be 'civilized' Indians).

Moreover, unlike other interactive maps that allow spatiotemporal movement, but remain monological in their content (i.e. formed by a single author), the cybercartographic maps discussed in this volume allow for dialogical (i.e. multi-perspective) construction, a construction that is iterative and ongoing rather than subject to closure (Pyne, 2014). This is important not solely as a means to reflect the dialogical practices of knowledge formation in Indigenous communities, but also as the participatory nature of Indigenous cultural life. It also reflects a commitment to capturing the diversity of possible experiences within the unevenness of the settler colonial mesh. In particular, the cybercartographic map presents the potential to map schools down to their most intimate, micro-level details. These details can be embedded in the map, located on a timeline and in relation to their specific spatial location. In this manner, the map loses its macro-sociological bias and opens to the singularity of human experience.

This offers some hope for mapping those forms of intimate colonialism that are so frequent in Indigenous boarding school testimony, such as the everyday forms of violence that communicate the denigration of one's culture, or devalue one's very self as an Indigenous person. Such experiences are not well represented by macro-level maps that place schools according to the archival record, but fail to capture any of their movement or day-to-day practices. By increasing user-contributed content—for example, by layering the map with links and hotspots that allow one to drill past the map's surface—promises more insight into the everyday life, but also the mood of residential schools. Whereas traditional maps are without affect, the cybercartographic map brings affect back into the picture, as individual contributions, or site-based remembrances, recall the feeling of a specific space at a specific time.

6.6 Conclusion

In this chapter, I have discussed some of the challenges faced in representing the unevenness of settler colonial destruction as it took place through assimilative boarding schools. In so doing, I have offered the model of the settler colonial mesh as a means to grapple with this complexity. Focusing on the micro-level of the settler colonial mesh, I further discuss the forms of intimate colonialism that appear in assimilative boarding school testimony and on occasion produce confusion in listeners who are tempted to treat these as redemptive or benevolent moments within assimilative education. I argue that such readings of the settler colonial past miss how seduction and discipline, the carrot and the stick, coexist within assimilative institutions. Finally, I conclude by reflecting on how these representational challenges confront the world of map-making, arguing that contemporary cybercartographic approaches to map-making have potential to better reflect the settler colonial mesh and highlight such moments of intimate colonialism, as experienced by survivors of these schools.

References

Abtahi, H., Webb, P., 2008. The Genocide Convention: The Travaux Preparatoires (2 Vol). Brill Academic Publishing, Leiden.

Adams, D.W., 1995. Education for Extinction: American Indians and the Boarding School Experience, 1875–1928. University Press of Kansas, Lawrence.

Bays, E., 2009. Indian Residential Schools: Another Picture. Baico Publishing, Ottawa.

Bourdieu, P., 1990. In Other Words: Essays Toward a Reflexive Sociology. Stanford University Press, Stanford.

Bourdieu, P., Wacquant, L., 1992. An Invitation to Reflexive Sociology. University of Chicago Press, Chicago.

Brean, J., 2015. Canada opposed concept of 'cultural genocide' in 1948 accord. National Post. June 8. https://nationalpost.com/news/canada/canada-threatened-to-abandon-1948-accord-if-un-didnt-remove-cultural-genocide-ban-records-reveal.

Browning, C., 1998. Ordinary Men: Reserve Police Battalion 101 and the Final Solution in Poland. Harper Perennial, New York.

Bush, C.N., 2014. Subsistence fades, capitalism deepens: the 'net of incorporation' and Diné livelihoods in the opening of the Navajo-Hopi land dispute, 1880–1970. American Behavioral Scientist 58 (1), 171–196.

Carney, R., 1971. Church-State and Northern Education: 1870–1961. Unpublished PhD diss. University of Alberta.

Cahill, C.D., 2011. Federal Mothers and Fathers: A Social History of the United States Indian Service, 1869–1933. University of North Carolina Press, Chapel Hill.

Callon, M., 1986. Some elements of a sociology of translation: domestication of the scallops and the fishermen of St Brieuc Bay. In: Law, J. (Ed.), Power, Action and Belief: A New Sociology of Knowledge. Routledge & Kegan Paul, London, pp. 196–223.

Chrisjohn, R., Young, S., 1997. The Circle Game: Shadows and Substance in the Indian Residential School Experience. Theytus Books, Penticton, BC.

Churchill, W., 2004. Kill the Indian, Save the Man. City Lights, San Francisco.

Clifton, R., Rubenstein, H., 2015. Debunking the half-truths and exagerations in the truth and reconciliation report. National Post. June 4. http://nationalpost.com/opinion/clifton-rubenstein-debunking-the-half-truths-and-exaggerations-in-the-truth-and-reconciliation-report.

Deleuze, G., Guattari, F., 2004. Trans. In: Massumi, B. (Ed.), A Thousand Plateaus. Continuum, London and New York.

Giesbrecht, B., 2017. TRC Report Political, Not Balanced. Winnipeg Free Press. March 31. https://www.winnipegfreepress.com/opinion/analysis/trc-report-political-not-balanced-417745543.html.

Glenn, C.L., 2011. American Indian/First Nations Schooling: From the Colonial Period to the Present. Palgrave, New York.

Hagan, W.T., 1985. The Indian Rights Association: The Herbert Welsh Years, 1882–1904. University of Arizona Press, Tucson.

Haney, C., Banks, W.C., Zimbardo, P.G., 1973. Interpersonal dynamics in a simulated prison. International Journal of Criminology and Penology 1, 69–97.

Jacobs, M.D., 2014. The habit of elimination: Indigenous child removal in settler colonial nations in the twentieth century. In: Woolford, A., Benvenuto, J., Hinton, A. (Eds.), Colonial Genocide in Indigenous North America. Duke University Press, Durham, pp. 189–207.

Jacobs, M.D., 2009. White Mother to a Dark Race: Settler Colonialism, Maternalism, and the Removal of Indigenous Children in the American West and Australia, 1880–1940. University of Nebraska Press, Lincoln.

Kirkup, K., 2017. Lynn Beyak removed from Senate committee over residential school comments. The Globe & Mail, 5 April. http://www.theglobeandmail.com/news/politics/beyak-removed-from-senate-committee-over-residential-school-comments/article34610016/.

Latour, B., 1987. Science in Action: How to Follow Scientists and Engineers through Society. Open University Press, Milton Keynes.

Lemkin, R., 2013. In: Frieze, D.L. (Ed.), Totally Unofficial: The Autobiography of Raphael Lemkin. Yale University Press, New Haven.

Lemkin, R., 1944. Axis Rule in Occupied Europe: Laws of Occupation, Analysis of Government, Proposals for Redress. Carnegie Endowment for International Peace, Washington, DC.

Macdonald, D.B., 2014. Genocide in the Indian residential schools: Canadian history through the lens of the UN genocide convention. In: Woolford, A., Benvenuto, J., Hinton, A. (Eds.), Colonial Genocide in Indigenous North America. Duke University Press, Durham, pp. 306–324.

Miller, J.R., 1996. Shingwauk's Vision: A History of Native Residential Schools. University of Toronto Press, Toronto.

Milloy, J.S., 1999. A National Crime: The Canadian Government and the Residential School System, 1879 to 1986. University of Manitoba Press, Winnipeg.

Nichols, R.L., 1998. Indians in the United States and Canada: A Comparative History. University of Nebraska Press, Lincoln.

Philp, K.R., 1977. John Collier's Crusade for Indian Reform, 1920–1954. University of Arizona Press, Tucson.

Powell, C., 2011. Barbaric Civilization: A Critical Sociology of Genocide. McGill-Queen's University Press, Montreal and Kingston.

Powell, C., 2007. What do genocides kill? A relational conception of genocide. Journal of Genocide Research 9 (4), 527–547.

Pyne, S., 2014. The role of experience in the iterative development of the Lake Huron treaty Atlas. In: Taylor, D.R.F., Lauriault, T.P. (Eds.), Developments in the Theory and Practice of Cybercartography. Elsevier, Amsterdam, pp. 245–262.

Reyhner, J., Eder, J., 2004. American Indian Education: A History. University of Oklahoma Press, Norman.

Short, D., 2010. Cultural genocide and indigenous peoples: a sociological approach. International Journal of Human Rights 14 (6), 833–848.

Stoler, L.A., 2002a. Carnal Knowledge and Imperial Power: Race and the Intimate in Colonial Rule. University of California Press, Berkeley and Los Angeles.

Stoler, L.A., 2002b. Colonial archives and the arts of governance. Archival Science 2, 87–109.

Stoler, L.A., 2001. Tense and tender ties: the politics of comparison in North American history and (post) colonial studies. Journal of American History 88 (3), 829–865.

Straus, S., 2006. The Order of Genocide: Race, Power, and War in Rwanda. Cornell University Press, Ithaca.

Szasz, M.C., 1999. Education and the American Indian: The Road to Self-Determination since 1928. University of New Mexico Press, Albuquerque.

Tinker, G., 1993. Missionary Conquest: The Gospel and Native American Cultural Genocide. Fortress Press, Minneapolis.

Titley, E.B., 1986. A Narrow Vision: Duncan Campbell Scott and the Administration of Indian Affairs in Canada. University of British Columbia Press, Vancouver.

Truth and Reconciliation Commission of Canada (TRC), 2015. Honouring the Truth, Reconciling for the Future: Summary of the Final Report of the Truth and Reconciliation Commission of Canada. Truth and Reconciliation Commission of Canada, Ottawa, Ontario.

Valverde, M., 2015. Chronotopes of Law: Jurisdiction, Scale and Governance. Routledge, London.

Wacquant, L., 2009. Punishing the Poor: The Neoliberal Government of Social Insecurity. Duke University Press, Durham.

Walker, F.A., 1874. The Indian Question. James R. Osgood and Company, Boston.

Wolfe, P., 2006. Settler colonialism and the elimination of the native. Journal of Genocide Research 8 (4), 387–409.

Woolford, A., 2015. This Benevolent Experiment: Indigenous Boarding Schools, Genocide, and Redress in Canada and the United States. University of Nebraska Press, Nebraska.

Woolford, A., 2014. Discipline, territory, and the colonial mesh: boarding/residential schools in the U.S. and Canada. In: Woolford, A., Benvenuto, J., Hinton, A. (Eds.), Colonial Genocide in Indigenous North America. Duke University Press, Durham, pp. 29–48.

Woolford, A., 2009. Ontological destruction: genocide and aboriginal peoples in Canada. Genocide Studies and Prevention: International Journal 4 (1), 81–97.

Woolford, A., Gacek, J., 2016. Genocidal carcerality and Indian residential schools in Canada. Punishment & Society 18 (4), 400–419.

Workhouses and residential schools: From institutional models to museums

Trina Cooper-Bolam*

Doctoral Candidate, Cultural Mediations and Geomatics and Cartographic Research Centre (GCRC), Carleton University, Ottawa, ON, Canada

*Corresponding author

7.1 Introduction: A reconciliatory museology

Can institutions that were once models for Indian residential schools and since musealized become models for a reconciliatory museology? That industrial schools were models for Canada's Indian residential school system is well documented. Their common provenance in the workhouse, however, is less so.

As the Canadian Museums Association initiates Truth and Reconciliation Commission (TRC)-driven efforts 'to undertake, in collaboration with Indigenous Peoples, a national review of Canada's museum policies and best practices to determine the level of compliance with the United Nations Declaration on the Rights of Indigenous Peoples' (UNDRIP; Truth and Reconciliation Commission of Canada (TRC), 2015 Call to Action 67), it is incumbent upon the museology profession to consider the implications to future museology of UNDRIP articles that do not overtly speak to museum practices and collections. For example, what are the obligations, if any, of state cultural heritage infrastructure with respect to providing effective mechanisms of prevention and redress for the actions listed in UNDRIP Article 8, Section 2, subsections 'a' through 'e'? These subsections articulate actions that constitute and result in cultural genocide. If the CMA and the larger museological community are to take seriously the TRC of Canada's Call to Action 67, a robust postrupture museological repertoire—one that extends the 'New Museology' of the 1990s and becomes an impetus to reconciliation—will need to be developed. In short, Canada will require a reconciliatory museology. This need is particularly urgent in light of the TRC revelation that, in forming and administering the Indian residential school system, the Canadian state and its partners are implicated in cultural genocide. My contention is that such a museology is already in formation, both in Canada, and in similar contexts of national historical reckoning.

This chapter investigates comparative emergent reconciliatory museologies at the Irish Workhouse Centre in Portumna, East Galway, Ireland, and The Workhouse (National Trust site) in Southwell, Nottinghamshire, UK. These and other workhouses are architectural, social and cultural precursors to Indian residential schools. As musealized sites, they can inform the emerging musealization of former Indian residential schools, and in particular the Shingwauk Indian Residential School in Sault Ste-Marie, Ontario, which is currently undergoing a museal transformation. In studying the synergies between these projects, I seek to contribute to both bridging a gap in our understanding of the provenance of Indian residential schooling in Canada and to the framing of a needed Canadian and even transcultural and transnational reconciliatory museology.

7.2 Rupture, difficult history and reconciliatory museology

A postrupture museological repertoire is one that both interprets and responds to a critical event. Here, 'critical event' is seen not in the sense of incident or discrete event, but rather as an occurrence, whether of momentary or long duration, resulting in societal rupture and a corresponding change of social action and trajectory. This usage follows Ruth Phillips' deployment of the concept of 'critical event', first theorized by Das (1996) in her analysis of the impacts of the critical event of Partition in India. The words rupture and event, however, often fail to describe phenomena as remembered/constructed over time. That a critical event causing a rupture occurred is perhaps as often a conclusion of a slow and oscillating progression of historical transmission, re-evaluation and conscientization, as it is of immediate and widespread condemnation.

However recognized, critical events lay the foundation for the construction of a postrupture society, one founded on a new identity and moral platform. Of the Great Famine, arguably Ireland's critical event, N.G. Ravichandra opined 'modern historians regard it as a dividing line in the Irish historical narrative, referring to the preceding period of Irish history as "pre-Famine"' (2013, p. 16). This assessment of Irish historiography, albeit from the unlikely source of a study of plant pathology (ibid), has become ubiquitous among public information websites such as Wikipedia ('Legacy Of The Great Irish Famine, 2018'), testifying to public perceptions of the Famine having caused a fundamental rupture in Irish history. As the centenary of the famine approached, Gráda (1992) recounted that the Taoiseach, and in particular Éamon de Valera, encouraged and funded historical research, reviving and deploying famine memory within a nationalist regime of Irish folkloric historicization. A period of revisionist and postrevisionist — if sparse — history-making, well chronicled in Gillissen's 2014 historiography, followed by the sesquicentenary of the Famine in 1995, which 'resulted in an outpouring of books on the subject', (Gillissen, 2014, p. 208; Tóibín, 1999) has stimulated official commemorations 'reflecting growing public interest in the Famine in recent years' (212). Against this backdrop, workhouse memory ebbs, a situation complicated and exacerbated by social ambivalence and intergenerational trauma.

Histories of Irish workhouses often layer onto those of the Famine, sometimes strengthening nationalistic arguments of British culpability for deaths seen as coincident with but not caused by the potato blight (Mitchel, 1876; O'Hegarty, 1952; Coogan, 2012). However, as a last respite of the poor, workhouses conversely stand as evidence of social welfare and caretaking by British authorities. Sustained social ambivalence dilutes collective mnemonic potency, such that intentional forgetting renders memories fallow. Correspondingly, collective amnesia or forgetting, refusals of an event's occurrence or importance, obstruct museological inquiry. Yet, historical reevaluation and attempts at synthesis remain a latent force, and together with museological methods of inquiry, are poised to intercept returns of the repressed. I argue then, that the remit of postrupture museological repertoire is 'difficult history' or 'difficult knowledge', whether vividly remembered and historicized or in uncanny, ambivalent and latent form.

Defining 'difficult knowledge,' Simon (2011) suggests that knowledge presented in a curatorial context can be deemed difficult when

> [i]t confronts visitors with significant challenges to their expectations and interpretive abilities—for example when historical narratives produce conflicting, complex, and uncertain conclusions; elicits the burden of 'negative emotions' [...] revulsion, grief, anger, and/or shame that histories can produce ... if they raise the possibility of complicity of one's country, culture, or systemic violence such as the seizure of aboriginal land, the slave trade, or genocide; or, evokes heightened anxiety that accompanies feelings of identification of the victims of violence, the perpetrators [...] or bystanders (194).

Simon's constructions of the 'terrible gift' and 'pedagogy of witness' are both more nuanced and ambitious than that of 'difficult knowledge', and reflect strategies that are conducive to postrupture museological inquiry. Such approaches are configured within a coherent commitment to prevent recurrence of the originating trauma and can initiate rituals and performances that implicate viewers and demand reciprocation: Simon's (2006) statement, '[T] here is no futurity (no break from the endless repetition of a violent past) without memories that are not your own but nevertheless claim you to a responsible memorial kinship and the corresponding thought such a problematic inheritance evokes' (203), offers motivation and context for his conception of a pedagogy that 'centers the demands of testament and supports the corresponding work of inheritance' (ibid). Postrupture museology is both a museal reckoning with and critique of the originating trauma or critical event, and a work of praxiological museology and institutional critique that observes how difficult pasts and their terrible gifts are musealized within the institution, which in the examples presented here, is also the original site of trauma. More concisely, postrupture museological inquiry is the critical practice of musealizing difficult history concomitant with actively critiquing and periodically intervening in its production. It is not the exclusive domain of artists commissioned by the museum as has been theorized as institutional critique/praxiological museology, but rather the responsibility of the assemblage of museal producers (See Marstine, 2017 for comprehensive coverage of critical art practice).

That neutral representations of difficult histories, such as mass atrocity, genocide and state violence, are ethically untenable I take as uncontested. While taking on memories not one's own (witnessing) and accepting the responsible moral kinship this compels is the responsibility of visitors, producing an ethical, transparent and critical mnemonic environment is correspondingly that of museal producers. A museological repertoire informed by rupture enables a process and mode of inquiry, one with healing and reconciliatory potential to be sure, but which may equally produce tentative, conflicted, deflective and hegemonic representations.

So, what must a reconciliatory museology do? First, it is not a museology as such but rather a process that deploys methods of museological inquiry to investigate a critical event toward redress and reconciliation. While the experience of phenomena that may come to be identified as having constituted a critical event may not result in immediate and widespread disapprobation and rupture, indeed the repressive responses of collective denial or amnesia and resulting smaller-scale social fragmentation along mnemonic lines may make the identification of the event itself tentative, the sense of its occurrence must be present. Next, a reconciliatory museology must reconstitute what has been lost. This is a preanalysis recovery, aggregation and documentation phase guided by a provisional ethics that defers judgements and prioritizes processes of collecting and witnessing in personal and cultural safety. In the subsequent phase of analysis and critique, a critical ethics emerges, which imposes curatorial judgements that shift collecting and display practices. It is in this phase that the investigative subject is constructed as a critical event and its musealization becomes an 'epitomizing event'. Ethnohistorian Raymond Fogelson defines epitomizing events as 'narratives [non-events in themselves] that condense, encapsulate, and dramatize longer-term historical processes', compelling inventions of such 'explanatory power that they spread rapidly through the group and soon take on an ethnohistorical reality of their own' (1989, p. 143).

Staged and deployed by museums, epitomizing events function as proxies for the historical processes they represent, perform and embody, yet are themselves necessarily abstractions, begetting secondary witnessing and prosthetic memory formation. A critical ethics at this juncture moves toward an ethics of conscientization, of seeing rupture in the present. For that reason, and as a final phase, a reconciliatory museology must seek redress in the present.

But in doing so, it must resist authorizing monologic and synchronic narratives. It must embrace complexity, polysemy and mess, and be receptive to dialogue and critique. So, while early phase exhibitions may be documentary in character, and middle phase critical, latter phase become dialogic and create contact zones between past events and the present, authenticity and prosthesis, and a range of subject positions and identifications.

For the purposes of discussing the reconciliatory potential of the sites that are the subject of this chapter, I will examine three vectors of musealization: site intervention and use, interpretation design, and animation and programming. The nature of this research is exploratory and this discussion is confined to the outputs of participatory observation research I conducted during site visits in April 2018. Such research demands an element of autoethnographic critique and while the scope of this neither can nor should extend to encompass autoethnography, a statement of my positionality is certainly warranted. I am a Canadian settler-scholar/practitioner in the field of critical museology. My praxis focuses on the collaborative (Indigenous/settler) production of exhibitions and commemorations through critical, emergent and experimental design and curatorial practices that recognize and reckon with difficult histories and promote justice, healing and decolonization. My approach, ethics and labour are a response to the terrible inheritance that I—as a Canadian—have received: the history of cultural genocide of Indigenous Peoples to which I have become conscientized, the stories of survivors to whom I bear an obligation of reciprocation, and the legacy that manifests in the present. It is not as an impartial observer that I visit the sites of this study, rather it is as a trained museum scholar and professional with specific decolonial commitments. Moreover, for the past 5 years I have been engaged in a project to musealize the former Shingwauk Indian Residential School at the behest of the Children of Shingwauk Alumni [Survivor] Association and am currently co-curating and co-designing one of its galleries; this is combined with my role as a doctoral research assistant on intersecting work under the Residential Schools Land Memory Mapping Project (see Chapters 2 and 8). The objective of my field research at workhouse museums was to determine what can be learned from them that may be applied to the Shingwauk context and, more broadly, to postrupture museology. This involves assessing to what degree these musealized sites admit of or identify with difficult pasts, and how they contribute to or foster reconciliation.

7.3 Workhouses and industrial schools in the United Kingdom and Ireland

First a few perspectives on workhouses:
For what, after all is a Workhouse?

Not merely the dumping ground for the waste human material of our cities and towns, but, as a competent authority has put it, "a home for imbeciles, a lying-in hospital for dissolute women, a winter resort for the casual labourer or summer beggar, a lodging-house for the tramps and vagrants, as well as a hospital for the sick. Within its four walls everything that is or has been of evil may be found in its symptoms or its effects, embodied in men and women whose souls are seared with the brand of vice, and whose standard of life is sometimes scarcely that of a brute. And there, too, are the children, daily exposed to the sights and sounds which cannot but have a deteriorating effect, and learning to look upon the Workhouse as a home to which in times of stress and struggle they will return in later life as surely as the homing pigeon returns to its loft *Day 1912, p. 170* (Fig. 7.3.1).

FIGURE 7.3.1   Southwell deadroom. *Photo courtesy of Iddon 2018.*

> The miserable inmates of union workhouses … were huddled together three tiers of beds in a room's height, like the scanty provision made in emigrant ships for those classes of men whose poverty alone brands them to a fate little preferable to the worst features of slavery **T.D. Barry,** *specialist in the 'superintendence and designing of workhouses', speaking to the Liverpool Architectural and Architectural Society in 1851, quoted in Wildman, 1974.*

The consequence of the dissolution of the monasteries from 1536, a primary source of poor relief, together with contemporaneous legislation to prohibit vagrancy, was placing the burden of providing for the poor on the British state. The 'Old Poor Laws' transferred this responsibility to parishes, which provided out relief from parish taxes. Despite the 'workhouse test', which threatened the 'able-bodied poor' with the prospect of houses of correction, they were, in actuality, provided the means to subsist in their own homes between periods of employment. By the 1800s, the increase in pauperism had virtually overwhelmed the existing relief system, necessitating the shift to more economical 'indoor relief'. The spectre of houses of correction became bricks and mortar workhouses. Parishes banded together forming unions to share the cost of poor relief, which was further decreased by the labour of workhouse inmates who received only bed and board for their efforts.

Following the 1834 *Poor Law Amendment Act*, the design and construction of union workhouses began in earnest (Higginbotham, 2012, p. 27). Architects set their skills to designing structures that would effectively segregate the various classes of inmates while facilitating the overseer's panoptical scrutiny.

A 1839 Report of the Poor Law Commissioners reveals that 'the style of building [was] intended to be of the cheapest description compatible with durability' (O'Connor, 1995, p. 80). Those constructed in Ireland were not deemed to merit wall plaster or proper floors of wood or stone (81). Places of extreme austerity and punishing accommodation, the workhouses were intended as a last respite of the poor, places of discomfort to be avoided at all costs. Indeed, a condition of admittance required entire families to enter the workhouse together, whereupon their members were classified and separated into distinct wards. According to popular historian John O'Connor, 'one tragic aspect of this separation was that during the famine times [in Ireland], parents often did not know that their children had died, husbands their wives, mothers their sons, daughters their fathers' (85). While the 'blameless' or 'helpless poor' (the elderly and infirm, mentally ill or incapacitated, some injured veterans of the Napoleonic War, others war widows) were segregated from the 'able-bodied poor', those judged tantamount to criminals, and occasionally afforded the luxury of extra tea, their accommodation was no less expressive of the desire of their keepers to turn them out on the street. Such was the only state provision for the elderly until the beginning of the 20th century. Abandoned women and their children were more frequently admitted than men (Thane, 1978). In the workhouse, able-bodied men and women were put to work rock breaking, oakum picking or maintaining the compound and gardens. Some unions had supplementary industries such as shoemaking, which could be learned by resident children. In most workhouses, children were given a rudimentary education, which soon was criticized as ill preparation for the life of industry they were expected to live upon reaching adulthood at 15 years of age.

In his 1859 article, 'Population, Crime, and Pauperism', Karl Marx opined, 'there must be something rotten in the very core of a social system which increases its wealth without diminishing its misery', emphasizing that the 'stationary million of English paupers' had only diminished by 26,233 between 1849 and 1858, despite the 'new start' to industry and commerce—with its industrial 'mania' and new avenues of trade and wealth (Marx n.p.). In Ireland, where the 1838 *Act for the Effectual Relief of the Destitute Poor* inaugurated a massive national workhouse construction program, workhouse conditions were contrived to be inferior to the already deplorable living condition of the Irish poor. This was challenging to the commissioners given the Irish poor of the early 1800s lived in a worse state than their English counterparts (Mahoney, 2016). Indeed, according to Mahoney, a particularly Irish abhorrence of confinement was a principal reason Poor Law Commissioners expected workhouses would succeed in thwarting the obtention of relief (Mahoney, 2016). Speaking to workhouse conditions during the Great Famine, Jonny Geber explains, 'for hundreds of thousands of Irish poor during the middle of the nineteenth century, the suffering of the Great Famine would be synonymous with first-hand experience of the much despised union workhouses', and that approximately 250,000 people died in workhouses between 1846 and 51 (2016, p. 102). Although the number of child deaths in the workhouses is not provided, Geber points out that over half of the victims of the Great Irish Famine (1845−52) were children, and that in

FIGURE 7.3.2   An empty cradle at the Irish Workhouse Centre. *Photo courtesy of Iddon 2018.*

the burial grounds adjacent to the former Kilkenny Union Workhouse, more than half of the skeletons were remains of children and adolescents (2016, p. 101—2). Geber notes:

> In the mass burial ground, there were remains present of 142 children belonging to the 2—5-year age class, of which most were about 3 years of age at the time of death A sharp rise in deaths of 3-year-olds is of particular interest, as it seems most likely to relate to the segregation policy between child and mother in the workhouse (2016, p. 112; Tables 3 and 4) (Fig. 7.3.2).

Mahoney (2016) attributes the popularization of Malthusian notions of population control as a reason for public apathy toward the plight of the Irish; and, while the Great Famine is claimed as a genocide by some, perhaps most vociferously by Irish historian Coogan (2012), the perpetuation of conditions in workhouses that would have been known to bring about death could meet some of the criteria for genocide stipulated in the United Nations Convention on the Prevention and Punishment of the Crime of Genocide.

Analyzing osteoarchaeological evidence from the former cemetery of the Manorhamilton Workhouse, researchers Tom Rogers, Linda Fibiger, L.G. Lynch and Declan Moore found 'the individuals buried here were under considerable physiological stresses, some evidently from early childhood [...] In addition, the double and triple burials of children suggest that they may have been dying in high numbers during some period of use of the cemetery' (2006, p. 96). The workhouse diet almost universally consisted mainly of potatoes, bread, meal and buttermilk or soured milk. Gruel made of meal, milk and water was a staple. During the potato blight in Ireland, potatoes were replaced with imported Indian meal, and according to Rogers et al. there was little evidence of green vegetables or fruit in their diet (2006, p. 95). Moreover, 'there was significant skeletal evidence of nutritional deficiencies, certainly from early childhood', along with evidence of other physical hardships (2006, p. 97) (Fig. 7.3.3).

FIGURE 7.3.3   I made gruel according to the Reverend Belcher's recipe, which I obtained while at The (Southwell) Workhouse. It was pitifully thin. *Photo courtesy of Graham Iddon.*

7.4 Constructions of the child at risk

Calls for the establishment of 'pauper schools' for the training of workhouse children came quickly on the heels of the institution of workhouses. Writing in 1838, Assistant Poor-Law Commissioner James Phillips Kay argued, 'it is difficult to perceive how the dependence of the orphan, bastard, and deserted children, and the children of idiots, helpless cripples, and of widows relieved in the Union workhouses, could cease, if no exertion were made to prepare them to earn their livelihood by skilled labour' (16). With the separation of children from their parents, Kay observed that skills typically taught in the home remained unlearned in the workhouse, citing the example of 'the child of a labourer reared beneath its parent's roof [who is] early trained to labour […] initiated in the duties of husbandry […] ploughing, harrowing, thrashing, milking, and the charge of horses' (16).

Rather than being an argument for a reversion to out relief, however, and after a series of comparative cost projections, Kay proposes a solution in 'County Schools of Industry', residential schools intended to put children to work, 'not to make a profit of their labour, but to accustom them to patient application to such appropriate work as will be most likely to fit them for the discharge of their duties of that station they will probably fill in after-life' (24). Of note, Kay urged that 'children should not be taught to consider themselves paupers', and that contact with 'adult paupers maintained in workhouses [including their own parents] from whose society the children could acquire nothing but evil' should be avoided (22). Lauding the success of the 'adoption of a similar system of industrial training to reclaim juvenile offenders' by the Children's Friend Society, Kay warns that without such instruction, 'the education of the pauper children can afford no effectual guarantee for their future independent subsistence by the wages of industry' (26).

Former poor law guardian and workhouse critic Day (1912) illuminates the deleterious effects on workhouse children of 'institutionalism'. It is a system of unremitting routine that saps children of their will, judgement and powers of decision and imprints upon them a preference of strict automation, rendering the challenge to adapt to the changing circumstances of working life, or indeed any life outside of the institution, insurmountable. 'Small wonder', Day opines, 'that a prejudice exists in the public mind against employing these children. Shop and factory owners will not receive them as apprentices because "they are not fit to associate with other employees", who in turn object to intercourse with "paupers"' (175). Day proposed a twofold solution to counter the effects of 'institutionalism', contamination and debasement of children by their institutionalized parents, and to redress their general lack of education and the missing 'give and take' of home life. Day's first strategy was to allow workhouse children to attend national schools, thereby integrating with the 'sons of the working classes" (175); her second, following Australia's example, to board out children with their own mothers (those judged fit and worthy) or with a foster mother (178). With statements such as, 'Australia thinks it worth her while to pay women to be good mothers […] in Ireland we are too poor to spend £15 or £18 a year upon the maintenance of healthy children, while we keep up a ruinously expensive Lunacy Authority, cheerfully disgorging £150 per annum to keep a suicidal maniac alive' (178), Day argued emphatically, anticipating her 'boarding out' strategy would be conflated with 'the crime' of out relief.

7.5 Toward industrial schooling

Day's arguments followed largely unsuccessful attempts to establish district schools for workhouse children. According to workhouses.org.uk/Ireland, '[b]oarding-out was introduced in 1862, initially in an attempt to reduce the mortality rates of those under five. The age limit for boarding-out was increased to 10 in 1866, and to 15 in 1898. In 1900, a total of 2,223 children were being boarded out or "out at nurse", although the majority of children remained in the workhouse' ('Poor law unions in Ireland, 2018'). Industrial Schools Acts were enacted in England (1857), Ireland (1868) and Ontario, Canada (1874). J. Walker and A. Glasner establish the English origins of Canadian 'training schools', where 'it was discovered at an early stage in the history of the reformatory movement [that] there was an urgent need to do something for the numerous street urchins who gathered in the lanes and alleys of the large cities and who, if not cared for, would sooner or later join the ranks of criminal offenders' (1964, 344). Showing little regard for impoverished and institutionalized children, Kate Kenny reveals 'as far back as the 1920s, government reports show that senior ministers, including one future Taoiseach (prime minister), refer [to them as] no great acquisition to the community [as persons whose] "highest aim is to live at the expense of taxpayers," and from whom Ireland would experience a "decided gain if they all took it into their heads to emigrate"' (2016, 948). Kenny compellingly constructs these children as 'abject-boundary', an excluded, distasteful 'other', enabling and even legitimizing the perpetuation of violence against them (2016, 939).

Their 'at-risk' status legitimized a system that was later found to have victimized them. Indeed, 'although ostensibly set up to care for children [w]e now know that many [Industrial Schools] were violent and neglectful institutions, with excessive beatings and sexual abuse commonplace' (Kenny, 2016, 939). The road to Indian Residential Schools, which have been characterized in a like manner, is often thought to be exclusively paved by way of Indian industrial schools in the United States, with specific reference to the influence of Captain Richard Pratt in Carlisle, Pennsylvania (See also Chapters 5 and 6). Discussing Canada's imperial inheritance, Milloy (1999) situates Indian residential schools within a late-Victorian empire-wide endeavour of social reform, one of 'heroic proportions and divine ordination' metamorphosing the various tribes of 'savages' of the British realm into civilized subjects (6).

Indeed, Milloy highlights the nationalistic rhetoric of Confederation-era Canada, wherein Prime Minister Sir John A. Macdonald, 'dreamed of discharging his benevolent duty', fulfilling a 'sacred trust' and 'national duty' to redeem and emancipate Indigenous Peoples, such that they become productive, peaceable and self-sustaining citizens, rather than 'dependants' on government funds (1999, p. 6–7). Of course, this construal of Indigenous Peoples is a profound misinterpretation on the part of the government of its nation-to-nation treaty obligations, negotiated by First Nations to secure a portion of revenues from the transfer of land and natural resources extracted therein. As I argue, the Indian residential school system was equally a project of (1) early intervention, designed to neutralize the threat of 'Indian insurrection' and criminality and (2) poor relief.

7.6 Indian Residential Schools in Canada

Industrial boarding schools, and later 'residential schools', were part of a government policy of aggressive assimilation — found by Canada's TRC to have constituted a cultural genocide. Over 150,000 Aboriginal children were forcibly removed from their homes to attend these church-run schools designed to, 'kill the Indian in the child' (Truth and Reconciliation Commission of Canada (TRC), 2016 TRC.ca; see also Chapter 1). The underfunded and poorly monitored system, which operated from 1831 to 1996, gave rise to neglect and abuse of every kind. High death rates at the schools (as high as 75% in the File Hills School over its 16 years of operation) were documented by medical examiners and known to Indian Affairs administrators as early as 1907 (Milloy, 1999, p. 97). Despite this, the system expanded, becoming compulsory for Indian children in 1920, and later for Inuit and Métis as well (Titley, 2004, 90). The intergenerational effects of Indian residential schools are ongoing and continue to negatively impact individuals, families and entire communities.

The Indian residential school system, I argue, constitutes a critical event in Canada. However, the rupture to which I refer in contextualizing this work in postrupture museology was not triggered by the critical event itself, but by the TRC process through which the majority of Canadians came to know that it had transpired. Thus, the disruptive power of the critical event of the Indian residential school system was only realized in translation, in the abstracted and distantiated form of the TRC.

7.7 Three sites of difficult history musealized: Portumna, Southwell, Shingwauk

As previously mentioned, the focus of my site-based research at the *Irish Workhouse Centre* (Portumna, Ireland) and *The Workhouse* (Southwell, UK) was to investigate and analyze phenomena within the overlapping categories of site intervention and use, interpretation design, and animation and programming. Specifically, I sought evidence of critical historical recovery, reclamation and transformation, as a means of realizing social and spatial justice. Moreover, I wanted to see if the remediations of place that these workhouses underwent were intended to promote healing and reconciliation in addition to tourism and education. The following questions emerged:

1. Were these motivations even present and if so, who needed healing and who reconciliation?
2. Was a desire for decolonization a motive for the restoration of the Portumna workhouse? And, is it a focus of the Irish Workhouse Centre administration?
3. Was any degree of functional coherence maintained or realized in the rehabilitation of these buildings? Phrased differently, do they currently perform critically and ethically informed functions consistent with their original program of poor relief? And if so, how?
4. What techniques of communication and translation do they implement? That is to say, how do they translate historical or cultural concepts and experiences, including experiences of trauma that have few contemporary counterparts and are thus, in the main, incommunicable to visitors?
5. How do they communicate across cultural divides?

7.8 Site intervention and use

My fieldwork could not have and indeed did not furnish answers to all of the above questions. Nevertheless, these places offered important insights toward a reconciliatory museology, which are discussed below.

7.8.1 Portumna

The former Portumna workhouse lay derelict and mostly covered in ivy for decades. In 1999, South East Galway Integrated Rural Development (IRD) Company, a not-for-profit local development company, approached the then-Western Health Board, owners of the complex, with a view to examining ways to conserve and reuse the site for the benefit (including economic benefit through labour and tourism) of the area. A masterplan was commissioned and in tandem, South East Galway IRD commenced work on conserving the buildings. In 2011, the workhouse was opened to the public as the Irish Workhouse Centre. The Centre outlines the following objectives in the introduction to its website under the heading 'Conservation, Re-Use, Restore':

- tell the story of the Irish workhouse as an institution, which operated from the early 1840s to the early 1920s,
- showcase a conservation and redevelopment work in progress,
- attract visitors to the area,
- provide employment and
- provide space for community events, projects and other appropriate uses (Conservation, 2018) (Fig. 7.8.1.1).

Despite the finding of the Commission to Inquire into child abuse that widespread abuse had occurred within the Catholic Church—operated, government-funded system of residential 'Reformatory and Industrial Schools', its recommendations did not include provisions related to redress via historical representation, museums and former institutional sites. In the absence of such impetus for reclamation and remediation of former industrial school sites through musealization, and given the trajectory of pauper children from workhouse to industrial school, I sought to go back further in history. I was looking to see if the project of the Irish Workhouse Centre could be seen to embody an unrecognized need to reckon both with the trauma workhouse sites introduced into the 19th-century Irish landscape, and the role these sites played in laying a foundation for the compounded trauma of Irish industrial schools.

Portumna curator Elizabeth Carter (Personal communication, 2018) affirms the strong and enduring presence of traumatic memory in Ireland as it relates to workhouse history. This view is consistent with that of Wildman (1974) who opined, 'the chief barrier to appreciation of the workhouse as a significant architectural species is, of course, its unpleasant associations' (291) and who predicted that, 'if these gradually recede into folk-memory, it becomes possible for us to understand more clearly how the new unions became, in the middle of the last century, "objects of civic pride and self-satisfaction, rather like swimming baths or libraries today"' (291)—a possibility consigned to perhaps a very distant future. George Wilkinson, English architect of all 130 Union workhouse buildings, himself entertained 'the belief that, when the present calamitous period (the famine) has passed, the Irish poor-houses will, at

FIGURE 7.8.1.1   Dormitory at the Irish workhouse centre. *Photo courtesy of Iddon 2018.*

no distant time, be found, with regard to their general building arrangements, a very superior class of public institutions of the kind' (1847, quoted in O'Connor, 1995, p. 93). This belief has certainly not borne out to date. Moreover, Carter identifies an underlying inhibitor to both cultivating relationships with communities and to inspiring patronage/visitorship in a 'deep-seated unease with conventional museums and spaces of authority' (personal communication, 2018). She explains, 'these experiences are background-specific; for example, people who have had negative school experiences or industrial school experience, or those from largely rural communities, will all be of a demographic more likely to feel that a museum space is not a "safe" space' (Carter, 2018).

Beginning with the opening by the British state of the National Gallery of Ireland in 1854, followed by the Natural History Museum and National Museum of Ireland, museums took their place within a series of institutions that imposed British cultural values on the Irish. Commenting on O'Connor's 1995 claim that 'the workhouse was the most feared and hated institution ever established in Ireland' (13), historian Peter Gray (2012) asserts, 'the penitentiaries and asylums of the nineteenth century were surely as much feared, and perhaps with more reason; the record of the industrial schools and Magdalene asylums has more recently attracted the appalled attention of Irish society' (22). As embodied material legacies of British cultural imperialism and rule, compounded with pervasive negative associations with public institutions, particularly those of confinement, the perception of museums as unsafe and inherently *institutional* spaces in the Irish imagination can hardly be questioned. While museums of the ilk aforementioned may not be indigenous to the Irish context, Carter asserts art and music are integral to Irish constructions of culture and identity, providing a promising entry point to would-be visitors. Arts-based programming contributes to the distinctly Irish museum typology in formation at the Irish Workhouse Centre. Its twofold strategy to mitigate site-based associations with traumatic memory and difficult history includes (1) prioritizing the relationship of the Centre with the local and visiting community and (2) adopting a caretaking attitude with respect to both the local visiting public and the site itself (Personal communication, 2018).

7.8.2 Southwell

The National Trust acquired the Southwell Workhouse ostensibly to preserve this Grade II listed building, which had come on the real estate market in the late 1990s and to further its social history mandate. This was consistent with its general purposes under Section 4.1 of the National Trust Acts 1907–71 (Post-Order, 2005) to promote 'the permanent preservation for the benefit of the nation of lands and tenements (including buildings) of beauty or historic interest and as regards lands for the preservation (so far as practicable) of their natural aspect features and animal and plant life' (4). During a guided tour of the Southwell Workhouse on 23 April 2018, the National Trust tour guide opined that it had been acquired by the Trust to diversify its attractions, most of which were historic houses. The reason for acquisition given by Kim Drabyk, Volunteer and Visitor Experience Manager, was to promote the Trust's 'social history mandate' (2018). During the time of writing, the Trust had yet to respond to my inquiry on the rationale for acquisition of the Southwell Workhouse or its specific mandate.

While the museal objectives of The Workhouse are not explicitly articulated on its website or in its promotional or information literature, the following may be inferred from those sources:

- Promote awareness of the history and heritage of Poor Law policy and Victorian workhouses
- Help visitors to connect with the experiences of Southwell Workhouse residents
- Promote local, regional and national tourism
- Connect to local history and community
- Promote educational experiences for children and youth
- Promote ongoing site-based conservation
- Create sustainability for this Grade II listed building

National Trust 'Spirit of Place' statements set the tone for historical and site-based interpretation. For the Southwell Workhouse, a property known formally as the Thurgarton Hundred Workhouse,

[t]he spirit of the place is a bleak one. Not a prison, the building and the confined spaces in and around it speak of a basic regimented existence for those within; an atmosphere of dejection and lost dignity pervades. The purely utilitarian nature of every element is repeated again and again throughout the building, yet there is symmetry and a sense of intent and design permeating the structure and detail which is solid quality. This building was built to last, a solution to the high cost of out relief for the poor, a means of providing a sanctuary for the infirm and for separating out the morally inferior of the Thurgarton Hundred. The rural location is a stark reminder of this fact; glimpses of Southwell Minister from the upper windows serve to reinforce this, beckoning poignantly towards a better world beyond (qtd. in Nottingham County Council and Newark & Sherwood District Council, 2012)

Because the Southwell Workhouse operated as a site of social service provision continuously since its erection in 1824, and has only within a few decades ceased to lodge people on social assistance, it retains social stigma. Closure and a period of disuse or of different use would have created a distantiating effect, as is the case with the Shingwauk Schools and indeed with the Portumna Workhouse. According to Volunteer and Visitor Experience Manager Kim Drabyk, the stigma that attaches to the site is mitigated by the National Trust brand (Personal communication, 2018).

7.8.3 Shingwauk

The Shingwauk Indian Industrial School or 'Shingwauk Home' established in 1875, was one of Canada's earliest residential schools. Shingwaukonse, 'The Pine' (1773–1854), was an Anishinaabe Chief, War Chief of the Ojibwe, leader in the War of 1812, and signatory to the Robinson Huron Treaty of 1850. He envisaged and promoted the development of 'teaching wigwams', schools offering a European-based education and training program to the Ojibway (Shingwauk and Agawa, 1992). In funding and founding the school, Shingwauk's sons Augustin (1800–90), and Buhkwujjenene (1811–1900), found a willing ally in the young Rev. E.F. Wilson (1844–1915). Wilson worked with the younger Shingwauks to found the first Shingwauk Home in 1875. By 1892–93, disillusioned with the school system he had helped to build and disheartened with an administration that failed to recognize the competence of Indigenous Peoples, Wilson resigned from his post as Principal of the Shingwauk Home. The school was subsequently absorbed into the chronically underfunded nationwide Indian Residential School System.

In 1970, the Shingwauk Indian Residential School closed, and in the subsequent year Algoma University College, later Algoma University, took possession of the building and the campus, expanding dramatically over the intervening 40 years. Shingwauk Hall was erected in 1934–35 as part of a modernization scheme concurrent with the school's transition from an industrial to a residential program; although both were residential in the sense that Indigenous children were boarded. In 2018, the main floor of Shingwauk Hall was musealized and opened to the public as the Reclaiming Shingwauk Hall Exhibition, while retaining its use and function as part of the larger Algoma University complex. The exhibition's mandate, 'healing and reconciliation through education', builds on decades of Residential School Survivor-driven 'sharing, healing, and learning' work, based at the Shingwauk Residential Schools Centre (SRSC) at Algoma University (See also Chapter 4).

7.9 Interpretation design

7.9.1 Portumna

The physical spaces of the Irish Workhouse Centre have been variously fallowed, conserved and restored. Staging is kept to a minimum. A prominent room of the main building, formerly the Guardian's Board Room, has been restored and fitted with modern amenities, now serving as the Centre office, resource room and gift shop. A few other rooms on the same building's ground and lower levels offer meeting and educational program space, modern toilets and an art gallery. On the second level appears an exhibition room fitted with displays of artefacts and interpretative panels. The Centre does not make use of digital media, which may be the result of both a lack of funds, and the view that digital media constitute an undesirable or inappropriate form of interpretation. Rather, guides bring visitors on tours of the various spaces (from those representing the most conservation intervention to the least) and provide live interpretation. The various spaces are sparsely populated with interpretative panels, making live interpretation critical to the Centre's didactic offering, as my colleague Graham Iddon discovered. For example, when he toured the site with staff archaeologist Mary Healy, she pointed out that the children's dormitories were fitted with standard height windows on one side and well above eye-height windows on the other. The purpose was to prohibit children from seeing their parents, whose outdoor work yards were situated just one storey below. This information provided valuable insight into the psychology of the workhouse scheme, yet was not available on panels or through other means (Personal communication, 2018).

The Centre suffers from a lack of artefacts, having only a very modest collection comprising pottery and other vessels, and small implements. Centre staff have considered putting out an open call for artefacts. Though acquisitions were of interest, no ethical conflicts or impediments to collecting were identified. A citizen having donated an artefact in an effort to promote its preservation suggests at least one potential motivation for would-be donors. According to Carter (personal communication, 2018), artefacts will fill in blanks of the story and bridge the gap between policy, its implementation, and what actually occurred in the Portumna Workhouse. Replicas and contextual artefacts are not used in the Centre's displays. The lack of such items reinforces the presence of historical absences.

7.9.2 Southwell

Visitors to The Workhouse are encouraged to take first a guided tour, and subsequently a self-guided tour which conducts them through the complex in a particularized and directed manner. Live interpretation provided by the volunteer guide of the guided component traces the contours of the Poor Law and of the workhouse system as a backdrop to a more specific and anecdotal history of the Southwell site. The tour involves a visit to the grounds and focuses on spaces that surround the workhouse; while the self-guided tour and its encounters with live animators takes place within the workhouse enclosure. Most of the tour sites (workhouse exterior and outbuildings) had been restored while a few, such as the mortuary, have been minimally conserved. At the time of my visit, the only fully modernized building was Firbeck House, an infirmary built on the site of the Workhouse in 1871, and partially converted to house a café and washroom facilities. Disparities between narratives delivered by the tour guide, animators, and through interpretive panels and materials were marked. In many ways, the site also spoke for itself, becoming its own interpretative agent. For example, the guided tour 'script' and its delivery was seemingly designed to entertain, going so far as to paint a flattering picture of the institution and its history, which may be equally attributed to the guide as to the direction of the administration. Peppered with amusing anecdotes (i.e. workhouse children being brought to the circus) and observations, the tour script contradicted other site-based information sources.

On the self-guided tour, visitors are instructed to view a video after which they are provided with a guidebook to help them navigate the tour pathway. The organization of the workhouse itself, designed to restrict the movements of inmates, enforces the prescribed visiting sequence. Live animators depicting adult male and female inmates, the master and matron, ensure that visitors do not deviate from the tour pathway. The effect is of being herded, which promotes, through embodiment, an experience of the control exercised over the lives of workhouse inmates. Such circulation management, however, comes at the expense of a fulsome visitor encounter. It is not until visitors arrive in the cellars that the horrors of the workhouse are made manifest. While the areas above grade are bright and cheerful, there is no disguising the cold, dank and gloomy aspect of the cellars where women worked for long hours preparing food, often ankle deep in freezing water. Here, the site defies interpretative revision.

While most workhouses across the United Kingdom were gradually absorbed in the National Health Service becoming hospitals, the Southwell Workhouse retained its social service function, closing only in the 1980s. As is popularly expressed, a greater number of Britons are descended from 'the poor', many of whom were residents of workhouses, rather than from solvent classes. Indeed, some of the guides working at the Southwell Workhouse are descendants of its inmates. Local history information-gathering occurs through the initiative of descendants, and, in fact, teams of volunteer researchers are responsible for much of the historical research that informs the site's anecdotal interpretation.

7.9.3 Shingwauk

Before delving into a discussion of interpretation at Shingwauk, I remind the reader that I am implicated in it as its cocurator and codesigner, the advantage of which is that I can provide insight into decision-making. The corresponding disadvantage, however, is that I cannot provide the critique of a disinterested observer.

The Reclaiming Shingwauk Hall exhibition is currently comprised of three galleries, with a fourth in development. The galleries occupy the main entrance and transversal hallways of Shingwauk Hall, with testimony-based East—West galleries and a contextual and densely historical North—South gallery. Together the three galleries strive to communicate the history of the site and its administration in a manner that foregrounds the experiences and stories of the students/survivors who spent their childhood and teenage years there. While the didactic 'history' gallery, 'From Teaching Wigwam … to Residential School' is text-dominant, the 'story' galleries are visually rich, offering layered archival and contemporary photographs spanning 110 years of experience, showing a continuity of site use into the present.

The exhibition, while ostensibly an effort to reclaim and decolonize a former space of colonization and assimilation, concerns itself far less with issues of heritage preservation and authenticity of the built environment than with its transformation into a culturally safe space of Indigenous articulation and reclamation. In Shingwauk Hall, former student thoroughfares—now gallery spaces and Algoma University thoroughfares—have become a mere canvas for interpretative space: 'substrate' for content that communicates context as a means of reckoning with the paradox of having to preserve the integrities of place in order to enunciate a place-based history, while at the same time neutralizing their mnemonic and other potencies. Becoming substrate, however, did not suppress all of the site's features. Rather, they were concealed and revealed at the discretion of the exhibition's producers.

Post 1970, the east and west wings of Shingwauk Hall were connected to newly constructed buildings and the former Indian residential school became absorbed into a sprawling complex. From the interior, the extremities of Shingwauk Hall could no longer be easily discerned. The exhibition, bounded within the footprint of the original building's interior, returns to it some of its interpretative agency, allowing it to speak for itself. Unlike the Southwell Workhouse, which in its materiality resists remedial interpretation, the historical configuration of Shingwauk Hall is intentionally revealed in the Exhibition's design. The pastiche of doors that admit Algoma University faculty and staff to offices directly off the hallways galleries were reconsidered and replaced with found and refinished originals, which enhanced the studied anachronism of the galleries themselves. Such reassertions of character-defining elements of place instantiated the residential school in a manner that supports the overall effort of historical recovery and representation but does not undermine its reclamative aspirations. In combination with visual argumentation and stories through archival and contemporary photographs, testimony and a didactic contextual history that draws on and displays textual artefacts, site attributes support layered interpretation.

7.10 Animation and programming

7.10.1 Portumna

The Irish Workhouse Centre has a temporary exhibition gallery (mentioned on its website in the section on site interpretation) with which to augment its museal offerings. At the time of my visit, *Dark Shadows,* an exhibition of sculptures on the Irish Famine by Kieran Tuohy, was on display. While the exhibition attracted visitors for whom the Great Famine is indelibly etched in Irish history, memory and identity, it did little to further efforts to promote reckoning with the distinct and unique history and legacy of Irish workhouses. That being said and in counterargument, such exhibitions entice visitors to the Centre, which fulfils a social function beyond musealizations and enables visitors to encounter Irish workhouse history. The Centre is both integrated with and provides support to East Galway community social and employment services, such that a continuity of the social function of 'the workhouse' as respite of the poor, aged and infirm is preserved. Employment at the Irish Workhouse Centre is tied, in part, to Ireland's social benefit (welfare) system through the Rural Social and Tus Schemes. These programs support rural seasonal supplemental employment and development of both community employment infrastructure and resources. The Centre provides work placements for individuals with special needs, for clients of East Galway Mental Health Services, and offers volunteer placements for retired individuals. Increasing its relevance to the community is a major goal of the Centre, which also provides placements for Heritage and Tourism students and is a site of pedagogy for postgraduate local history programs. Social justice is enacted through spatial justice: the deployment of the Centre as a means to support the mental, social and economic health of the local Portumna community and its residents at risk.

In terms of Centre-specific programming, curator Elizabeth Carter is attempting to use art as a new entry point and contact zone. Children's art programming at the site offers an unintentional remediation of place and creates positive associations with a new generation of descendants. Indeed, Carter asserts that art practice, in the form of embroidery, was a part of daily life for female residents of the workhouse—a history Carter aspires to bring to the forefront and a heritage practice she hopes will inspire new art production. For her, art is the key to relevance to the local community and a way to connect to the Irish public (Personal communication, 2018) (Fig. 7.10.1.1).

FIGURE 7.10.1.1   The (Southwell) Workhouse. *Photo courtesy of Iddon 2018.*

7.10.2 Southwell

Not 700 m southwest of the Southwell Workhouse sits the Southwell Court Care Home, specializing in dementia, old age and sensory impairment care, and bearing an uncanny resemblance to the workhouse in architecture and function. In addition, care facilities for impaired children border the Workhouse's north-easterly perimeter. Their timelines overlap with a historically discordant or anachronistic exhibition on the top floor of the Workhouse, which shows how its more recent inmates, those from the 1960s and 1970s, made it their home. At the time of my visit, a dialogic and participatory community textile exhibition, entitled *Struggle for Suffrage: Workhouse Women and the Vote,* occupied the various spaces of the Workhouse, while an exhibition of works from the National Portraits Gallery, entitled *Votes for Women: Faces of Change,* was being mounted in a temporarily sealed off room. The exhibition program appears to augment the Workhouse's social history education function, simultaneously distantiating it from its more recent history and bridging the historical divide through contemporary art practice and community-based interpretation. Thus, the Workhouse appears as a nexus of reflection and interpretation of historic social services and related functions within a larger complex and continuity of social services packaged in mimetic architectural form. It is a remediation certainly, but less a reclamation.

7.10.3 Shingwauk

Shingwauk Kinoomaage Gamig (Shingwauk University), an Anishinaabe Institute, is colocated with Algoma University, temporarily occupying the former Principals' residence as it awaits the completion of its new facility, currently in construction on the footprint of long since demolished industrial school era outbuildings. The site, which was purchased in trust for Anishinaabe education, is today used for Anishinaabe education. It also stands as a dual testament to the enduring legacy of the foresight of Chief Shingwauk and to the trauma generations of Indigenous students suffered as a result of the Indian residential school system's deviation from Shingwauk's vision of the 'teaching wigwam'.

While the onsite SRSC offers tours of the new exhibition galleries, specific exhibition-based programming has yet to be developed and the Centre's primary efforts are focused on both implementing the fourth gallery, which is now in development and obtaining funding for a subsequent phase that includes significant art-based programming. Having only just launched its exhibition in August 2018, Shingwauk is in a very early phase of musealization, which makes the study of places with similar curatorial contexts essential.

7.11 Discussion and conclusion: Toward a reconciliatory museology

In Canada, the physical remains of the built heritage of Indian industrial school sites along with archival documents and a limited number of photographs constitute the material landscape of this difficult history. The project of historical recovery of the Indian industrial school era is both necessary for ethical and robust postrupture museology and complicated by the paucity of testimony. No living survivors of the 19th century industrial schooling remain and very few recorded survivor testimonies relate to this period. These conditions are similar to the curatorial environments of the two workhouse museums.

Common inhibitors of museological development offer another parallel. Like the Shingwauk site, the Irish and English workhouses are sites of trauma requiring remediation to succeed as places of historical interpretation and meaning-making. Despite this, all three are situated in close proximity to groups of descendants of those incarcerated and recognize them as source communities. Archaeological and architectural interventions have not yet been undertaken at the Shingwauk site. However, the Irish and English workhouses have both been the subject of such interventions, which can provide useful models for the Shingwauk site going into the future.

In my experience of working through interpretive strategies with survivors, I have heard them repeatedly express a desire to (1) simultaneously reclaim, occupy and overwrite places where they experienced trauma as children and in doing so change them *and* (2) create a window into the past such that they may be interpreted as they presented in the moment of trauma—in effect, to freeze them in time. Thus, the drive for healing and conquest contends with that to preserve the evidence of trauma, in which the potential for its recurrence inheres. An ethical, museal response prioritizes healing in the physical environment through reclamation, while it creates the means to manifest historical environments and evidence virtually, as digital overlays made visible or invisible on visitor demand. This is high-stakes exhibitionary, entailing Hegelian notions of negation, transcendence and preservation.

One would think this is well-trodden ground in Canada, that Canada would by this time have well-developed museological infrastructure built on the reclaimed sites of the 139 Indian residential schools. This is far from the case. Reclaiming Shingwauk Hall is the only such reclamation project and the remnants of the built environment of the residential schooling landscape have more often been characterized as detritus (Carr, 2009)—mere rock and rubble—rather than being fertile ground for reconciliatory museology.

Surprisingly, or perhaps unsurprisingly, one encounters a similar museal environment in Ireland and the United Kingdom with respect to the history and legacy of workhouses. There are no industrial school museums in Ireland or the United Kingdom, although a few exist in the United States, where a former Indian residential school, Carlisle Indian School, now occupied by the United States Army Way College, offers a digital resource centre similar to the SRSC (See Chapters 4 and 5). As such, further investigation of institutional and even museal continuities may be directed southward. Of course, there has been no TRC process in the United States, and while the Ryan Commission revealed a difficult history, it made no recommendations for remedy that would set museologists on a reclamative and reconciliatory trajectory.

Museology at the Irish Workhouse Centre is driven by the professions of history, art history and archaeology, and although this work is rooted in the local and is practiced by source communities, it is not driven by them. Archaeology orients historical inquiry at the Centre, which is consistent with trends in Irish historical production. This privileges the land as a keeper of memory (artefact-based) over people as living memory keepers (oral history-based). Curator Elizabeth Carter stressed that the remains of centuries of past societies persist in Ireland, suggesting a consciousness of deep history, heterotopia/crony and palimpsest that has a more recent historical basis in Canada. Rather than drawing on source communities for historical data and for direction, the Centre staff seeks to develop ties with, and to become relevant to, the local community. An apprehension toward museums, similar to that described by Carter, manifests in Canada as a historically rooted and justified suspicion of ethnographic (and other) museums by First Peoples. In terms of speaking across cultural divides, it is evident that only part of the story of Indian Residential Schools can be told effectively in museums. Moreover, should Shingwauk become a model for museological development, source communities of Survivors will lead the charge, and reclaimed and musealized former residential schools will express *their* priorities.

The (Southwell) Workhouse, arguably the best funded and most visited of the three institutions, appears to operate both as a novel excursion alternative for the heritage house-going set and a place of education for school groups. It is clear that all three places as musealized embody a reckoning with a difficult past, albeit to different degrees. What is considered difficult is considered as such differently; what constitutes the aspect of difficulty is relative. Because the rupture is differently constituted and understood, the sutures are different, creating distinct reconciliatory possibilities.

Challenges to musealization persist, among them the lacks of material investment, of infrastructure, ownership/responsibility, of public will, and the prohibitive distance of the some of the sites themselves. Paucity of artefacts compound these challenges, demanding either absence (as in Portumna) or prosthesis through reproduction and the introduction of contextual artefacts (as done at Shingwauk). No matter how the sites are musealized, represented and interpreted, animated and programmed, they are done so with the understanding that their subject is both a difficult and increasingly distant past, one demanding insertions of necessary inauthenticities and aids to interpretation in order to communicate and indeed translate their stories to contemporary audiences.

Yet, the past persists and communicates in the embodied and emplaced memory of the site itself, and perhaps also through the performative embodiment of historical visitors in contemporary acts of visitation. Yet visitors are never passive and receptive vessels for museal ambitions. We come to it with our own lenses, which are necessarily personal and different from those of others. We form a part of the learning, history/memory formation process, which while a weak form of redress, initiates subjective acts of reckoning. Reckoning, which must come before redress, and which is also discernible at all three sites, is perhaps the first step in a reconciliatory museology.

References

Carr, G., 2009. Atopoi of the modern: revisiting the place of the Indian residential school. ESC: English Studies in Canada 35 (1), 109–135.

Carter, E., April 15, 2018. Personal Communication.

Conservation, Re-use, Restore, the Irish Workhouse Centre, Portumna, Galway, Ireland | South East Galway IRD, 2018. Irishworkhousecentre.Ie. http://irishworkhousecentre.ie/conservation-and-redevelopment/intro/.

Coogan, T.P., 2012. The Famine Plot: England's Role in Ireland's Greatest Tragedy. St. Martin's Press.

Das, V., 1996. Critical Events: an Anthropological Perspective on Contemporary India. Oxford University Press, USA.

Day, S.R., 1912. The workhouse child. The Irish Review 2 (16), 169–179.

Drabyk, K., April 23, 2018. Personal Communication.

Geber, J., 2016. Mortality among institutionalised children during the Great Famine in Ireland: bioarchaeological contextualisation of non-adult mortality rates in the Kilkenny Union Workhouse, 1846–1851. Continuity and Change 31.1, 101–126.

Gillissen, C., 2014. Charles Trevelyan, John Mitchel and the historiography of the great famine. Revue Française de Civilisation Britannique. French Journal of British Studies 19, XIX-2, 195-212.

Gráda, C.Ó., 1992. Making history in Ireland in the 1940s and 1950s: the saga of the great famine. The Irish Review 87–107.

Gray, P., 2012. Conceiving and constructing the Irish workhouse, 1836–45. Irish Historical Studies 38 (149), 22–35.

Higginbotham, P., 2012. Voices from the Workhouse. The History Press.

Iddon, G., April 16, 2018. Personal Communication.

Irishworkhousecentre, 2018. Conservation, Re-use, Restore, the Irish Workhouse Centre, Portumna, Galway, Ireland | South East Galway IRD. As described in. Irishworkhousecentre.Ie. http://irishworkhousecentre.ie/conservation-and-redevelopment/intro/.

Kay, J.P., 1838. On the establishment of county or district schools, for the training of the pauper children maintained in union workhouses. Part I. Journal of the Statistical Society of London 1 (1), 14–27.

Kenny, K., 2016. Organizations and violence: the child as abject-boundary in Ireland's Industrial Schools. Organization Studies 37 (7), 939–961.

Legacy of the Great Irish Famine, 2018. En.Wikipedia.Org. https://en.wikipedia.org/wiki/Legacy_of_the_Great_Irish_Famine.

Let Us Lay These Ghosts to Rest as author with Tóibín C, July 10, 1999. The Guardian, Guardian News and Media. www.theguardian.com/books/1999/jul/10/historybooks.books.

Mahoney, P., 2016. Grim Bastilles of Despair: The Poor Law Union Workhouses in Ireland. Quinnipac University Press.

Marstine, J., 2017. Critical Practice: Artists, Museums, Ethics. Routledge.

Marx, K., 1859. Population, crime and pauperism. New York Times 16.

Milloy, J.S., 1999. A National Crime: the Canadian Government and the Residential School System, 1879 to 1986. vol. 11. University of Manitoba Press.

Mitchel, J., 1876. The Last Conquest of Ireland (Perhaps). Cameron, Ferguson.

Nottingham County Council and Newark & Sherwood District Council, 2012. Southwell Landscape Setting Study. Nottingham County Council and Newark & Sherwood District Council.

O'Connor, J., 1995. The Workhouses of Ireland: The Fate of Ireland's Poor. Anvil Books.

O'Hegarty, P.S., 1952. A History of Ireland under the Union, 1801–1922. Methuen, London.

Poor Law Unions in Ireland, 2018. Workhouses.Org.Uk. http://www.workhouses.org.uk/Ireland/.

Ravichandra, N.G., 2013. Fundamentals of Plant Pathology. PHI Learning Pvt. Ltd.

Rogers, T., et al., 2006. Two glimpses of nineteenth-century institutional burial practice in Ireland: a report on the excavation of burials from Manorhamilton workhouse, Co. Leitrim, and St. Brigid's hospital, Ballinasloe, Co. Galway. The Journal of Irish Archaeology 93–104.

Shingwauk, Agawa, J., 1992. From Teaching Wigwam to Shingwauk University. Woodland Printers, Sault Ste. Marie, ON.

Simon, R.I., 2006. The terrible gift: museums and the possibility of hope without consolation. Museum Management and Curatorship 21 (3), 187–204.

Simon, R.I., 2011. Afterword: the turn to pedagogy: a needed conversation on the practice of curating difficult knowledge. Curating Difficult Knowledge, Palgrave Macmillan, London, pp. 193–209.

Thane, P., 1978. Women and the Poor Law in Victorian and Edwardian England. History Workshop, no. 6, JSTOR, pp. 29–51. JSTOR. www.jstor.org/stable/4288190.

Titley, E.B., 2004. A Narrow Vision: Duncan Campbell Scott and the Administration of Indian Affairs in Canada. UBC Press.

Truth and Reconciliation Commission of Canada (TRC), 2016. Trc.Ca. http://www.trc.ca/websites/trcinstitution/index.php?p=39.

Truth and Reconciliation Commission of Canada, 2015. The Final Report of the Truth and Reconciliation Commission of Canada. Canada's Residential Schools: Reconciliation, vol. 6. McGill-Queen's University Press.

Walker, J., Glasner, A., 1964. The process of juvenile detention: the training school act, the child welfare act. Osgoode Hall Law Journal 3, 343.

Wildman, R., 1974. Appendix: workhouse architecture. In: The Workhouse: A Social History. St. Martin's Press, Inc., New York.

Wildman, R., 2003. Appendix: workhouse architecture. In: The Workhouse: A Social History, vol. 592. Random House, p. 296, 291.

CHAPTER

8

Talk, templates and developing a geospatial archives tradition: Stories in the making of the Residential Schools Land Memory Atlas

Stephanie Pyne*,1, Melissa Castron2, Kevin Palendat3

1Postdoctoral Research Fellow Geomatics and Cartographic Research Centre (GCRC), Carleton University, Ottawa, ON, Canada; 2Department of History (Archival Studies), University of Manitoba, Winnipeg, MB, Canada; 3MA Candidate, Department of History (Archival Studies), University of Manitoba, Winnipeg, MB, Canada

*Corresponding author

8.1 Introduction

As Bourdieu (1992) intimated in his Paris workshop, presentations that focus on disseminating knowledge about finished products or final results may obscure significant aspects of the products or results, from issues related to their generation to research process insights (see Chapter 1). In a similar fashion a variety of research processes are shaping the development of various components of the Residential Schools Land Memory Atlas (some of which have been discussed in Chapter 2) in a manner that makes them constitutive of the Atlas itself. This chapter reflects on the relationship between relational approaches and a relational space performative approach to cartography, in particular, Cybercartography; where the common thread of process and its reflexive documentation is shared by each approach. In addition to this interlinking theme, an explicit concern with the concept of community and the concomitant practice of community building is also shared by new approaches to cartography and archival studies. These commonalities reflect the intersectional relationship that is necessary to create a geospatial archive, with the map acting as a portal to information, knowledge and perspectives relating to residential schools, and the principal of using place as an organizing factor.

The turn to technology taken by cartography offers new possibilities for organizing digital archival collections according to geospatial principles. Critical archival studies take an inclusive approach to digital archives database management; it seeks ways to engage the broader community in the creation and use of these databases (Christen, 2011; Flinn et al., 2009; Shilton and Srinivasan, 2007), in addition to employing ethical, decolonizing approaches to understanding and interpreting these processes (Jean-Bastian, 2013; Christen, 2011; Nesmith, 2006). A brief review of the spatial turn to 'relational space' in critical geography will provide the basis for understanding the kind of space that many critical cartographers are mapping, and why their approaches to mapping are so different from the conventional notions of cartography that many are used to.

Following the next section on new approaches in cartography, which informed the research assistant orientation phase and our ongoing atlas development work, we illustrate the links between theory and practice in the early to mid stages of work to conceptualize and develop the map modules referred to in the project proposal (see Chapter 2): First, by sharing some of the results from our literature review on new approaches to archival studies; and second, by sharing insights and observations from our atlas development research processes.

8.2 Relational space, performance and community approaches to cartography

In this new age, maps are increasingly being viewed in a performative and nonrepresentational manner (del Casino and Hanna, 2006). This means seeing maps as being continually in the making. Every time a map user 'uses' a map, they are remaking it according to their own contextual needs, abilities and interpretations. As a result of work in critical cartography, instead of being used solely as a means to exert authority over territory and people's understanding of it, mapping is increasingly being thought of as a means to contribute to 'human [and world] flourishing' (Sen, 1999) through the enhancement of agency and the inclusion of multiple — often previously silenced or underattended to — perspectives and ways. The performative turn in cartography includes an emphasis on emergent knowledge, on rethinking and redoing mapping, often with a focus on journeys and experience. Critical cartographers generally resist rationalism (Turnbull, 2000) and question conventional approaches to space.

How does thinking about space relationally differ from conventional conceptions of space? Nancy Ettlinger summarizes the contrast well:

Unlike 'location', which is a Cartesian matter of latitudinal/longitudinal fixes and related patterns, spatiality concerns processes that occur across space and over time, and are integrally related to social relations – not by cause and effect (e.g. action has spatial effect) but rather by being inextricably bound up in one another. Thus understanding social relations requires understanding processes of the space economy and vice versa (Ettlinger, 2004, p. 29–30).

Perhaps the most important aspect of the relational space perspective is its nonessentialist rejection of binary thinking, especially with respect to the global and the local, space and place. According to this view, space is as real as place (Massey, 2004) and not some abstract, ethereal entity. Further, place is not a territorially predefined entity that somehow determines the identities of its inhabitants (Amin, 2004), but is rather an entity that derives its identity through the negotiations of a heterogeneous set of inhabitants, each of which has an evolving and unique set of proximate network relations, in addition to a unique set of more distant network relations (Amin, 2004; Massey, 2004).

Reflecting a relational space approach to 'identity', Massey concludes, 'identities are forged in and through relations (which include non-relations, absences and hiatuses). In consequence they are not rooted or static, but mutable ongoing productions' (2004, p. 5). In her nonessentialist construal of individuals, Massey maintains, 'We do not have our beings and then go out and interact'. This view acknowledges the way that people – as relational individuals – are always 'in the making', which opens up possibilities for considering personal transformation and growth as a function of relationships. Paradoxically, it would seem that it is through a relational perspective that we are able to appreciate the uniqueness of places and people. Erik Swyngedouw has argued for the ontological priority not of the global nor of the local, but of process where "scale is not and can never be the starting point for sociospatial theory", and (1997, p. 141), '[a]ccording ontological priority to a process based view takes the focus away from both the "global" and the "local" as the starting point for descriptive and explanatory analysis' (1997, p. 141). Instead, 'scalar spatial configurations' are produced through processes that involve sociospatial dynamics: 'The theoretical and political priority, therefore, never resides in a particular geographical scale, but rather in the process through which particular scales become reconstituted' (Swyngedouw, 1997, p. 141). This process-based and relational view of space is contrasted with approaches to geography of scale that assign 'motive, force, and action to pregiven geographical configurations and their interaction rather than to struggles between individuals and social groups through whose actions and their nested articulations become produced as temporary standoffs in a perpetual transformative sociospatial power struggle' (Swyngedouw, 1997, p. 141). Despite the fact that Swyngedouw focuses on struggles and does not mention more positive, mutually supportive and co-operative interactions, his process view of the scales of relational space, which focuses on interactions between people, leads us to an analytical focus on the ways in which relationships and interactions reflect scale and define spaces.

The relational space perspective is a process-oriented view that sees places as being inherently political nodes of negotiation or contestation that are constituted through encounters. The basic ontology of an emotionally mediated relational space view extends the relational space ontology to include a concern with understanding the dynamics of the actual encounters. It assumes (1) the basic ontological fact of people as thinking-feeling creatures, (2) that people's thoughts and feelings interact in ways that are subject to standards of moral rationality and (3) that those thoughts and feelings shape the moral conduct and outcomes of negotiations and contestations that comprise relational places. In this context, it is important to have a language or conceptual framework with which to understand this human geographical microspace (Thrift, 2004); and in this regard, people like Kwan (2007) have begun to try mapping these microspaces of experience.

In their guest editorial to the Environment and Planning A (special issue on performance), Latham and Conradson (2003) acknowledge the history of geography's interest in studying practices and experience, pointing to people like Hagerstrand (1975, 1976, 1982), and others from 'the more interesting strands of humanistic geography (Buttimer, 1976; Buttimer and Seamon, 1980; Ley and Samuels, 1978; Tuan, 1977)' (1901). At the same time, they point out two significant turns in geography that have led to the renewed interest in practice: First, 'a more general willingness to explore culturally inflected styles of analysis and interpretation [which] has helped move questions of social practice closer to the intellectual centre of the discipline'. Second, the renewed interest in social practice 'in combination with an engagement with a diverse collection of social-theoretic traditions' have contributed to the prevailing social construction of knowledge view of reality (Latham and Conradson, 2003, p. 1901—02). Together these turns focus on space as a relational entity and provide a supportive environment for the emergence of performative approaches to cartography. A performative perspective of maps as processes, including those related to their making and their use, 'goes beyond seeing the map as an immutable and neutral physical object' (Pyne and Taylor, 2012, p. 95) and rests on two central assumptions according to David Turnbull: 'One is that meaning, understanding and knowledge are based in embodied practices. The other is that the performance of knowledge practices and their attendant knowledge spaces and artefacts simultaneously structure and shape our sociocultural world in a process of coproduction. We make our world in the process of moving through and knowing it' (2007, p. 142).

In 'Maps, mapping, modernity: Art and cartography in the twentieth century', Denis Cosgrove suggests 'an examination of art and cartography in the twentieth century should focus on mapping practices rather than on maps as such' (2005, p. 35). Del Casino and Hanna (2006) approach the process of mapping from a performative perspective that includes the practices involved in 'making' the map and the intellectual—emotional acts of interpreting maps each time they are perceived. For them, a map is more than just ink on paper or a digital display on the computer screen: It is the processes of meaning-making that comprise its making and remaking. The approach to maps as practices is consistent with the 'ontogenetic' conception of mapping practices put forward by Kitchin and Dodge (2007) who 'propose a radical departure in ontological thinking concerning maps: a shift from ontology (how things are) to ontogenesis (how things become), or from the nature of maps to the practices of mapping [...] they are not ontologically secure representations, but, rather, a set of unfolding practices' (Kitchin, 2008, p. 213). In contrast with the view that maps are the results of carefully laid out plans and theories, several critical cartographers take an emergent approach to mapping that emphasizes their exploratory nature.

David Turnbull (2007, p. 142) recommends a hodological approach to mapping, where '[i]n geography, hodology is the study of paths, in philosophy, the study of interconnected ideas, and in neuroscience, the study of the patterns of connections in the white matter of the brain'. A hodological approach to the map-making process sees mapping both as trail making, in the sense of knowledge creation, and as trail following, in the sense of tracking the emerging knowledge that results from a series of knowledge-sharing interactions. Emphasizing the trail-making function of maps is relevant to critical Indigenous cartographies (Pyne, 2013, Chapter 9). Kitchin and Dodge (2007), Turnbull (2007) and others cast the critical cartographic project as one that involves 'rethinking knowing and mapping — where the key questions relate to the similarities and differences in the ways space, time and movement are performed and to how those similarities and differences are handled' (Turnbull, 2007, p. 141). Along the same theme, Turnbull advocates creating 'a third space, a space in which the possibilities of agonistic pluralism can occur based on a performative rethinking of knowing and mapping' (2007, p. 141–42).

For Kitchin and Dodge, '[m]aps emerge in process through a diverse set of practices. Given that practices are an ongoing series of events, it follows that maps are constantly in a state of becoming; they are ontogenetic (emergent) in nature. Maps have no ontological security, they are of-the-moment; transitory, fleeting, contingent, relational and context-dependent. They are never fully formed and their work is never complete' (2007, p. 340). For Crampton (2009), maps can be both 'part' of performance the performance itself, and Cosgrove (2008) sees maps as cultural practices. Both authors discuss community mapping in their discussions of performative cartography, revealing the interconnectedness of these themes.

The wake of the digital age has seen an increase in community mapping projects, which can provide fruitful contexts for collaborative work involving different perspectives. An important purpose of community mapping is to 'ground truth' policy by involving community input into sustainability and other issues. There are many issues related to the democratization of cartography. These include concerns regarding the need to educate people about maps and mapping techniques if they are going to be engaged in cartographic practices (Johnson et al., 2006). In 'Participatory mapping as a tool for capturing local perspectives on cultural landscape: A case study of Ostlänken', Chia Jung Wu and Karolina Isaksson contrast participatory and scientific maps: 'Participatory maps are distinguished from scientific maps[...] to emphasize that the contents of the maps derive from a local perspective rather than a scientific perspective' (2008, p. 7). With this said, taking a broad view of community, it is important to neither reify nor exclude the scientific perspective from participatory mapping exercises.

There is a variety of literature concerned with new approaches to archival processes and cartographic records, which a similar shift in thinking to that referred to above. In this regard, Podolsky Norland (2004) makes special mention of Brian Harley's contributions to revisioning cartography in 'Deconstructing the Map' (1989), and Hugh Taylor's contributions to revisioning archival studies and practice, noting that although the thinking of each of these foundational thinkers is quite similar, their thought developed independently of one another in two distinctive, yet related disciplines. In an attempt to contribute to a view of provenance that reflects this new thinking Podolosky Norland offers the following characterization of 'secondary provenance': 'A document is more than its subject content and the context of its original creation. Throughout its life cycle, it continually evolves, acquiring additional meanings and layers, even after crossing the archival threshold. As such, archivists need to read documents against the grain to search for the deeper contexts of their meaning' (1989, p. 147).

In our discussions and email dialogues about this article, it was pointed out that adopting the concept of 'multiple provenance', which extends to public participation in provenance, would be the more useful and appropriate approach (e.g. https://www.nunatsiaqonline.ca/stories/article/65674archivists_seeking_names_of_arctic_inuit_in_historic_photos/). Part of the orientation phase for the research assistants from the Archival Studies program at University of Manitoba involved reviewing critical cartography literature reflecting relational space, performance and community participation themes. Early work also involved collective work toward a rich bibliography organized around emergent themes and incorporating the implicit cartographic strategy of zooming in from the general to the specific. With this in mind, research assistants were provided with an orientation regarding the nature of Cybercartography, and the evolution of the atlas development work, which provided the launch point for the Residential Schools Land Memory Mapping Project (RSLMMP) (see Chapter 2 for a summary of this background). Consistent with the granting agency institutional ethics specifications, the research assistants also completed the online ethics training required by the grant (see Chapter 10), and we began to engage in a community-oriented approach to group research via audio and/or video calls, meetings, email and sometimes text and Facebook messages, and in person meetings. The literature we reviewed next ranged in terms of topic and type from discussions of community/relational approaches to archives and issues in digital archives, which have counterparts in the critical GIS and digital mapping literature, and which we documented in a Zotero data base and in a variety of spreadsheets.

8.3 Relationality and community in new approaches to archives

Throughout our literature review period, we were concerned with informing both the formal and the content dimensions of the Residential Schools Land Memory Atlas, which is the central output of the RSLMMP. The review was aided by suggestions made by Greg Bak, a project collaborator; and we attended (at different times) the MA courses in Archival Studies and Digital History led by Greg and Tom Nesmith, which covered material relevant to our investigation. For example, Paradigm (http://www.paradigm.ac.uk/workbook/index.html), a digital project related to personal archives creation; the InterPARES 2 Project (http://www.interpares.org/ip2/ip2_index.cfm), in which the Geomatics and Cartographic Research Centre has contributed policy relating to archival practices and geospatial information in a digital world (see, for example, Lauriault and Craig, 2008; Lauriault et al., 2008); the Bodleian Electronic Archives and Manuscripts (BEAM) (https://www.bodleian.ox.ac.uk/beam/about/projects/futurearch-project), which is a platform designed to handle born-digital and manuscript material; Preservica or AIMS born-digital collections: an inter-institutional model for stewardship (https://preservica.com/blog/aims-born-digital-collections-an-inter-institutional-model-for-stewardship), with a division for culture and heritage [more]; and the web magazine ARIADNE (http://www.ariadne.ac.uk/issue50/hilton-thompson/), which includes a variety of articles concerning archival practice in a digital world.

In our review, we encountered literature informing our understanding of the history and context of the 'digital world', and were particularly interested in the common roots and issues that faced both archival and cartographic practices. For example, commenting on the theme of community in the digital world, Mahoney (2005) inserts human agency into the history of computing noting that the computer is in no way a natural phenomenon, but a tool developed by humans for specific purposes and utilized to reach an intended goal, which is shaped by a sociopolitical context. Mahoney refers to those responsible for the development of computers as 'communities of computing' and points to a multiplicity of histories including commercial and military examples. This places the computer (or 'computings' (124) in a category of emergent technology that involves the borrowing and adaptation of software templates across knowledge communities.

Software is highlighted as the core aspect of the history of computing: 'software as a model, software as experience, software as medium of thought and action, software as environment within which people work and live' (Mahoney, 2005, p. 128). What we want computers to do and what we can make them do is a narrative with significant failures, most of which remain hidden behind more substantial documentation of successes. It is therefore important to develop a reflexive approach to the systematic archiving of both hardware and software in order to retrieve the 'dynamic artifact' (Mahoney, 2005, p. 130). This speaks to the focus in the RSLMMP on archiving events, performances, and other phenomena that constitute the processes in the iterative development of a cybercartographic atlas. Mahoney calls for greater critical awareness of the computer as a tool and medium, especially in the context of the humanities, as computing used by this community is usually borrowed from other communities due to a lack of internal development (for example, tools developed for use in a corporate environment are routinely used in academic contexts). What has stifled this development is the inability of participants in the field to come up with problems that are simplified enough to be effectively translated to computer programming, a form of technology just as the pencil and paper are forms of technology: 'The historian inherits the problems of software maintenance: the farther the program lies from its creators, the more difficult it is to discern its architecture and the design decisions that inform it' (Mahoney, 2005, p. 131).

Carroll et al. (2011) discuss the challenges within archives as the shift from analogue to digital materials has occurred, and focus on the Salman Rushdie Collection at Emory University as a case study. The collection reflects the need for a new approach and prompted the archivists responsible to develop a holistic approach that would consider the contents of the collection as a whole, with the goal of integrating analogue and digital while carefully considering the needs of both the donor and researchers. The article outlines the development of appropriate tools for synchronized searches of emulated environments such as Rushdie's personal computer, as well as item-level, database-driven searches, and the finding aid. It is also a means of contextualizing 'the processes, workflows, and products that comprise Emory's born-digital archives program' within the team itself — a collaborative group including technologists, librarians and archivists (2011, p. 62); and a useful introduction to the many challenges connected to archiving materials in a broad range of formats that extend beyond preservation and into the realm of accessibility.

New Skills for a Digital Era (Pearce-Moses and Davis, 2006) is the edited proceedings of a colloquium sponsored by the National Archives and Records Administration, the Society of American Archivists and the Arizona State Library, Archives and Public Records and includes reports on case studies. In general, these proceedings confirm an increase in digitized systems that support all aspects of the work of information professionals, and comment on the digital era, in which archivists and other records professionals 'must be able to manage electronic collections, including the ability to select, acquire, describe, organize, reference, and preserve these digital works' (viii). Digital works include analogue materials that have been digitized, and born digital materials, which must be kept 'permanently alive' (viii). Cases studies cover major archival functions (such as acquisition, processing, preservation, reference and access) and are presented by archivists and other information professionals from American universities and other national and state-level organizations. Each section deals with the same issue — the new skills that information professionals need to develop to meet the needs of a changing records landscape — and provides both theoretical and practical information to address this challenge (x). While some information may be outdated due to the rapid evolution of digital records and how we use them, there are both theoretical and practical elements that remain relevant today, for example the value of communication and collaboration, which implies community (Pearce-Moses and Davis, 2006, p. 34).

Cooban (2017) highlights the need for collaborative approaches to archives using Web 2.0 tools and technology in a literature review that presents the involvement of archives with Wikipedia through the lens of 'Archives 2.0', which refers to 'an approach to archival practice that promotes openness and flexibility', and access to and knowledge of collections of interest to the public (259). In relation to participatory approaches, Cooban highlights Duff's and Harris (2002) call for descriptive systems that support the potential inclusion of user voices, as well as Eveleigh's (2014) observations on crowdsourcing as a form of participatory archiving, which has its counterpart in volunteered geographic information as a form of participatory digital mapping. Cooban contrasts processes that allow user contributions, but where archivists are fully in control, with Wikipedia, where 'engagement can place archivists in the role of "facilitator" rather than "gatekeeper", sharing their authority rather than seeking to defend it' (2017, p. 263). The evidence collected by Cooban suggests that engagement with Wikipedia is well aligned with archival ethics and Archives 2.0, although a critical approach is necessary for an appropriate understanding of the relationship between archives and Wikipedia, as well as the role of the archivist as an editor of content in a collaborative project. In particular, Cooban highlights the tension between Wikipedia editors and archivists using the encyclopedia as a tool to direct potential users to their collections. In the past archivists have simply inserted links to the finding aids associated with their holdings, using it as a tool rather than as a platform for engagement with a wider audience or for their own expertise (267). What Wikipedians encourage is more descriptive content that will exist in connection with but separate from authoritative archival websites and documents, thereby providing an open space that is not under the direct control of an institution (267—68).

Nesmith (1999) discusses archival theory in relation to key characteristics overlooked in Trevor Livelton's book Archival Theory, Records, and the Public (1996) by applying science historian Paul Thagard's notion that a "fuzzy" or speculative approach to theory can yield a more accurate interpretation of complexity and nuance (137). Nesmith illustrates this approach through the archival concept of the origin or provenance of a record, stating that archivists "focus on certain aspects of [the origin of a record] which seem meaningful to them and omit much else which may not be knowable" (141). In addition, he defines the role of archivists as a creative one, in contrast with the view of archivists as the passive receivers of records. Just as each interaction between records and researchers becomes an additional layer of context, so too the archivist leaves evidence through their functions and process, thereby creating the "recordness" of the record through their own conceptualization of its role (144.). Nesmith suggests that both the record and archival functions represent ongoing mediations of understanding. The ghosts of archival theory, represented by this speculative approach, should be addressed (or spoken to as Derrida suggests) as a means of also preserving the role of archivist as a visible force that shapes public records publicly (149).

In her consideration of hypertext author and professor, Michael Joyce, Stollar Peters (2006) examines active digital preservation strategies with respect to private collections, which differ significantly from institutional, government or corporate records and can be characterized by the idiosyncratic record-keeping strategies and practices of an individual. Similar to the previous examples from Emory University and the National Library of Australia, this article focuses on the strategies developed to handle the ingest and preservation of born-digital records and various physical components and containers. The level of detail devoted to the process of arrangement extends beyond the previous two examples, where Stollar Peters notes that the team used both traditional archival arrangement and on-demand item-level arrangement to adequately handle the digital objects in the collection. According to Stollar Peters, a utilitarian approach to digital preservation (collecting as much metadata as possible, while leaving the records relatively untouched) best facilitates scholarly use (2006, p. 32). The conclusions and recommendations listed are potentially useful to approach any type of project that relies on preservation of digital materials to remain viable. This includes advocating the use of automated, open-source tools, specialized training and knowledge (though training staff or creating hybrid positions, as well as collaborative projects), the adapting traditional archival functions to suit electronic records on a theoretical and practical level, and clear policy and procedure in place in advance of new projects.

When it comes to considerations of archives in a community context, cultural celebrations — folk traditions at the core of a community identity — reveal the 'heart of the community itself' (Bastian, 2013, p. 122). The paradigm of a cultural archive offers the theory that an annual celebration (conceptualized as a cultural community expression) can be considered a living archive made up of events within the celebration that constitute the archive's records, which will be both fixed (traditional variety) and mobile (dance, oral performance, costumes, folklore); as the celebration and the community are one entity (122). Using the St. Thomas carnival as an example, Bastian demonstrates that the history of the place is retold and reinterpreted annually. Focussing on the 'basic and fundamental issue of locating cultural expressions beneath a wide and all-encompassing umbrella of records and archives', Bastian asks, 'how can the traces and signifiers of cultures and traditions fit within an archival structure' (123), especially if they exist in a wide range of formats not normally accommodated within the traditional western archival theory.

According to Bastian, the 'postcolonial archives' is 'both a potential framework for locating, preserving, and maintain cultural signifiers of memory and identity, and a mental map for thinking about traditional and non-traditional records in unconventional ways' (2013, p. 124). Bastian presents examples of archival records being read against- or with-the-grain through a postcolonial lens, which has baggage and well as limitations, and discusses the formation of archives by postcolonial governments and associated challenges (2013, p. 125–126). From Bastian's performative perspective, archives can exist as both a state of mind and physical place (127). A postcolonial archive supports the theoretical reformulation and rediscovery of narratives that counter master narratives, and the practical restructuring of archival processes to create the space for marginalized actors, voices, and small communities through the prioritization of cultural expression that extends beyond text-based records (127). Bastian includes examples that range from monuments to textiles, as well as an 'archive of place' reflected in the story of the Noongar who won a native title to their land through the use of text-based colonial records, which served to strengthen the oral records already in their possession within the framework of the Australian legal system (128). New technologies in particular can support the continued focus on a plurality of narratives within the archives (2013, p. 129–130). The cultural archive exists in the same space as the traditional archive, as a part of the extant archival continuum, 'supporting collective memory and communal identity as it embraces an inclusive societal provenance that considers all the elements that are essential for the full societal record' (2013, p. 130). The use of archival power is thus employed for social as well as cultural justice.

Flinn et al. (2009) examine examples in the United Kingdom, situated within an international milieu, of archives created by and for communities marginalized by the dominant narrative espoused by heritage institutions. The definition of independent community archives includes, but is not confined to, the collections of various materials, both created and accumulated, gathered primarily by community members who then retain some level of control over said collections. This definition includes collections that are independent or that receive support from mainstream heritage institutions, however they argue that the 'defining characteristic of community archives is the active participation of a community in documenting and making accessible the history of their particular group and/or locality *on their own terms*' (2009, p. 73). These archives also fill a gap within the collections of mainstream institutions, as well as address certain issues of representation within such institutions.

The growth of community archives in the United Kingdom reflects growing interest within historically marginalized groups, such as the black community, to challenge dominant narratives, fill gaps in public awareness of these histories, and document significant themes such as the impact of migration, place and belonging, the impact of deindustrialization and other changes that have affected these communities (2009, p. 74). The availability of funding and information communication technologies has enabled significant growth and 'the formation of virtual communities and allowed geographically distributed individuals to focus on and collaborate around the heritage of a shaded identification, or specific geographical location' (74). The cases studies included in the article include London-based organizations that archive materials that document African, Caribbean, Asian and the Black LGBT experience in the United Kingdom (75). A large part of their research agenda is the recognition and understanding the role of archives in the formation and maintenance of evolving identities, while also expanding the notion of what encompasses identity in the United Kingdom (77). The authors adopted an ethnographic approach 'employing an open participatory observation method to collect "thick description" data', with an emphasis on insider and bottom-up perspectives (77), noting that archival ethnography is gaining purchase in the field.

The rise of community archives is directly linked to social and political movements and as such, archival activists are constructing community archives and heritage as a tool to empowerment and social change, and in the pursuit of social justice. This movement also reflects the need for critical assessment of how processes of reconciliation can be supported through democratized heritage in the form of mainstream and/or community-based initiatives (84).

According to Shilton and Srinivasan (2007), the power of the archivist is grounded in the functions of appraisal, acquisition, arrangement and description, as they choose which records to preserve and how narratives will be framed in relation to these materials (88). With this in mind, they argue that memory institutions have created gaps within archives due to a privileging of certain actors and voices within they societies in which they work. In addition to this, archives have also made efforts to acquire collections documenting marginalized communities which, being created and managed by nonmembers, create 'archives *about* rather than *of* the communities', distorting their narratives through a loss of context and descriptive metadata, while upholding a 'marginalized' and 'mainstream' binary (89). However, by identifying archival interest in preserving contextual and evidential value in relation to digital records, the authors identify a similar strategy that would employ the same 'archival principles of appraisal arrangement and description', and extend to actively incorporating participation from traditionally marginalized communities, 'which simultaneously empowers these communities to preserve their own narratives and allows archivists to move closer to the goal of representative collections' (90–91).

An appropriate framework for archival practice emphasizes the collaborative nature of participatory archiving. For example, archivists should work alongside rather than on behalf of communities to appraise materials that are culturally relevant, including nontraditional materials such as performance (2007, p. 92–93). Similarly, arrangement and description schemes might be adapted to more appropriately and/or accurately reflect contextual value provided by and for the community (94–95). Citing projects that incorporate the participatory archiving model, the authors acknowledge this approach is labour-intensive and involves multiple actors, with success hinged on the ability to respect a plurality of perspectives and priorities. However, by exploring the practical application of this approach they are optimistic about the outcome, which is intended to provide solutions for and encourage other organizations to adopt the same strategies (100).

8.4 Reflections on the iterative development of a geospatial archive: Putting theory into practice

A broad approach to community guides the RSLMMP, where community extends to the team that has been working since 2015 on various aspects of the collaborative project to create the Residential Schools Land Memory Atlas under the RSLMMP (funded by the Social Sciences and Humanities Research Council of Canada, 2015–20 and referred to in Chapter 2 and elsewhere), and where Mahoney's (2005) understanding of 'communities of computing' can be applied to the project's technological dimension. It was proposed that the RSLMMP would put commemoration into action through a range of participatory mapping activities focused on Residential School buildings and sites. It would do this by further developing the Residential Schools component of the online Lake Huron Treaty Atlas (Caquard et al. 2009; Taylor and Pyne, 2010; Brauen et al. 2011; Pyne and Taylor, 2012; Pyne, 2013, 2014) through the expansion of the Residential Schools Map into a new map series, including a cybercartographic digital archives map, nine cybercartographic maps based on experience and memory, a participant network map, and new Residential Schools Land Memory Atlas to host the series. This chapter charts some of the initial research processes and decisions in the atlas development process in order to contribute to a better understanding of the evolution of a research project with emergence as one key methodological component. Upon receiving the grant, it was necessary to revise the project outline to reflect a lower amount of funds that requested. While we maintained the proposed methodology, which involved gathering and mapping archival records and engaging in participatory mapping, we reduced the number of school case study sites from nine to six.

Work on the project began with introductions and orientations, talks with collaborator Greg Bak about critical approaches to archival studies, participation in Master's courses in archival studies, literature review, a focus on background map development, exploratory work gathering digital records from different categories; and map module and schema development, which involved prototype creation, sharing, revision and an emergent vision for a final prototype version. As the orientation phase and literature review were underway, in a group meeting, I introduced the idea of a continuum of online map interactivity, which I had begun to develop several years earlier as a start to understanding participatory online mapping. At one end of the continuum was the 'informative' online participatory map, which involved the map user having an interactive relationship as an information seeker with the map; whereas 'active' online participatory maps involved the participation of the map user in the creation of the map, in addition to being an information seeker (Fig. 8.4.1).

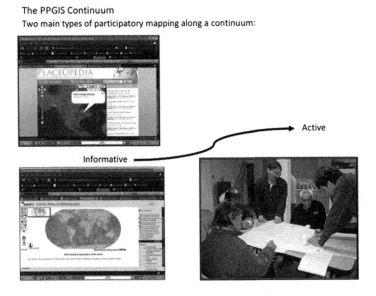

FIGURE 8.4.1 Early diagram illustrating simple participatory mapping continuum.

After hearing my thoughts on a continuum approach to online participatory mapping, Kevin Palendat suggested that we read 'Archivists and Changing Social Information Spaces: A Continuum Approach to Recordkeeping and Archiving in Online Cultures' (Upward, McKemmish and Reed, 2011), which considers the idea of a continuum approach to information and reflects similar themes to those discussed in the critical cartographic, for example the notion that maps and information in general are in a constant state of becoming, emergence, or ontogenesis (Kitchin and Dodge, 2007; Turnbull (2007).

Upward, McKemmish and Reed (2011) describe the records continuum approach developed in the late 1950s by Australian archivists, which integrated historic and current record-keeping and archiving processes and adopted a view of archival functionality as a multidimensional and pluralistic. Although significant changes in technology have impacted records and other archival materials, the approach remains relevant, especially with respect to the continuum approach to information it posits. This approach involves the 'infinite expansion of recorded information in a modern information ecology in which there can be no end products. Any event in the lifespan of an information object can be relevant to its archival management; "creation" recursively occurs in places of situated action' (Upward, McKemmish and Reed, 2011, p. 199).

Mirroring Turnbull's (2007) discussion of hodology or trail-making dimensions ranging from neurobiology to Indigenous trail-mapping, Upward, McKemmish and Reed (2011), identify three areas of concern along the information continuum: 'continuum thinking as a form of consciousness; [transformations in] archival functionality and professional record-keeping practice to better engage with complexity and plurality; [and] the need for strategies, tactics, and structure [to] address the pluralization and massive complexification of the infinitely expanding and indefinitely divisible continuum of recorded information that is engulfing us' (Upward et al. 2011, p. 200). The continuum is a metaview, such that 'all is archive', where a 'continuum consciousness' broadens the understanding of records. Rather than static objects, they might be conceptualized 'always in a process of becoming' (McKemmish in Upward, McKemmish and Reed, 2011, p. 202).

The continuum approach to information and archival practice is considered in light of a case study by Sue McKemmish, which involved research into the resettling of the relationship between Indigenous Australians and the archival community as part of a continuum that supports decolonization. McKemmish describes this process as 'a collaborative, co-creative journey involving Indigenous and archival communities' which 'would involve respectful and carefully negotiated partnerships' between these communities, with a 'sharing of governance and control, allowing multiple voices to contribute to the decision-making about current and historical recordkeeping, and joint exploration of how archival functionality and professional recordkeeping practice could be pluralized so that they can embrace and better support multiple ways of knowing, recordkeeping and archiving, and multiple forms of records' (Upward, McKemmish and Reed, 2011, p. 218). Reed examines WikiLeaks through a continuum lens with the understanding that this online platform challenges access approaches traditionally employed in archives and suggests solutions based in a continuum-style practical application (2011, p. 232).

McKemmish in Upward, McKemmish and Reed (2011) promote a 'rhizoid (weed-like) thinking' that mirrors the digital networks that have developed around and with record-keeping activity, and expand through both space and time (235). With this vision in mind, they introduce concepts that involve a new understanding of 'information objects' as being part of a continuum of recordkeeping events rather than being finite products of action, where collective and coordinated action across multiple platforms is essential, and can be supported through online cultures. This creates space for plurality:

> 'As the case study relating to Indigenous communities illustrates, within continuum consciousness, technologies and social networking can be employed to implement participatory recordkeeping and archival models (globally and locally), and negotiate appraisal by records co-creators. Meta-metadata schemes can deal with multiple and parallel provenance, and related rights and responsibilities management in current and historical recordkeeping settings. Shared, collaborative recordkeeping and archival spaces can be configured to respect the rights in records and protocols of all parties involved, allowing contested views and multiple perspectives to coexist, and providing for differentiated access' (*2011, p. 236*).

Emergent structures are necessary to manage this kind of complexity. In this regard, plurality is important, as well as developing strategies that are creative and agile, respecting rights to access through appropriate stewardship (Upward et al., p. 236–237). In this respect, Cybercartography is primarily concerned with the presentation and geo-archiving of multiple perspectives in multiple ways (Pyne, 2013; Taylor, 2003, 2005; Taylor and Pyne, 2010; Taylor and Lauriault, 2014; Pyne and Taylor, 2012).

Our subsequent review and discussion of this paper in relation to the discussion of the continuum of interactivity in an online mapping context resulted in a practical revisioning for the Residential Schools modules; thereby providing a significant example of the interplay between theory and practice, in addition to demonstrating that archival theory and critical cartographic practice can indeed work together. Whereas our original approach involved separating archival and historical information from information we expected to gather via participatory mapping exercises, after reviewing the literature and discussion, we decided to do away with the distinction between archival and participatory 'content' and to create school specific modules that could incorporate both types of information. In addition to our literature review concerned with system level issues, we also devoted time to exploring the archival, historical and academic literature relating to residential schools, paying special attention to the cybercartographic principle of place as an organizing factor for information.

The themes revealed in the excerpts from our email communications below were discussed in our meetings, and provide one example of the 'talk' we were engaged in that involved transforming theory into the practice of developing a cybercartographic map module template for the case study schools:

> I have been meaning to send this along for the last day or so. I have attached an article that, to my mind, dovetails incredibly well with our project and may be used to inform (or at least reinforce) our design principles as they relate to archives of decolonization/reconciliation. I should say as a point of clarification, but also as a means of tantalization, that the attached article represents but one intervention in a very large body of literature treating the "continuum" as an archival ontology (email communication from Kevin to group, 9 March 2016).
>
> Hi everyone,
> I am looking forward to reading through this. This paper should be relevant to further informing the decision to dispense with the archival sources module (as being distinctive from the RS Map Modules, which were originally intended to display "data" relating to the participatory mapping).

We are now going forward with a one map module per school approach where each map module will "house" the continuum of "data" relating to the school. Our thematic lens and main organizing feature is "place" (place/space). An important attribute of 'place' is scale. In terms of schema development, which partially corresponds to the concept of 'finding aid' (email communication from Stephanie to group, 10 March 2016).

Despite our interest in underlying theoretical aspects, our research was importantly aimed at designing and developing the map modules, which included consideration of background map design and scope and nature of content to be included in the modules. As we began reviewing newspaper articles, theses, Indian Affairs Annual reports, the SRSC's holdings in the Algoma University online archives, the NCTR, LAC and other sources, we began to realize that some of the Residential Schools' buildings had undergone changes over the years. This observation led us to focus on information that could inform a background map that reflected the evolution over time of the Residential Schools' buildings, with the intention of later 'mapping' the content we were gathering to inform the background map as the first examples of map module content. For example in the case of Assiniboia Residential School, it began as a Children's Home in Winnipeg (190x), transitioned to a Veteran's Home (195x–195x), was repurposed as an Indian Residential School (1958–72), and is currently the Canadian Centre for Child Protection. Chris Calesso researched newspaper articles related to three of these phases and prepared a spreadsheet with some relevant metadata (see Fig. 8.4.2)

With respect to research on the timeline evolving background map for Shingwauk Residential Schools, we video recorded a meeting with Trina Cooper-Bolam, the PhD research assistant on the project, in which she guided us through her research process with archival images, plans and text to inform her design of the Shingwauk Industrial Home over time. The video was shared amongst research team members to give them a better idea of the rationale behind the background map we were working on designing for the Shingwauk Schools Module (Fig. 8.4.3).

	Children's Home (1915-1945)	Veteran's Home (1945-1958)	Assiniboia Indian Residential School (1958-1973)
General History/ Timeline	May 31, 1915- "To Erect $65,000 Children's Home" (Tribune, page 9)- Description of site for the Children's Home	Jan 5, 1945- "Children's Home Sold as Veterans' Hospital" (Tribune, Page 9)- Sale of building to Veterans Affairs as convalescent home	April 22, 1958- "Indian School; Not Girls' Home" (WFP, Page 1, continued on page 5)- Announcement of use of Home as residential school in September/ for students that can't be accomodated elsewhere
	June 5, 1915- "Will Build New Children's Home" (Tribune, page 7)- Location of children's home/ possibly three buildings (although it is unclear)- main building, infants' building, boys' building	Sept 9, 1955- "Veterans Home To Go When Hospital Ready" (WFP, page 3)- Transfer of residents to Deer Lodge and home transfered to Crown Assests Disposal Corporation/ property of 6.9 acres	April 23, 1958- "River Heights Indian School Open Sept. 1" (Tribune, Page 19)- Announcement of plan to open a residential school at the Acadmey Road location in September/ 100+ from aboriginal communities in MB
	May 15, 1918- "Estimates Total Million and Half" (WFP, page 5)- Mention of building of four room school on the property	April 8, 1958- "More About Home (Continued from Page 1)" (WFP)- Transfer of Veterans' Home	June 16, 1958- "Indian School To Open Sept. 2" (WFP, Page 1)- Announcement of opening of Assiniboia as school and hostel/ Grades 8-10, 110 students, 12 taking higher education elsewhere/ pricipal Father O. Robidoux
	Oct 5, 1932- "Childrens' Home Wards kept happy" (Tribune)- Mention of buildings being built in 1915, separate school building and surgery/ hospital room	April 22, 1958- "Indian School; Not Girls' Home" (WFP, Page 1, continued on page 5)- Announcement of use of Home as residential school in September/ for students that can't be accomodated elsewhere	Sept 2, 1958- "Indian Children Romp to School in Winnipeg" (WFP, page 2)- Opening of school
	Jan 5, 1945- "Children's Home Sold as Veterans' Hospital" (Tribune, Page 9)- Sale of building to Veterans Affairs as convalescent home	April 23, 1958- "River Heights Indian School Open Sept. 1" (Tribune, Page 19)- Announcement of plan to open a residential school at the Acadmey Road location in September/ 100+ from aboriginal communities in MB	Sept 11, 1958- "School Borrows To Open" (Tribune, Page 23)- Opening of school
			Sept 20, 1967- "5 Canadian Indians Enrolled at U Of M" (WFP, page 37)- Mention of Assiniboia being used as a hostel with students going to Winnipeg high schools- part of attempted integration in schools to improve education
			Jan 15, 1971- "Cross Lake children to attend school here" (Tribune, page 5)- 50 students will attend high school in Winnipeg and stay at the Assiniboia residence for Indian

FIGURE 8.4.2 Screenshot of a portion of the Assiniboia Residential School Newspaper stories spreadsheet.

FIGURE 8.4.3   Screenshot from research video showing the digital plan for the Shingwauk Industrial Home prepared by Trina Cooper-Bolam based on information including the archival images of the Home (also shown in the screenshot).

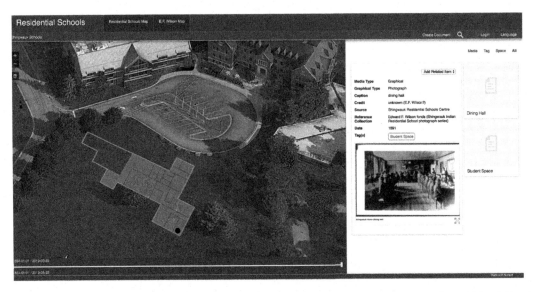

FIGURE 8.4.4   Screenshot showing the Shingwauk Schools Map of the in-development Residential Schools Land Memory Atlas with the first phase of the Shingwauk Industrial home shown in green and the dining hall area highlighted in purple in the central map window, with an archival photograph of dining hall in the right side panel.

Based on Trina's research, we designed a prototype cybercartographic module with the first two phases of the Shingwauk Industrial Home sketched out (Fig. 8.4.4).

We introduced the idea of mapping theses as we began to geo-transcribe theses relating to Residential Schools. We began this process by creating a spreadsheet that includes quotations relevant to our investigation of the evolution of the Shingwauk and Assiniboia Residential Schools (Fig. 8.4.5).

We began to add the text from our spreadsheet based on 'White Man's Burden' to the pilot Shingwauk Schools Module as part of the process of testing the Records Schema we had also

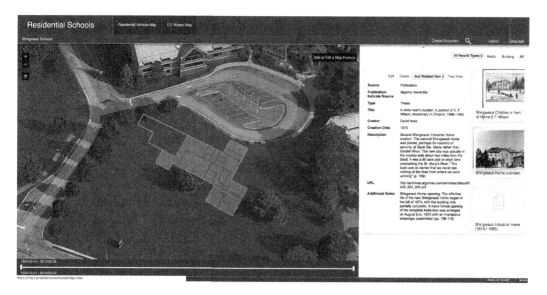

place	start date	end date	longitude (if town, city, region etc)	latitude (if town, city, region etc)	room or space (if in building)	passage	page	brief description / caption for passage
First Shingwauk Industrial Home	1873-09-22	1873-09-28	-84.24950071	46.51559492	n/a	On September 22nd, 1873 the Shingwauk Home at Garden River was completed and opened — 12 days after F.D. Fauquier was elected as the first Bishop of the new missionary Diocese of Algoma.	108	First Shingwauk Industrial Home opens
						Only six days later the school went up in flames and all was lost except, happily, lives. Both Wilson and the Rev. E.H. Capp, a later historian of Sault Ste. Marie indicate that arson may have been involved. Neither indicates the suspected party.	108	First Shingwauk Home is destroyed in fire
Second Shingwauk Industrial Home	1874	1895	n/a	n/a	building polygon for phase one of Shingwauk Home	The second Shingwauk Home was placed, perhaps for reasons of security, at Sault Ste. Marie rather then Garden River. The new site was actually in the country-side about two miles from the Sault. It was a 90 acre plot of virgin land overlooking the St. Mary's River. "The bush was so dense that we could see nothing of the river from where we were working"	109	Second Shingwauk Industrial Home location
						The effective life of the new Shingwauk Home began in the fall of 1874, with the building only partially complete. A more formal opening of the complete institution was arranged on August 2nd, 1875 with an impressive entourage assembled.	109-110	Shingwauk Home opening

FIGURE 8.4.5 Screenshot of a portion of the Geo-transcribing Residential Schools theses spreadsheet.

FIGURE 8.4.6 Screenshot showing the Shingwauk Schools Map of the in-development Residential Schools Land Memory Atlas with the first phase of the Shingwauk Industrial Home shown in green, with a building-relevant thesis quotation and thumbnails of archival photographs showing the Home in the right side panel.

developed to accompany the time-lined interactive map background intended to reflect the evolution of the Shingwauk Industrial Home building; in addition to adding test media primarily in the form of digital archival images associated with the building at different stages of development (Fig. 8.4.6).

8.5 Discussion and conclusion

We began our research inquiries with questions about different approaches to archival systems, including issues related to interoperability, and compared Archives including the NCTR, SRSC and LAC in our exploration of the nature and scope of Residential Schools related materials.

Our inquiries into the archival studies literature informed us about basic categories (acquisition, processing, preservation, reference and access) to consider, which helped shape our approach to the development of the RSLMMA. Noting intersections along several themes between critical cartography and archival studies discourses, including the common concern with inclusiveness, both in terms of modes of information and modes of communication and interactivity, we turned our attention to more practical design issues.

We also examined issues related to temporal scope and discussed the types of survivor stories that were relevant to place and that could be mapped. For example, many survivors lived far away from the schools they attended. In relation to this, there was some interest for a map showing the home communities of residential schools students. As we progressed through our literature and content reviews, we also began discussing the design of our schemas, or information input forms, which would contain the content metadata and reflect a continuum approach to information. While the brief summary of atlas development research is nowhere near complete, in the tradition of Bourdieu (1992) with his high speed tour, the examples provide a taste of the processes behind what will eventually be launched as the Residential Schools Land Memory Atlas. This Atlas contains two reflexivity map modules (referred to in Chapter 2), which are intended to geo-document atlas development processes.

It became apparent early on in the project — and even before — that we would be involving broad concepts of community and participation that involved the research team. Combined with the concept of emergence in the context or project design and development, this has meant that the proposed cybercartographic digital archives map is in actuality a digital archives component of all of the map modules being developed in the RSLMMP. The research process during the first half of the project involved talk to a great extent: One on one phone meetings, Group Google chat meetings, face-to-face meetings and email communications. We also discussed project intersections and participatory/archival intersections. Amidst the research was RA training and development, which included the emergence of a teaching and research approach that proceeds in a consultative fashion and involves transdisciplinary group learning and concentration contributing in a collaborative way to a common output.

Reflexivity in project design and development is important for understanding cybercartographic atlas output, especially since cybercartographic atlases are developed in an iterative manner, with multiple forms of contributions, and share the open-ended, emergent, always-in-the-making character of critical approaches to both archives (Upward, McKemmish and Reed, 2011) and cartography (Kitchin and Dodge, 2007; Turnbull, 2007). Together over the years, the research team has included a number of research assistants in discussions, email dialogues, spreadsheet and map module development going well beyond the example presented in this Chapter. A fuller account of these processes will be given in future publications following the expected launch of the Residential Schools Land Memory Atlas in 2020–21. Whereas this chapter is primarily concerned with presenting a glimpse of our work with archival records, Chapter 9 focuses more on a participatory mapping event with Residential Schools survivors, which has come to form the heart of the Assiniboia Residential School map component of the RSLMMP, and which also involved the active participation of research assistants.

References

Amin, A., 2004. Regions abound: towards a new politics of place. Geografiska Annaler 86B, 33–44.

Bourdieu, P., 1992. The practice of reflexive sociology. In: Bourdieu, P., Wacquant, L. (Eds.), An Invitation to Reflexive Sociology. University of Chicago Press, Chicago, pp. 217–253.

Brauen, G.S., Pyne, A., Hayes, J.P., Fiset, Taylor, D.R.F., 2011. Transdisciplinary participation using an open source cybercartographic toolkit: The Atlas of the Lake Huron Treaty Relationship process. Geomatica 65 (1), 27–45.

Buttimer, A., 1974. Values in Geography. Association of American Geographers, Washington.

Buttimer, A., Seamon, D. (Eds.), 1980. The Human Experience of Space and Place. Croom Helm, London.

Carroll, L., Farr, E., Hornsby, P., Ranker, B., 2011. A comprehensive approach to born-digital archives. Archivaria 72, 61–92.

Caquard, S., Pyne, S., Igloliorte, H., Mierins, K., Hayes, A., Taylor, D.R.F., 2009. A "living" atlas for geospatial storytelling: The cybercartographic Atlas of indigenous perspectives and knowledge of the Great Lakes region. Cartographica 44 (2), 83–100.

Del Casino, V.J., Hanna, S.P., 2006. Beyond the 'binaries': a methodological intervention for interrogating maps as representational practices. ACME: An International E-Journal for Critical Geographies 4 (1), 34–56.

Christen, K., 2011. Opening archives: respectful repatriation. American Archivist 74 (1), 185–210.

Cooban, G., 2017. Should archivists edit Wikipedia, and if so how? Archives and Records 38 (2), 257–272.

Cosgrove, D., 2005. Maps, mapping, modernity: art and cartography in the twentieth century. Imago Mundi 57 (1), 35–54.

Cosgrove, D., 2008. Geography and Vision: Seeing, Imagining, and Representing the World. I. B. Taurus, London and New York.

Crampton, J.W., 2009. Cartography: performative, participatory, political. Progress in Human Geography 33 (6), 840–848.

Duff, W., Harris, V., 2002. Stories and names: archival description as narrating records and constructing meanings. Archival Science 2 (3–4), 263–285.

Ettlinger, N., 2004. Toward a critical theory of untidy geographies: the spatiality of emotions in consumption and production. Feminist Economics 10 (3), 21–54.

Eveleigh, A., 2014. Crowding out the archivist? Locating crowdsourcing within the broader landscape of participatory archives. In: Ridge, M. (Ed.), Crowdsourcing Our Cultural Heritage. Ashgate, London, pp. 211–229.

Flinn, A., Stevens, M., Shepherd, E., 2009. Whose memories, whose archives? Independent community archives, autonomy and the mainstream. Archival Science 9 (1–2), 71–86.

Hagerstrand, T., 1975. Space, time and human conditions. In: Karlqvist, A., Lundquist, L., Snickars, F. (Eds.), Dynamic Allocation of Urban Space. Saxon House, Farnborough, Hants.

Hagerstrand, T., 1976. Geography and the study of interaction between nature and society. Geoforum 7, 329–334.

Hagerstrand, T., 1982. Diagram, path and project. Tijdschrift voor Economische en Sociale Geografie 73, 323–339.

Jean-Bastian, J., 2013. The records of memory, the archives of identity: celebrations, texts and archival sensibilities. Archival Science 13 (2–3), 121–131.

Johnson, J., Louis, R., Pramono, H., 2006. Facing the future: encouraging critical cartographic literacies in indigenous communities. ACME: An International E-Journal for Critical Geographies 4 (1), 80–98.

Kitchin, R., 2008. The practices of mapping. Cartographica 43 (3), 211–216.

Kitchin, R., Dodge, M., 2007. Rethinking maps. Progress in Human Geography 31 (3), 331–344.

Kwan, M., 2007. Affecting geospatial technologies: toward a feminist politics of emotion. The Professional Geographer 59 (1), 22–34.

Latham, A., Conradson, D., 2003. Guest editorial. Environment & Planning A 35, 1901–1906.

Lauriault, T.P., Craig, B., 2008. General study 10 preservation practices in scientific data portals: Part 1 – case and general studies in the artistic, scientific and governmental sectors focus task force. In: Duranti, L., Preston, R. (Eds.), InterPARES 2: Experiential, Interactive and Dynamic Records, Rome, Italy, Associazione Nazionale Archivistica Italiana. http://www.interpares.org/ip2/display_file.cfm?doc=ip2_book_part_1_focus_task_force.pdf.

Lauriault, T.P., Hacket, Y., Taylor, D.R.F., 2008. InterPARES 2, case study 06, cybercartographic atlas of Antarctica: Part 1 – case and general studies in the artistic, scientific and governmental sectors focus task force. In: Duranti, L., Preston, R. (Eds.), InterPARES 2: Experiential, Interactive and Dynamic Records, Associazione Nazionale

Archivistica Italiana, Rome, Italy, pp. 24—38. http://www.interpares.org/ip2/display_file.cfm?doc=ip2_book_part_1_focus_task_force.pdf.

Ley, D., Samuels, M., 1978. Humanistic Geography: Prospects and Problems. Croom Helm, London.

Livelton, M., 1996. Archival Theory, Records and the Public. The Society of American Archivists and Scarecrow Press Inc., Lanham, Maryland and Oxford.

Mahoney, M., 2005. The histories of computing(s). Interdisciplinary Science Reviews 30 (2), 119—135.

Massey, D., 2004. Geographies of responsibility. Geografiska Annaler 86B, 5—18.

Nesmith, T., 1999. Still fuzzy, but more accurate: some thoughts on the 'ghosts' of archival theory. Archivaria 47, 136—150.

Nesmith, T., 2006. The concept of societal provenance an records of nineteenth-century Aboriginal-European relations in Western Canada: implications for archival theory and practice. Archival Science 6 (3—4), 351—360.

Pearce-Moses, R., Davis, S.E. (Eds.), 2006. New Skills for a Digital Era, Colloquium Proceedings. National Archives and Records Administration, Society of American Archivists, Arizona State Library. Archives and Public Records. http://files.archivists.org/pubs/proceedings/NewSkillsForADigitalEra.pdf.

Podolsky Norland, L., 2004. The concept of "secondary provenance": Re-interpreting Ac ko mok ki's map as evolving text. Archivaria 58, 147—159.

Pyne, S., 2013. Sound of the Drum, Energy of the Dance: Making the Lake Huron Treaty Atlas the Anishinaabe Way. Unpublished PhD Dissertation. Carleton University, Ottawa. https://curve.carleton.ca/392a68ac-086c-4470-976d-101d4e96f9f7.

Pyne, S., Taylor, D.R.F., 2012. Mapping indigenous perspectives in the making of the cybercartographic atlas of the Lake Huron treaty relationship process. Cartographica, Special Issue on Indigenous Cartography and Counter Mapping 47 (2), 92—104.

Pyne, S., 2014. The Role of Experience in the Iterative Development of the Lake Huron Treaty Atlas. In: Taylor, D.R.F., Lauriault, T. (Eds.), Developments in the Theory and Practice of Cybercartography: Applications and Indigenous Mapping. Elsevier, Amsterdam (Chapter 17).

Sen, A., 1999. Development as Freedom. Oxford University Press, Oxford.

Shilton, K., Srinivasan, R., 2007. Participatory appraisal and arrangement for multicultural collections. Archivaria 63, 87—101.

Stollar Peters, C., 2006. When not all papers are paper: a case study in digital archivy, Provenance. Journal of the Society of Georgia Archivists 24 (1), 22—34.

Swyngedouw, E., 1997. Neither global nor local. Glocalisation and the politics of scale. In: Cox, K. (Ed.), Spaces of Globalization: Reasserting the Power of the Local. Guilford Press, New York, pp. 137—166.

Taylor, D.R.F., 2003. The concept of Cybercartography. In: Peterson, M.P. (Ed.), Maps and the Internet. Elsevier, Amsterdam, pp. 405—420.

Cybercartography: theory and practice. In: Taylor, D.R.F. (Ed.), 2005. Modern Cartography Series, vol. 4. Elsevier, Amsterdam.

Taylor, D.R.F., Lauriault, T. (Eds.), 2014. Developments in the Theory and Practice of Cybercartography: Applications and Indigenous Mapping. Elsevier, Amsterdam.

Taylor, D.R.F., Pyne, S., 2010. The history and development of the theory and practice of Cybercartography. International Journal of Digital Earth 3 (1), 1—14.

Thrift, N., 2004. Intensities of feeling: towards a spatial politics of affect. Geografiska Annaler 86B, 57—78.

Tuan, Y.-F., 1977. Space and Place: The Perspective of Experience. Edward Arnold, London.

Turnbull, D., 2000. Masons, Tricksters, and Cartographers: Comparative Studies in the Sociology of Scientific and Indigenous Knowledge. Harwood Academic, Amsterdam.

Turnbull, D., 2007. Maps, narratives, and trails: performativity, hodology, and distributed knowledges in complex adaptive systems: an approach to emergent mapping. Geographical Research 45, 140—149.

Upward, F., McKemmish, S., Reed, S., 2011. Archivists and changing social information spaces: A continuum approach to recordkeeping and archiving in online cultures. Archivaria 72, 197—237.

Wu, C.J., Isaksson, K., 2008. Participatory Mapping as a Tool for Capturing Local Perspectives on Cultural Landscape: A Case Study of Ostlänken. Report for School of Architecture and the Built Environment Urban Planning and Environment, Stockholm. Available at: www.includemistra.org.

9

Site-based storytelling, cybercartographic mapping and the Assiniboia Indian Residential School Reunion

*Stephanie Pyne**

Postdoctoral Research Fellow, Geomatics and Cartographic Research Centre (GCRC), Carleton University, Ottawa, ON, Canada
*Corresponding author

9.1 Introduction: Community-based research, mapping and site-based storytelling

There is a variety of both communities and approaches to cartography. The wake of the digital age has seen an increase in community mapping projects, which provide fruitful contexts for collaborative work involving different perspectives. An important purpose of community mapping is to 'ground truth' policy by including community perspectives in various ways on maps, in addition to 'scientific' and other perspectives. For example, Cidell (2008) is concerned with the role of critical cartography in the context of local knowledge and environmental issues in her discussion of a case in which scientific contour maps of airport noise fail to adequately account for and represent perceptions of excessive noise experienced by residents in proximity to the airport. While Cidell applauds the increased emphasis on public participation GIS, including volunteered geographic information and lay people making their own maps, she points out the need for more study of how average people do and can read maps critically. According to Cidell's observations, the public is increasingly aware of the limitations of maps to convey all perspectives unless specific efforts to include alternative perspective are made. Cidell's study demonstrates that community members can deconstruct maps by challenging their authority and pointing out their assumptions.

There are many issues related to the democratization when it comes to cartography and community. These include concerns regarding the need to educate people about maps and mapping techniques if they are going to be optimally engaged in—especially digital—cartographic practices (Johnson et al., 2006), and inclusiveness with respect to styles of knowledge and communication Young (2000). For example, Liu and Palen (2010) discuss emergent neogeographic practices involved in the design and creation of crisis map mashups and the importance of cartographic literacy, which can be enhanced by combining professional and participatory geotechnologies (69), and Kim (2015) provides a review of 'exuberant experimentation' in several disciplines and arenas, from geographers and community development to activists, artists and new media innovators (Kim, 2015, p. 215). Clearly linking participation and democracy as key means and ends of her interest in mapping, Kim is concerned with developing the potential of a critical approach to mapping in a landscape and urban planning context. In another study that integrates the quantitative and qualitative dimensions, Gilmore and Young (2012) take a broad approach to ethnobiological research, leading with an emphasis on community empowerment via participatory mapping.

In her examination of the usefulness of critical cartography to digital journalism and environmental knowledge production, Salovaara (2016) discusses the use of 'computer-assisted cartography as part of environmental knowledge production' via the case of InfoAmazonia, a project involving cartographic approaches to journalistic practice in the development of 'a digitally created map-space', in which 'journalistic practice can be viewed as dynamic, performative interactions between journalists, ecosystems, space, and species' (827). According to Salovaara, geo-visual representations of information in InfoAmazonia 'enhances embodiment in the experience of the information' (827). Crisis response and management is another important area for mapping, which is very useful to communities for both contributing to and obtaining information from map-based digital websites such as Ushahidi in times of crisis (Palen and Liu, 2007; Palen and Vieweg, 2008; Palen et al., 2009; Liu and Palen,

2010). Focussing more on issues at the intersection of semiotics and community participation in mapping in a sub-Saharan African village context, Burini (2012) explores a choreographic approach to creating cartographic tools for use in cooperative projects in environmental governance and sustainability and invokes semiotic concepts to interpret the form and function of map symbols and related features.

The critical cartography literature concerned with community or participatory mapping is full of conceptual transformational thinking that intersects closely with work in indigenous cartographies. Turnbull (2007) has stated that the best way to engage in a critical approach to mapping—especially in an intercultural context—is to engage incommensurable perspectives in a dialogical tension with one another in a nondominatory manner (Pickles, 2004), where Western epistemologies and ontologies do not take priority or precedence over indigenous understandings and ways of approaching the world. Woodward and Lewis (2000) refer to the lack of a cross-cultural consensus on the meanings of 'map' and 'cartography' and point out the Eurocentric assumption that Indigenous peoples did not make maps, an assumption that was based on their inability to find evidence resembling their own conceptions of cartography (1).

Getting at the question of the ontologies of mapping, Renee Pualani Louis cites Rundstrom (1995), who in turn draws on Connerton's (1989) distinction between incorporating and inscribing cultures: Incorporating cultures traditionally emphasize oral communication and other performance-based modes (e.g. dance, painting) in transmitting all sorts of meaningful information. The actions, lasting hours or days, carry greater meaning than any object they produce. In contrast, inscribing cultures hold and fix meaningful information years after humans have stopped informing and typically must do so by means of some object (e.g. maps, GIS). Storage is crucial and leads to stasis and fixity (Rundstrom, 1995, p. 51, quoted in Louis, 2004, p. 9). According to Louis, such an incorporating culture can be found in Hawaiian cartography. Like the Maori, pre-contact Hawaiians had a clear understanding of the world they lived in and communicated their perception of the world orally (Kelly, 1999, p. 1). Hawaiian cartographers privilege process by incorporating their understanding of their island setting into their mo'olelo (stories), oli (chant), 'ö lelo no'eau (proverbs), hula (dance), mele (song) and their mo'o kü 'auhau (genealogy). This is a form of cartography categorized by Woodward and Lewis (1998) as 'performance or spiritual cartography' and may 'take the form of a nonmaterial oral, visual, or kinesthetic social act [in order] to define or explain spatial knowledge or practice' (4) (2004, p. 9).

It is becoming increasingly apparent through deconstructive analysis of colonial maps and related research that Indigenous peoples around the world engaged in mapping practices in a manner that displayed flexibility in their ability to adapt to the colonial mapping mindset. While Europeans had their own ideas about how mapping should be done, Indigenous peoples knew about 'what' should be mapped, and 'where' in their own lands, in addition to having their own understandings of mapping processes. This observation unsettles the conventional view of Indigenous peoples as consistently helpless victims in the context of settler—indigenous relations. A good example of indigenous agency in mapping is Beyersdorff's (2007) account of the Andean-Spanish encounter between 1577 and 1586, which proposes that the Andean way of mapping 'led the way' in a joint mapping venture. Although the result was a map that reflected the mediation of Spanish cartographic technology, its contents and style were also mediated by practices of the local indigenous culture, including a pre-contact Andean mapping methodology known as 'amojonamiento' (130), and Bernstein (2016) summarizes mapping practices involved in American state building as 'a reflection of both Euro and Native American mapping practices and typified the syncretic nature of how the American West was mapped' (626). In a contemporary context, Harris and Hazen (2006) discuss the benefits of incorporating indigenous perspectives into conservation-oriented counter-mapping initiatives in their consideration of a project led by Leakey (2002) that incorporated indigenous knowledge in the design of a semipermeable 'fence' to replace the primarily electric fencing surrounding Lake Nakuru National Park in Kenya.

Irene Hirt reflexively describes her work with epistemological inclusiveness in a Mapuche counter-mapping and participatory mapping project in southern Chile from 2004 to 2006. Hirt notes how 'dreams and dreaming practices are integrated into knowledge-building processes in many Indigenous societies, and therefore represent a source of geographical and cartographic information' (2012, p. 105). Mapping dreams and dreaming practices expands the cartographic limits of 'indigenous territorial dimensions—such as the sacred and the spiritual, as well as the presence of nonhuman actors' (Hirt, 2012, p. 105). Hirt's work is evidence of nonindigenous cartographic researchers taking dreams seriously as basis for knowledge and incorporating them into their work.

Reflecting a concern with both the conceptual and the practical arms of critical cartography, Sletto (2015) appeals to the ontological instability of maps and the performative nature of mapping practices (see also Chapter 8) in his ethnographic study of a participatory mapping project in Yukpa indigenous territory in the Sierra de Perija, Venezuela, to reinscribe maps with indigenous perspectives. Margaret Pearce and Renee Pearce and Louis (2007) echo others (Johnson et al., 2006; Palmer, 2009, 2012) in their concern with optimizing the benefits of GIS technologies for Indigenous peoples with their attention to how cartographic language can incorporate both 'indigenous and nonindigenous conventions in the same map' (107). Hunt and Stevenson (2017) explore digital counter-mapping, which overlays indigenous maps onto 'conventional' maps and 'guerrilla mapping techniques' that include indigenous names being affixed over settler street names (372).

Reid, Naqvib and Waldichuk (2014) favour a place-based focus in their approach to incorporating walking tours into the educational curriculum. In their view, 'place-based education has emerged as a pedagogical approach [that] counters overly general, exclusive, and economistic views of place, and emphasizes engagement with local community [...] providing an ideal medium for engagement, interpretation, and interaction, allowing personal connections, and the perception of complex place relationships which can be masked by theoretical abstractions' (2015, p. 63). Site-based storytelling lends itself naturally to talk of mapping, which is useful not only for documenting details relating to places and spaces but also information related to movement and travels between these entities. Yet, mapping in this new age may be more than this.

Deconstructive approaches to map interpretation question the motives and context of the map maker (Harley, 1988, 1989); reconstructive approaches include making maps through participatory processes and community collaboration (Caquard, 2011; Corner, 1999; Cosgrove, 2005; Craig et al., 2002; Crampton, 2001, 2009; Crampton and Krieger, 2006; Elwood and Ghose, 2004; Fox et al., 2005; Harmon, 2003; Kitchin, 2008; Kitchin and Dodge, 2007; Pearce, 2008; Pearce and Louis, 2007; Pyne, 2013; Turnbull, 2007; Taylor and Lauriault,

2014). Today's cartography extends to participatory collaborations with individuals from a variety of knowledge communities. It can be 'on the ground' and involve art, experience and immediacy, on one hand, and involve technologies, on the other (Carver, 2003; Craig et al., 2002; Irwin et al., 2009; Parker, 2006). There is increasing practical and theoretical attention to the intersections between art and cartography (Caquard and Taylor, 2005; Harmon, 2003; Irwin et al., 2009), and there is growing attention to mapping experience, emotions and indigenous perspectives (Caquard, 2011; Harmon, 2003; Hirt, 2012; Louis 2007; Louis et al., 2012; Pyne, 2013; Taylor and Lauriault, 2014). Collaborative projects to map traditional knowledge and place names, and/or to tell interactive stories through maps, provide examples of initiatives with potential to contribute to the empowerment of indigenous communities (Laidler et al., 2010; Pyne, 2013, 2014; Pyne and Taylor, 2012; Tobias, 2000). An understanding of maps and atlases as narrative vehicles in relation to the land (or place) is common to both deconstructivist and reconstructivist cartographic projects, in addition to their concern with reflexive patterns in cartographic design, development and use (Brauen et al., 2011; Pyne, 2013).

In his editorial introduction to a special issue on cartography and narrative in the *Cartographic Journal*, Kenneth Field notes the motivational power of maps as stories and their ability to link people passionately (and actively) to place:

> At the heart of [...] all good maps is a story and as a map reader we become a very important actor in the interaction with the map [...] Story telling is the very essence of good map-making and good cartographers have forever been successfully telling stories. In fact, good storytellers have made very good use of maps as vehicles for their work. Whether abstract or pictorial, good maps convey a sense of place regardless of whether you are using it to uncover some basic factual information or purely to gaze at for enjoyment (2014, p. 99).

Connected to their narrative capacity to engage and motivate, maps and atlases contribute to enhanced awareness and focus attention by using 'the language of cartography to graphically structure their meaning and to concentrate people's attention on the aspect of importance' (Field, 2014, p. 99). Field remarks on the variety that exists in today's public online mapping environment in terms of people's stories related to places despite a fairly universal online map format, process or 'language':

> The maps may tell very personal stories of someone's movement, perhaps tracing some historical journey or they may explore a single theme and allow people to interactively follow the path or the 'top ten' of something or other. These are really just an alternative way of telling a story through the use of maps but they've become popular since we now have the tools to create very personalized maps ... to tell our own stories and bring our own sense of place to the spaces that are important to us. These maps live online and can be shared widely. What they do is allow people to narrate ... to bring a story to life in ways that may once have been the preserve of professional cartographers. We may cringe at some of the maps and, really, some are not much more than geolocated holiday snaps, but the ability for people to create their own stories and accounts of things that are important to them has never been more prevalent. People are chronicling, commentating, documenting and creating their own spatial annals (Field, 2014, p. 99–100).

Despite the newfound ability of these public mapping platforms to mediate the expression and communication of 'sense of place', Field emphasizes how the new world of cartography is far more expansive than one that simply provides 'people with the ability to map their holiday photos on a web map' (2014, p. 100) and applauds Sébastien Caquard and William

Cartwright for their edited special issue, which extends to theoretical, ontological and methodological aspects of narrative approaches to cartography and contributes to the advancement of critical thought (2014, p. 100). In their introduction to the special issue, Caquard and Cartwright identify two primary categories of relationship between cartography and narrative. First, 'maps have been used to represent the spatio-temporal structures of stories and their relationships with places' (2014, p. 101) and second, expanded approaches to cartography have given rise to notions of maps as narratives, and efforts to reflexively tell the stories behind map- and atlas-making processes (2014, p. 101). Cybercartographic atlas projects reflect both types of relationship between cartography and narrative, allowing individuals to present their perspectives and understandings in multiple ways in relation to places and providing reflexive interpretations of cybercartographic atlas-making processes. The cybercartographic Lake Huron Treaty Atlas project, which lays the narrative groundwork for further development of its Residential Schools mapping component, provides a good example of an innovative cartographic practice in this regard.

The critical perspectives of community, indigenous mapping and narrative reflected in the approaches summarized above are relevant to the site-based storytelling at the Assiniboia Indian Residential Schools (AIRS) Reunion reported on in this chapter and help to interpret the site-based storytelling in this case study. The chapter continues by tracing the early stages of the work to create the Assiniboia Residential Schools Map component of the Residential Schools Land Memory Mapping Project (RSLMMP), including research relationships, stories of transdisciplinary and inter-project knowledge sharing and emergence (9.2). To provide an idea of the AIRS Project, which intersects with the RSLMMP, the AIRS project is summarized, including the AIRS June 2017 Reunion from which a variety of media was gathered to document the stories and reflections of survivors and others (9.3). Following this description, a summary of the media coverage included in the prototype Assiniboia School Module Mapping provides an idea of the attention given to the reunion and demonstrates the value of mapping this information (9.4), and a brief reflection is provided regarding the processes involved in mapping media related to the AIRS Reunion (9.5). The chapter concludes with a discussion of transdisciplinary research and emergence in the relationship between community-based participatory work and the work with archival records, which have come together post reunion to result in a more firm and focused design and development plan for the Assiniboia Residential School map component.

9.2 Building transdisciplinary research strength: Early development of the Assiniboia Residential Schools module

As outlined in Chapter 2, the proposed RSLMMP was envisioned to put commemoration into action through a range of participatory mapping activities focused on Residential School buildings and sites, and through innovative approaches to creating a relational geospatial archive. The project proposal included AIRS as one of the case study schools because it was the focus of an AIRS survivor-led participatory action research project directed by Andrew Woolford (see Chapter 6). Andrew became a co-investigator on the RSLMMP after contacting the GCRC team via an email to me about a different Residential Schools project, the Embodying Empathy project, which was to involve an inclusive approach to developing a prototype virtual Residential School (http://www.katherinestarzyk.com/empathy). Despite several potential project intersections between the RSLMMP and the Embodying Empathy project, we instead collaborated on work to integrate Andrew's AIRS Remembering Assiniboia Project, which was funded by the Social Sciences and Humanities Research Council of Canada through the Manitoba Research Alliance grant: Partnering for Change—Community-Based Solutions for Aboriginal and Inner-city Poverty, and the RSLMMP, funded by a Social Sciences and Humanities Research Council of Canada Insight Grant. This included archival research relating to Assiniboia Residential School intersecting to inform development of a related map module. The first stage of research consisted of a survey of Library and Archives Canada (LAC) as well as NRTRC materials on Assiniboia, and materials from other repositories, in addition to materials such as newspaper articles (see Chapter 8).

The initial research period was followed by a sharing circle with members of the AIRS survivors to solicit their views, the most important outcomes of a survivor-led project, and how such a project should be implemented. I attended this sharing circle and used the opportunity to listen and learn and to share my work in an introductory way. Survivors mentioned a desire to map the home communities of individual survivors. This reminded me of a comment from a Nishnawbe Aski Nation (NAN) outreach worker at the 2017 Shingwauk Residential School Reunion who mentioned that he found it difficult to know which Residential School materials to bring to communities, since students attending Shingwauk came from across the nation. The survivors also mentioned uses for mapping in connecting with other alumni, pointing to a potential networking function for a digital map, in addition to being a vehicle for sharing information with others, including younger generations.

I attended a second AIRS project breakfast months later, where I noted Theodore (Ted) Fontaine's interest in reaching local indigenous intergenerational survivors. Ted is known for his advocacy work in Residential Schools reconciliation and for the publication of *Broken Circle: The Dark Legacy of Indian Residential Schools* (Fontaine, 2010). Approximately 1 year after receiving feedback from the survivors concerning the direction they wanted to take, another breakfast was hosted (1) to receive their feedback in person on how well the researcher had followed their instructions and (2) to present them with the data from the research activities of the preceding year. During this time, I worked with the research assistants on schema and module development for the Assiniboia School module, following the model that was developing in our approach to building the Shingwauk School Module, which included attention to the background map and side panel. In addition, attending the meetings, gave me a better idea of how to intersect the participatory component of the RSLMMP with the AIRSS project activities planned for the reunion, the initiative that emerged from the survivor meetings.

9.3 The Assiniboia Indian Residential School survivor association (AIRSSA) project: On the ground with survivor-led commemoration in place

The AIRS was opened in 1958 in the former Julia Clark School on Academy Avenue in River Heights. It operated until 1973. Initially, it served as a Residential School, primarily for upper level students who had previously attended Residential Schools closer to their home reserves (Era 1). Later, it served as a dorm from which students were sent to attend other schools (Era 2). Working with survivors from both Eras of the AIRS (including Theodore Fontaine, Dorothy Crate, David Rundle, Daniel Highway, Carrie Perreault and Joe Malcolm), a participatory process was initiated with survivors who co-designed and directed the research project to ensure that outcomes met the needs and interests of the AIRS survivor community, including ownership and control by community members, consistency with indigenous epistemologies and values, and accessibility to and facilitation of the participation of survivors. An initial informal meeting was held with four AIRS survivors to discuss whether or not they felt there was enough interest among their members (graduates from the school, who meet on a semi-regular basis) to support a participatory research project. Their response was enthusiastic and they stressed in particular their concerns about the erasure of the school from public memory and the need for commemoration of its existence.

The first stage of research consisted of a survey of LAC as well as NRTRC materials on Assiniboia. It was followed by a sharing circle with members of the AIRS survivors to solicit their views on what the most important outcomes of such a project should be and how they wanted the project to be implemented. Breakfast was provided, and we worked to generate ideas and to identify research priorities. Although the purpose of the meeting was not to revisit any traumatic experiences at the schools, there was no guarantee such memories would not arise, and therefore a cultural support worker and Elder was present at this meeting. Potential research activities discussed at the consultation meeting included oral history interviews with survivors. Potential research outcomes included efforts to erect a commemorative plaque/monument at the AIRS site; media-based awareness raising about the school; placement of the school and information about the school in the Residential Schools Land Memory Atlas at the Geomatics and Cartographic Research Centre; public tours of the school with educational materials developed by survivors; a co-authored manuscript written for public readership that features both academic historiography, survivor memoir and other related contributions (artistic, creative nonfiction, etc.).

Approximately 1 year after receiving feedback from the survivors concerning the direction they wanted to take, another breakfast was hosted (1) to receive their feedback in person on how well the researcher had followed their instructions and (2) to present them with the data from the research activities of the preceding year. This provided an opportunity for community members to suggest new research activities or outcomes to be pursued, to request that we revise our activities and outcomes and to learn about our early findings. The most significant initiative that arose from the sharing circle was the idea to apply for funds and organize an Assiniboia Residential School Reunion for the following year. In addition, the survivor group became known as the 'Assiniboia Reunion and Commemorative Gathering Governing Council' as per its instructions. The governing council requested that the research team assist in the following activities: (1) planning a 'reunion and commemorative gathering' at the former Assiniboia school site that contributes to the further commemoration of Assiniboia; (2) facilitating the design and build of a commemorative marker and interpretive display for the former school site and (3) organizing and editing a book of survivor remembrances of their time at Assiniboia. We planned for several future planning meetings, which we felt were necessary to ensure that the project continued to be participant-led and driven. In addition to planning for the reunion and commemorative event, the Governing Council would have oversight over the development of the edited volume.

As the meetings progressed, survivors expressed their desire for the research team to seek funds to facilitate a reunion event that would serve to better educate people about the AIRS while also serving to (re)connect multiple generations of survivors. The AIRS survivors envisioned a 2-day gathering for survivors, intergenerational survivors and neighbourhood residents at the original Assiniboia school site. The Canadian Centre for Child Protection, which is housed in the former AIRS classroom building, offered to make classrooms available for these days, and the city would provide access to the grounds for the event. Multiple educational and commemorative activities were organized for the gathering, and the Governing Council decided upon all final plans.

As the plans became more detailed and the date for the reunion neared, the Governing Council meetings yielded the decision to engage interested reunion participants in recorded remembrance interviews and site-based storytelling related to Assiniboia in the context of the following activities:

(a) tours for local school groups and others to foster education about Assiniboia and the residential schooling experience;

(b) a group artwork project to create a commemorative monument for the school site;

(c) archival and informational display boards about the school;

(d) public talks by survivors, former staff and scholars about Assiniboia;

(e) site-specific storytelling and remembrances about areas of the school and

(f) interviews with survivors to enhance and add to what has already been captured through Truth and Reconciliation Commission testimony; and, reconciliation activities, including a feast shared by long-time River Heights residents, Assiniboia survivors and intergenerational survivors.

Many of these and other activities would be video and audio recorded (with appropriate permission forms signed) and incorporated into the Assiniboia Residential School map of the Residential Schools Land Memory Atlas. The Council planned to reach out to media, as well as local documentary makers, to ensure that the 2-day event would contribute to knowledge dissemination and education about Residential Schools, their continuing legacy and the prospects for reconciliation.

9.4 Mapping media coverage in the Assiniboia Residential Schools Reunion

The 23−24 June 2017 reunion was advertised and reported on locally and nationally both to the general public and to the Church, Indigenous and university communities. At the local level, UM News Today (2017) provided background about the school and the AIRS project in its announcement to University of Manitoba subscribers and others; and the Winnipeg Free Press (Sanders, 2017) featured a recent photograph of Ted Fontaine pointing to a photograph of one of the school's graduating classes (see Fig. 9.4.1) along with reflections by Fontaine that included contrasting his relatively positive experiences at the Assiniboia Residential (High) School with his prior negative experiences at Fort Alexander Residential School. Informing

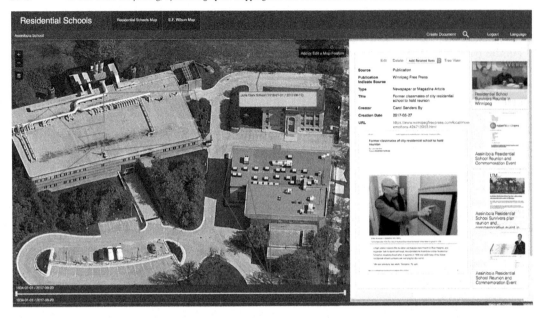

FIGURE 9.4.1   Screenshot showing the Assiniboia Schools Map of the in-development Residential Schools Land Memory Atlas with the school building (Julia Clark School) outlined in red in the central map window, with thumbnails of news media announcing the reunion in the right side panel, and featuring a screenshot of the photograph of Ted Fontaine (Sanders, 2017).

church communities, both Rupert's Land News online and the Winnipeg Presbytery provided the reunion program with a timetable of events, and City Councillor, John Orlikow, invited his constituents and others to attend. At the national level, Global News included a video interview with Ted Fontaine and Andrew Woolford on the importance of the reunion event, which was noted by Fontaine as being a demonstration of the ongoing resilience of Residential Schools survivors. Of interest to the map research we were engaged in, comments were also made about the Assiniboia Residential School site in the heart of Winnipeg. At the end of the interview, Ted announced that there would be a sunrise ceremony and smudge to begin the reunion events on the school site on Saturday, June 24. Windspeaker included a brief backgrounder on the school and the reunion project with its announcement as did the First Nations in BC Knowledge Network.

In terms of event coverage, APTN sent a crew to cover the reunion and produced both a video story that featured interviews with survivors, including Ted Fontaine (Robinson-Desjarlais, 2017), and CBC News (2017) included both a video and a written story based on interviews of event attendees (Fig. 9.4.2).

Reflecting a more comprehensive approach, JUST TV (https://www.thebnc.ca/just-tv), an indigenous youth film company in Winnipeg, produced Remembering Assiniboia, which included interviews and footage of reunions activities, including performances, testimonials and survivor-led school room and grounds 'tours' (Figs. 9.4.3 and 9.4.4).

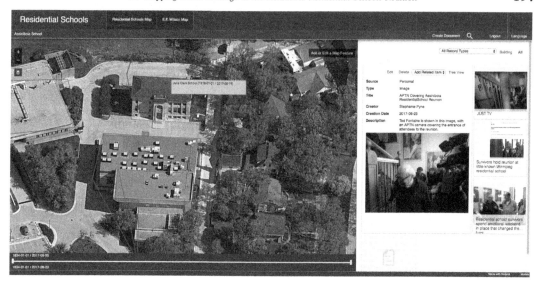

FIGURE 9.4.2 Screenshot showing the Assiniboia Schools Map of the in-development Residential Schools Land Memory Atlas with the school building (Julia Clark School) outlined in red in the central map window, with thumbnails of news media covering the reunion in the right side panel and featuring a photograph of Ted Fontaine with an APTN news camera trained on him. *Photo courtesy of author.*

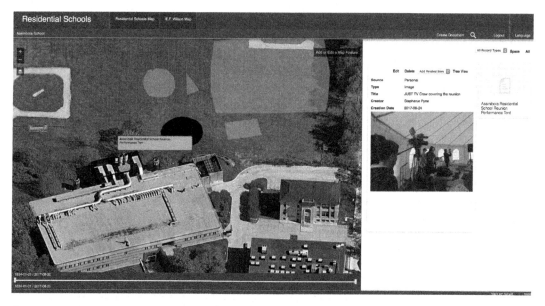

FIGURE 9.4.3 Screenshot showing the Assiniboia Schools Map of the in-development Residential Schools Land Memory Atlas with the Performance Tent (blue circle) in the central map window with a photograph of the JUST TV crew covering the events and activities in the right side panel.

FIGURE 9.4.4   Screenshot of scene from Remembering Assiniboia showing reunion attendees gathering outside the former Assiniboia Indian Residential School (JUST TV, 2017).

9.5 Mapping media related to the AIRS Reunion

Developing the Assiniboia School Map Module involves a variety of processes and includes intersections with the AIRS project, which extends to the gathering and processing media. While we had begun to get a better understanding of the school building, site and other aspects through our research with archival records and historical and academic publications (see Chapter 8), photo, video and audio documenting the events and activities at the reunion would be our first chance to get on the ground itself, and to meet the reunion attendees (Fig. 9.5.1).

Master's student Chris Calesso from the University of Manitoba assisted me with video recording, and Sarah Story accompanied Chris with her professional audio recording setup as he videoed survivors giving narrated tours through the rooms of the school on June 23 and as activities and events were happening in the various tents set up for the reunion on June 24. This left me free to video and photo document the other tours. In addition, Andrew Woolford had tasked a number of research assistants to gather audio media of informal interviews with survivors and other reunion attendees in the storytelling and food and beverage tents.

The news coverage provided by local and national distributors involved gathering 1 or 2 days of footage and boiling it down into relatively brief multimedia 'stories', whereas the JUST TV crew took a more comprehensive approach, both devoting larger teams to document the reunion, and producing a longer and more comprehensive video documentary of the reunion. Demonstrating the umbrella nature of the cybercartographic approach to documenting the reunion, our team was concerned with mapping the documenters of the reunion as well as the reunion itself, which represents a broader focus, which at the same time incorporates the focus of the documenters. In addition, our focus on later processing and mapping the media would necessitate a more detailed approach to our video and audio documentation as we followed survivors and the groups they were leading from room to room, floor to floor on the first day of narrated school tours. Rather than selecting aspects

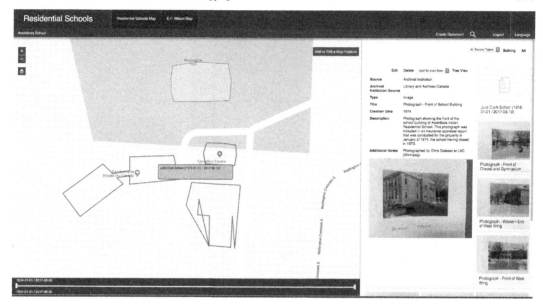

FIGURE 9.5.1 Screenshot showing the Assiniboia Schools Map of the in-development Residential Schools Land Memory Atlas with the school building (Julia Clark School) outlined in red in the central map window and photographs taken by Chris Calesso of archival photographs of the school in the right side panel.

of our total footage and distilling them into a story, our intention was to incorporate 'all' of our footage in the form of clips related to spaces.

Much work remains to be done in terms of media processing, including synching the professional audio recorded by Sarah with the video footage obtained by Chris, creating clips related to spaces and developing the background map, which is currently the default Google Maps background. For now we have sketched interactive polygons onto the Google Map background, where the polygon for the dormitory building was informed plans photographed by Chris during a visit to the Winnipeg branch of LAC (Fig. 9.5.2).

During my work transcribing the audio interviews, I noted many instances of stories being told that related to specific places and journeys. We have started to experiment with mapping survivor stories, which offer many possibilities for mapping broader historical geographies. For example, in transcribing the story of Ted Fontaine, I found myself following him on a map as he described an impromptu bus trip that he and his friends took to explore the Assiniboia school site as construction workers were putting the finishing touches on what would become the dormitory building for the school residents. In the process of experimenting with mapping Ted's story, I found it necessary to do some research on the likely bus route south from the Long Plain First Nation region to Winnipeg, and then through the city to the school. Ted's mention of the Winnipeg bus terminal they arrived at left me stumped when I searched for the location. In addition to the very interesting and revealing remarks in Ted's recantation of his 'journey to Assiniboia Residential School', my interest into the nature of Winnipeg and the surrounding region in the late 1950s was piqued, and the answers to my questions in this regard were necessary for me to map Ted's story.

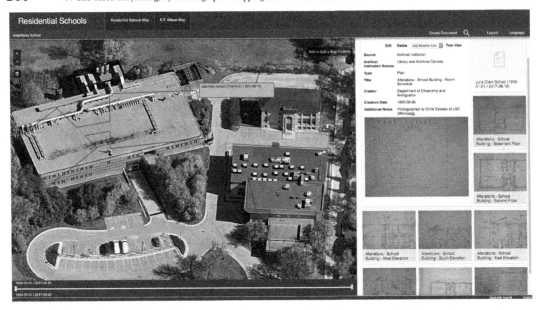

FIGURE 9.5.2 Screenshot showing the Assiniboia Schools Map of the in-development Residential Schools Land Memory Atlas with the school building (Julia Clark School) outlined in red in the central map window and photographs taken by Chris Calesso of archival photographs of the school in the right side panel.

9.6 Conclusion: Theory and practice in reflexive development

Our aims in mapping the Assiniboia Reunion are largely documentary and journalistic. In this respect, we share the opinion of Inka Salovaara that geo-visual representations of information '[enhance] embodiment in the experience of the information' (827), and agree with Annette Kim that 'exuberant experimentation' with interactive mapping is necessary. Allowing for the emergence of new ideas and approaches and developing the module in an iterative way, while incorporating an indigenous ethic of not wasting, is allowing us to reflexively fold the results of our archival and historical research (see Chapter 8) into our research to process and map the multimedia results of our participatory 'coverage' of the 2017 Assiniboia Reunion.

Turnbull (2007) has stated that the best way to engage in a critical approach to mapping—especially in an intercultural context—is to engage incommensurable perspectives in a dialogical tension with one another in a nondominatory manner (Pickles, 2004), where Western epistemologies and ontologies do not take priority or precedence over indigenous understandings and ways of approaching the world (Pickles, 2004). Our cybercartographic approach to developing the Assiniboia School Map Module involves bringing multiple perspectives together and is focused on place as an organizing factor for information. Reflecting critical approaches to space, cartography and archival studies, we are getting closer to producing a map module that reflects an emerging ongoing process of map production and use and a continuum approach to both information and interactivity.

Reflexivity in our ongoing mapping work is reflected in several ways. For example, our approach to multimedia documentation of the reunion event was influenced by literature on micromapping, where, for example, the walking tours of Reid, Naqvib and Waldichuk (2014) and our awareness that site-based storytelling lends itself naturally to mapping. This is reflective of the reflexive relationship between deconstruction and reconstruction/theory and practice where deconstructive approaches to map interpretation question the motives and context of the map maker (Harley, 1988, 1989); reconstructive approaches include making maps through participatory processes and community collaboration (Caquard, 2011; Corner, 1999; Cosgrove, 2005; Craig et al., 2002; Crampton, 2001, 2009; Crampton and Krieger, 2006; Elwood and Ghose, 2004; Fox et al., 2005; Harmon, 2003; Kitchin, 2008; Kitchin and Dodge, 2007; Pearce, 2008; Pearce and Louis, 2007; Pyne, 2013; Turnbull, 2007; Taylor and Lauriault, 2014). An understanding of maps and atlases as narrative vehicles in relation to the land (or place) is common to both deconstructivist and reconstructivist cartographic projects, in addition to reflexive patterns in cartographic design, development and use (Brauen et al., 2011; Pyne, 2013), which include education.

One of the central strands in reconciliation processes associated with the Residential Schools Legacy, and reconciliation in general (see Chapter 1), is 'truth', in this case, the sharing of stories by Residential Schools survivors, many of whom are eager to educate future generations and promote healing and social harmony. Place is an important dimension of survivor experiences and their stories (Field, 2014). Connected to their narrative capacity to engage and motivate, maps and atlases contribute to enhanced awareness and focus attention by using 'the language of cartography to graphically structure their meaning and to concentrate people's attention on the aspect of importance' (Field, 2014, p. 99). Cybercartographic Atlas projects reflect both types of relationship between cartography and narrative, allowing individuals to present their perspectives and understandings in multiple ways in relation to places and providing reflexive interpretations of cybercartographic atlas-making processes.

The AIRS survivors project, which culminated in the 2017 reunion provided an ideal project for our mapping work to intersect with. In addition to the power and potential of the reunion itself as a reconciliation process, and the associated media coverage, the mapping work we are doing around this combined with the results of our archival and historical research, should help to amplify the good work already being done.

References

Bernstein, D., 2016. Negotiating nation: Native participation in the cartographic construction of the Trans-Mississippi West. Environment & Planning A 48 (4), 626–647.

Beyersdorff, M., 2007. Covering the earth: mapping the walkabout in Andean Pueblos de Indios. Latin American Research Review 42 (3), 129–160.

Brauen, G., Pyne, S., Hayes, A., Fiset, J.P., Taylor, D.R.F., 2011. Encouraging transdisciplinary participation using an open source cybercartographic toolkit: the Atlas of the Lake Huron Treaty relationship process. Geomatica 65 (1), 27–45.

Burini, F., 2012. Community Mapping for Intercultural Dialogue. EspacesTemps.net. Available from: https://www.espacestemps.net/articles/community-mapping-for-intercultural-dialogue/.

Caquard, S., 2011. Cartography 1: mapping narrative cartography. Progress in Human Geography 37 (1), 135—144.

Caquard, S., Taylor, 2005. Art, maps and Cybercartography: stimulating reflexivity among map-users. In: Taylor, D.R.F. (Ed.), Cybercartography: Theory and Practice, Modern Cartography Series, vol. 4. Elsevier, Amsterdam, pp. 285—307 (Chapter 12).

Carver, S., 2003. The future of participatory approaches using geographic information: developing a research agenda for the 21st century. Urban and Regional Information Systems Association (URISA) Journal 15. APA I 61-72.

CBC News, June 23, 2017. Survivors Hold Reunion at Little-Known Winnipeg Residential School. https://www.cbc.ca/news/indigenous/winnipeg-survivors-reunion-assiniboia-residential-school-1.4176057.

Cidell, J., 2008. Challenging the contours: critical cartography, local knowledge, and the public. Environment & Planning A 40 (5), 1202—1218.

Connerton, P., 1989. How Societies Remember. Cambridge University Press, New York.

Corner, J., 1999. The agency of mapping: speculation, critique and invention. In: Cosgrove, D. (Ed.), Mappings. Reaktion Books, London, pp. 213—252.

Cosgrove, D., 2005. Maps, mapping, modernity: art and cartography in the twentieth century. Imago Mundi 57 (1), 35—54.

Craig, W., Harris, T., Wiener, D., 2002. Community Participation and Geographic Information Systems. CRC Press, Boca Raton, FL.

Crampton, J., 2001. Maps as social constructions: power, communication and visualization. Progress in Human Geography 25 (2), 235—252.

Crampton, J.W., 2009. Cartography: performative, participatory, political. Progress in Human Geography 33 (6), 840—848.

Crampton, J., Krieger, J., 2006. An introduction to critical cartography. ACME: An International E-Journal for Critical Geographies 4 (1), 11—33.

Elwood, S., Ghose, R., 2004. PPGIS in community development planning: framing the organizational context. Cartographica 38 (3—4), 19—33.

Field, K., 2014. Editorial: the stories maps tell. The Cartographic Journal 51 (2), 99—100.

Fontaine, T., 2010. Broken Circle: The Dark Legacy of Indian Residential Schools. Heritage House Publishing Company, Victoria, Vancouver, Calgary.

Fox, J., Suryanata, K., Hershock, P., 2005. Mapping Communities: Ethics, Values, Practice. East-West Center, Honolulu, Hawaii.

Gilmore, M., Young, J., 2012. The use of participatory mapping in ethnobiological research, biocultural conservation, and community empowerment: a case study from the Peruvian Amazon. Journal of Ethnobiology 32 (1), 6—29.

Harley, B., 1988. Secrecy and silences: the hidden agenda of state cartography in early modern Europe. Imago Mundi 40, 57—76.

Harley, B., 1989. Deconstructing the map. Cartographica 26, 1—20.

Harmon, K., 2003. You Are Here: Personal Geographies and Other Maps of the Imagination. Princeton Architectural Press, New York, NY.

Harris, L., Hazen, H., 2006. Power of maps: (Counter) mapping for conservation. ACME: An International E-Journal for Critical Geographies 4 (1), 99—130.

Hirt, I., 2012. Mapping dreams/dreaming maps: bridging Indigenous and western geographical knowledge. Cartographica, Special Issue on Indigenous Cartography and Counter Mapping 47 (2), 105—120.

Hunt, D., Stevenson, S., 2017. Decolonizing geographies of power: indigenous digital counter-mapping practices on Turtle Island. Settler Colonial Studies 7 (3), 372—392.

Irwin, R., Bickel, B., Triggs, V., Springgay, S., Beer, R., Grauer, K., Xiong, G., Sameshima, P., 2009. The City of Richgate: A/r/tographic cartography as public pedagogy. Jade 28 (1), 61—70.

Johnson, J., Louis, R., Pramono, H., 2006. Facing the future: encouraging critical cartographic literacies in Indigenous communities. ACME: An International E-Journal for Critical Geographies 4 (1), 80—98.

JUST TV, 2017. Remembering Assiniboia. https://www.youtube.com/watch?v=iQDSQuf7_iA.

Kelly, J., 1999. Maori maps. Cartographica 36 (2), 1—30.

Kim, A., 2015. Critical cartography 2.0: from "participatory mapping" to authored visualizations of power and people. Landscape and Urban Planning 142, 215—225. https://www.sciencedirect.com/science/article/pii/S0169204615001504.

Kitchin, R., 2008. The practices of mapping. Cartographica 43 (3), 211—216.

Kitchin, R., Dodge, M., 2007. Rethinking maps. Progress in Human Geography 31 (3), 331–344.

Laidler, G.J., Elee, P., Ikummaq, T., Joamie, E., Aporta, C., 2010. Mapping sea-ice knowledge, use, and change in Nunavut, Canada (Cape Dorset, Igloolik, Pangnirtung). In: Krupnik, I., Aporta, C., Gearheard, S., Laidler, G.J., Kielsen-Holm, L. (Eds.), SIKU: Knowing Our Ice, Documenting Inuit Sea-Ice Knowledge and Use. Springer, Dordrecht.

Leakey, R., November 1, 2002. The Future of Conservation and Wildlife. Guy Stanton Ford Memorial Lecture, University of Minnesota, Minneapolis.

Liu, S., Palen, L., 2010. The new cartographers: crisis map mashups and the emergence of neogeographic practice. Cartography and Geographic Information Science 37 (1), 69–90. https://www.tandfonline.com/doi/abs/10.1559/152304010790588098.

Louis, R.P., 2004. Indigenous Hawaiian cartographer: in search of common ground. Cartographic Perspectives 48, 7–22.

Louis, R.P., Johnson, J.T., Pramono, A.H., 2012. Introduction: indigenous cartographies and counter-mapping. Cartographica, Special issue on Indigenous Cartography and Counter Mapping 44, 77–79.

Palen, L., Liu, S., 2007. Citizen communications in crisis: Anticipating a future of ICT-supported participation. In: CHI 2007: Proceedings of the SIGCHI Conference on Human Factors in Computing Systems, ACM, San Jose, California, pp. 727–736.

Palen, L., Vieweg, S., 2008. Emergent, widescale online interaction in unexpected events: Assistance, alliance and retreat. In: Proceedings of the ACM Conference on Computer Supported Cooperative Work (San Diego, CA, USA, November 08–12, 2008). CSCW'08. ACM, New York, NY, pp. 117–126.

Palen, L., Vieweg, S., Liu, S., Hughes, A., 2009. Crisis in a networked world: features of computer mediated communication in the April 16, 2007, Virginia tech event. Social Science Computer Review, Special Issue on e-Social Science 27 (5), 467–480.

Palmer, M., 2009. Engaging with indigital geographic information networks. Futures 41, 33–40.

Palmer, M., 2012. Theorizing indigital geographic information networks. Cartographica, Special Issue on Indigenous Cartography and Counter Mapping 47 (2), 80–91.

Parker, B., 2006. Constructing community through maps? Power and praxis in community mapping. The Professional Geographer 58 (4), 470–484.

Pearce, M.W., 2008. Framing the days: place and narrative in cartography. Cartography and Geographic Information Science 35, 17–32.

Pearce, M., Louis, R., 2007. Mapping indigenous depth of place. American Indian Culture & Research Journal 32 (3), 107–126.

Pickles, J., 2004. A History of Spaces: Cartographic Reason, Mapping and the Geocoded World. Routledge, London, New York.

Pyne, S., 2013. Sound of the Drum, Energy of the Dance: Making the Lake Huron Treaty Atlas the Anishinaabe Way. Unpublished PhD Dissertation. Carleton University, Ottawa. https://curve.carleton.ca/392a68ac-086c-4470-976d-101d4e96f9f7.

Pyne, S., 2014. The role of experience in the iterative development of the Lake Huron Treaty atlas. In: Taylor, D.R.F., Lauriault, T. (Eds.), Developments in the Theory and Practice of Cybercartography: Applications and Indigenous Mapping. Elsevier, Amsterdam (Chapter 17).

Pyne, S., Taylor, D.R.F., 2012. Mapping indigenous perspectives in the making of the cybercartographic atlas of the Lake Huron Treaty relationship process. Cartographica, Special Issue on Indigenous Cartography and Counter Mapping 47 (2), 92–104.

Reid, R., Naqvi, K., Waldichuk, T., 2014. Place-based curriculum: revealing student connections with community through walking tours. Western Geography 20–21, 63–69. http://kamino.tru.ca/experts/home/main/bio.html?id=knaqvi.

Reid, R., Naqvib, K., Waldichuk, T., 2016. Place-based curriculum: revealing student connections with community through walking tours. Western Geography 21–22, 63–69.

Robinson Desjarlais, S., June 26, 2017. Residential school survivors spend emotional weekend in place that changed their lives. APTN National News. https://aptnnews.ca/2017/06/26/residential-school-survivors-spend-emotional-weekend-in-place-that-changed-their-lives/.

Rundstrom, R.A., 1995. GIS, Indigenous peoples, and epistemological diversity. Cartography and Geographic Information Systems 22 (1), 45–57.

Salovaara, I., 2016. Participatory maps. Digital Journalism 4 (7), 827–837.
Sanders, C., May 27, 2017. Former Classmates of City Residential School to Hold Reunion. Winnipeg Free Press. https://www.winnipegfreepress.com/local/mixed-emotions-424713353.html.
Sletto, B., 2015. Inclusions, erasures and emergences in an indigenous landscape: participatory cartographies and the makings of affective place in the Sierra de Perijá, Venezuela. Environment and Planning D: Society and Space 33 (5), 925–944.
Taylor, D.R.F., Lauriault, T. (Eds.), 2014. Developments in the Theory and Practice of Cybercartography: Applications and Indigenous Mapping. Elsevier, Amsterdam.
Tobias, T.N., 2000. Chief Kerry's Moose: A Guidebook to Land Use and Occupancy Mapping, Research Design, and Data Collection. Union of BC Indian Chiefs: Ecotrust Canada, Vancouver.
Turnbull, D., 2007. Maps, narratives, and trails: performativity, hodology, and distributed knowledges in complex adaptive systems: an approach to emergent mapping. Geographical Research 45, 140–149.
UM News Today, 2017. Assiniboia Residential School Survivors Plan Reunion and Commemorative Event in Winnipeg. May 24. http://news.umanitoba.ca/assiniboia-residential-school-survivors-plan-reunion-and-commemorative-event-in-winnipeg/.
Woodward, D., Lewis, G.M., 1998. The History of Cartography: Cartography in the Traditional African, American, Arctic, Australian, and Pacific Societies, vol. 2. The University of Chicago Press, Chicago. Book 3.
Woodward, D., Lewis, G.M., 2000. Cartography in the Traditional African, American, Arctic, Australian, and Pacific Societies. University of Chicago Press, Baltimore.
Young, I., 2000. Inclusion and Democracy. Oxford University Press, Oxford, New York.

10

Bridging institutional and participatory ethics: A rationality of care perspective

Stephanie Pyne*

Postdoctoral Research Fellow, Geomatics and Cartographic Research Centre (GCRC), Carleton University, Ottawa, ON, Canada

*Corresponding author

10.1 Introduction

The study of ethical issues in research intersects with qualitative research, which emphasizes a social justice component. According to Norman Denzin:

> There has never been a greater need for a critical qualitative inquiry that matters in the public sphere. We live in the audit cultures of global neoliberalism. The politics of evidence that define the audit culture marginalize critical inquiry. Our challenge is to push back, to resist, to redefine the place of the academy, indigenous epistemologies and the public intellectual in these public spaces. This is a call for interpretive, critical, performative qualitative research that matters in the lives of those who daily experience social injustice. This is a call for inquiry that addresses inequities in the economy, education, employment, the environment, health, housing, food, and water, inquiry that embraces the global cry for peace and justice" (Denzin, 2017, p. 8).

Concerns with social justice and relevance in research are echoed by researchers in many disciplines, including geography with calls 'for research practices that more actively promote social and political change, theorize and practice geographies of care and responsibility, and re-examine the ethical and moral commitments that shape what we do as researchers and how we do it (see for example, Cloke, 2002; Massey, 2004; Valentine, 2005; Lawson, 2007)' (Elwood, 2007, pp. 329–330). Both participatory and community-based approaches to research and ethical frameworks associated with institutional research funding have emerged in this social justice context. In addition, there are particular ethical challenges associated with the creation and maintenance of both digital archives and digital atlases. For example, the concepts of both 'archive' and 'atlas' have broadened in this new age of Web 2.0 technologies, partly due to the reconceptualization of 'citizenship'. In this regard Goode (2010) 'suggests that cultural citizenship invokes questions of access, visibility and cultural recognition, as well as tensions between intra- and inter-cultural communication online'. To deal ethically with this situation, Goode argues for 'a reflexive and critical research agenda which accounts for the "attention economy" of the Internet and issues of cultural ethics online', and insists on the need to 'acknowledge ongoing tensions and contradictions between a postmodern "remix" ethic in which the Internet serves as an open cultural archive which citizens can freely access and rework, on the one hand, and claims for cultural authorship, sovereignty and protection, on the other (p. 527)'.

The administration of nationally funded research in both Canada and the United States is governed by relatively general ethics regimes that are intended to be interpreted in a consistent fashion across academic institution research ethics boards. This presents challenges for a vast range of participatory research projects with a correspondingly vast range of research contexts. Elwood (2007) contrasts institutional and participatory approaches to ethics and 'proposes participatory ethics as a framework from which we might begin to proactively engage some of the contradictions and gaps inherent in institutional ethical frameworks' (p. 329). In addition to the significance of Elwood's distinctions, her work is relevant to the discussion insofar as it is concerned with the distinctive assumptions on which these approaches are based. Participatory research is often also characterized as qualitative inquiry, and the best part of action research is participatory and ethical: 'Action Research (AR) understood as a collaborative research method generating both scientific knowledge and democratic social change' (Greenwood, 2007, quoted in Strumińska-Kutra, 2016, p. 864). Elwood's (2007) distinction between 'institutional' and 'participatory ethics' aligns with this distinction and serves as a useful tool to contextualize a discussion of a spectrum of issues, from recognition to anonymity.

Morton Ninomiya and Pollock (2017) acknowledge the importance of the development of ethical principles by academic institutions for research involving Indigenous people, yet caution that 'real-world' experiences and events lie beyond the guiding purview of many principles. This chapter considers a variety of issues in institutional and participatory research ethics, including those related to inclusion, and presents some examples illustrating the importance of practising an ethic of care in academic research. It concludes with some reflections on research ethics in relation to the Residential Schools Land Memory Mapping Project (RSLMMP).

10.2 Institutional and participatory research ethics

While the need for ethical frameworks in institutional funding contexts is understandable, it is also known that the nature and interpretation of ethical requirements vary depending on the specificities of individual research projects. This situation has resulted in a variety of challenges when it comes to doing collaborative and participatory research that involves both 'participatory' and 'institutional' ethical dimensions (Elwood, 2007). Many challenges are rooted in the disconnect between the 'hegemonically lateral' or democratic, distributed and emergent approaches typified by participatory (often qualitative) research, and the top-down 'fixed code' approaches of institutional ethics, which include 'rules defining everything from what to say (and not say) in recruiting research participants, to when and how to conceal participants' identities, who may or may not participate, and what activities can or cannot be undertaken' (Elwood, 2007, p. 330). Although Elwood recognizes the importance of ethics in the implementation of institutionally funded academic research, especially with respect to 'protecting research participants from harm', as she admits 'in practice, negotiating the requirements and processes of institutional ethics can be frustrating and difficult' (2007, p. 330).

According to Elwood, '[o]n one level, institutional ethics and participatory ethics are both guided by a shared commitment that research practices and interactions should protect and respect participants' health, safety, privacy, and physical and emotional wellbeing. But beyond this shared principle, they are quite different' (2007, p. 331).

Institutional ethics assume

(1) that ethical problems and risks can be identified before they occur,
(2) that ethical problems and risks can be identified outside the context of the research situation,
(3) that rules for ethical practices can be universal,
(4) lack of agency in the 'subject' (within these frameworks, the 'subject' who is to be protected from harm (or who might benefit from participation) is envisioned as an individual, constructed as inherently vulnerable and not an active agent in identifying and responding to ethical dilemmas in research)(p. 331).

These assumptions underlie a core set of practices 'through which institutional ethics seek to minimize harm and avoid or respond to ethical challenges tend to be fixed standards, rules, and codes, and the main actors involved in these processes are the researcher(s) and institutional participants such\as IRB members' (p. 331).

In contrast with the assumptions indicated for institutional ethics, Elwood identifies one central assumption of participatory ethics − 'that ethical problems and dilemmas are situational, specific to the relationships and interactions of a particular research context' (p. 331). This generally described assumption leads to practices 'through which participatory ethics might be negotiated', which are inconsistent with the fixed ethical codes, values, practices and assumptions of universality that characterize institutional ethics. These practices include

1. shared dialogue among a wide range of research participants (potentially university researchers, community researchers, and people affected by the research), before, during and after research;

2. inclusion of a wider range of subjectivities;
3. "movement beyond the notion of an atomistic individual who assesses risks and benefits to determine participation in a research project" and
4. "speak[ing] to the ways that identities, group membership (or exclusion), and power relations can shape ethical practices and challenges in the social situations of research" (p. 331).

Dickens and Butcher (2016) join Elwood in contrasting institutional and participatory ethics, which they see as 'a situated and relational stance orientated around an ethos of care and responsibility' (p. 528). A participatory ethics involves 'ongoing negotiation of the nature and scope of participation [...] and a commitment to take action in order to achieve affirmative social transformations among, and on behalf of, those participating' (Dickens and Butcher, 2016, p. 528); whereas institutional ethics tends more toward 'mitigation' (Bradley, 2007) as a guiding principle and includes a regulatory approach to ethics processes that reflects 'values associated with paternalism (Miller and Wertheimer, 2007; Skelton, 2008), ethnocentrism (Mistry and Berardi, 2012) and medical governance (Dyer and Demeritt, 2009), as well as the epistemological limitations imposed by an overreliance on predictability (Thrift, 2003)' (Dickens and Butcher, 2016, p. 528). In their discussion of a community-based knowledge creation and dissemination research project involving youth as research assistants in the production of autobiographical films, Dickens and Butcher argue for 'an ethical stance that accounts for the conditions under which participants become publicly visible through the research process' (p. 529).

Wood (2017) refers to community-based research (CBR) as the political pursuit of 'knowledge democracy', which 'requires the use of democratic modes of enquiry that include all stakeholders in knowledge production; likewise, it entails using knowledge to strengthen democracy by including everybody into democracy building (Stern, personal communication, 16 June 2017)' (1). CBR involves both the creation of knowledge products, and 'in itself is a process of learning and development which has the potential to change society through changing individuals and organizations' (1−2). In her critique of institutional review processes, Wood (2017) talks of 'working in partnership with communities, rather than doing research "on" them and reporting findings "about" the issues they face', noting the inadequacy of institutional ethical protocols for projects involving 'participatory, community-based qualitative methodologies' (1). While Zeynep et al. (2016) emphasize the importance of imagination and emergence in participatory action research with young people. These features are consistent with an ethics of care approach, which is both a good way to engage in both institutional and participatory ethics and to bridge intercultural incommensurabilities in terms of valued epistemologies and ontologies.

In her video lecture on the conceptual approaches to ethics underlying participatory projects subject to institutional ethics requirements, Dr. Rose Wiles outlines principlism and consequentialism as the underlying themes in institutional ethics, while participatory ethics often involve virtue ethics or ethics of care approaches (2015). Principlism involves applying general principles to specific cases and consequentialism involves a focus on outcome, which is reflected in the preoccupation with risk aversion in many institutional ethics regimes. In 'Engaged scholarship: Steering between the risks of paternalism, opportunism, and paralysis', Strumińska-Kutra (2016) reflects on three types of research-related relationships involving power tensions: (1) participatory inquiry and its cultural, institutional, and social environment; (2) the heterogeneous 'community' being studied, 'which itself is not homogeneous in terms of interests, values, and ability of their realization'; and (3) the researcher and 'the community'. Connecting these tensions to the three mutually exclusive approaches of pragmatism, critical theory, and constructivism, Strumińska-Kutra categorizes critical theory as a paternalistic approach, which alienates rather than invites participation. The practice of applying general principles to specific cases is far more prevalent in institutional ethics contexts where ethics boards rarely have contact with research participants; whereas participatory ethics presuppose the development of ongoing relationships between researchers and participants, and often involve virtue ethics or ethics of care approaches. Although Sara Ruddick acknowledges the existence of a wide variety of caring activities and relationships in her ethics of care approach, she insists that mothering is a 'symbol of care' that can contribute to thinking about development (1989, p. 46) and related areas such as research ethics. Maternal thinking involves thinking, strategy, calculation, reflection, communication and 'reflective feeling': 'Rather than separating reason from feeling, mothering makes reflective feeling one of the most difficult attainments of reason' (Ruddick, 1989, p. 70).

10.3 Discussion and conclusion: Community, research and ethics of care

Underlying the contrast between institutional and participatory approaches to ethics is the philosophical distinction between deontological, or principle-based, approaches to ethics and virtue-oriented approaches to what is often termed 'morality'. According to Christa Fouché and Laura Chubb, 'Individualistic, agenda-driven, and outcome-guided research characteristically defines the academic world. These characteristics can create a tension with values of collaboration, flexibility, and adaptability which underscore community-participatory research approaches (Glass and Kaufert, 2007)' (Fouché and Chubb, 2017, p. 26). In this context, 'ethical reviews require iterative approval and engagement processes due to continual re-negotiation of consent, project design, ownership, and dissemination between those regarded as community and the researchers' (Fouché and Chubb, 2017, p. 27). Underlying the iterative approval process are ongoing relationships between researchers and participants that involve a certain rationality of care.

Ruddick (1989) distinguishes between two aspects of care — 'caring about' and 'caring for' — where 'caring for' is in an important respect an intellectual discipline that is based on and incorporates 'caring about'. Whether we conceive of care as a virtue, a process, a practice or a discipline, care appears to involve several stages or components that can be understood and assessed in terms of their rationality. In the process of caring about things, we feel responsible for them, or committed to them; however, we must also be able to care for them. Caring for things successfully involves a range of reasonable approaches, thought processes, feelings, attitudes, actions, habits and virtues, that is, a 'distinctive discipline' of rationality (Ruddick, 1989).

Appealing to the process of mothering as a model for a rationality of care, Ruddick describes processes related to the three goals of protection, growth and social acceptance. Successfully promoting these goals requires characteristic cognitive and emotional skills and attitudes. A brief examination of some of the skills and attitudes involved in caring for children will provide a glimpse of the rational nature of caring for, which involves characteristic aspects/virtues or intellectual-emotional complexes, including reflective feelings.

Effective protection of children requires humility. Although mothers may wish to protect their children from all harms (which they have to think about and identify in relation to their children), they require humility in order to realize that they cannot control every aspect of their children's lives. As children grow and develop, they acquire a sense of independence that can be compromised by [process aspect] overprotection in the caring they receive from their mothers. In order to guard against overprotection, a mother must develop the 'mental habit or cognitive style, which [Ruddick refers to as] scrutinizing' when it comes to watching over her child: 'A mother never stops looking, but she must not look too much. Attentiveness to a creature who perseveres in its own being and at the same time is perpetually at risk is peculiarly demanding'(1989, p. 71).

Ruddick describes the cognitive capacity of holding, which is a 'fundamental attitude' toward the vulnerable, and 'means to minimize risk and to reconcile differences rather than to sharply accentuate them' (1989, p. 78). This capacity is important for any interaction involving care as it 'is a way of seeing with an eye toward maintaining the minimal harmony, material resources and skills necessary for sustaining a child in safety. It is an attitude elicited by the work of "world-protection", world-preservation, world-repair [...] the invisible weaving of a frayed and threadbare family life' (1989, pp. 78–79).

Mothering involves a welcoming attitude to change, since as time goes on children grow and develop their abilities to do and to be. This is also the case with participatory research projects aimed at enhancing community and individual empowerment. A welcoming attitude to change is one of 'the most exigent intellectual demands on those who foster growth (1989, p. 89). It is also, moreover, an attitude that could be relevant to thinking about development in general: "Maternal experience with change and the kind of learning it provokes will help us to understand the changing natures of all peoples and communities. It is not only children who change, grow, and need help in growing. We all might grow—as opposed to simply getting older—if we could learn how"' (1989, p. 90).

Concrete thinking is 'called forth by and enables the work of fostering growth' (1989, p. 93). In contrast with abstract thinking, which seeks to 'simplify, generalize, and sharply define', concrete thinking involves relishing complexity, tolerating ambiguity and multiplying options (1989, p. 93). Ruddick does not discount the appropriateness of abstract thinking for 'any disciplined thinking', including scientific, mathematical, philosophical and even maternal thinking. However, concrete thinking is taken to be especially useful in making sense of open (changing and developing) structures (1989, p. 97).

By telling stories from a compassionate perspective, a mother teaches her child in a concrete manner about the nature of compassion and instils in that child the ability to be compassionate both toward others and toward himself: 'Children who learn about themselves through compassionate stories may develop a maternal generosity toward their lives, learning from their mothers the capacity to appreciate the complex humanness of their plight, to forgive themselves as they have been forgiven (1989, pp. 99–100)'. This capacity is especially important in the residential schools research ethics context, where transparency in process is of utmost importance.

Proper trust is the virtue associated with the training of conscience and is described as 'one of the most difficult maternal virtues' (1989, p. 119) where '[i]deally, proper trust is prepared by maternal protectiveness and nurturance [...] Clearly, to identify proper trust as a virtue is not to identify an achievement but an ongoing and difficult struggle (1989, p. 119). Attentive love is a discipline that combines the cognitive capacity of attention with the virtue of love, "which knits together maternal thinking"' (1989, p. 119). This virtue could be seen to have a parallel in a broader context where goodwill between people is necessary for productive relationships where the mother learns to ask:

> 'What are you going through?' and to wait to hear the answer rather than giving it. She learns to ask again and keep listening if she cannot make sense of what she hears or can barely tolerate the child she has understood. Attention is akin to the capacity for empathy, the ability to suffer or celebrate with another as if in the other's experience you know and find yourself. However, the idea of empathy, as it is popularly understood, underestimates the importance of knowing the other without finding yourself in her. A mother really looks at her child, tries to see him accurately rather than herself in him. 'The difficulty is to keep attention fixed on the real situation'—or on the real children (1989, p. 121).

As Elwood notes, space must be made in institutional ethics for emergence and imagination, which requires strengthening relationships and ties of trust. In addition to such significant issues as the 'ethics of recognition' (Dickens and Butcher, 2016; Bak, 2016), the 'ethics of confidentiality', and 'the ethics of informed consent', it is important to consider the 'ethics of trust'. Archives are increasingly becoming more participatory and relational, for example, Project Naming (http://www.bac-lac.gc.ca/eng/discover/aboriginal-heritage/project-naming/Pages/introduction.aspx) and Co-Lab (https://co-lab.bac-lac.gc.ca/eng/), both initiatives of Library and Archives Canada geared toward enhancing public engagement in the digital world and both democratizing and enriching its record appraisal processes; and ongoing work at the Shingwauk Residential Schools Centre and the National Centre for Truth and Reconciliation.

With respect to the conventional relationship between Archives and the public, Greg Bak highlights Hickerson's characterization of the virtue of trust:

> Archivists are seen as trusted agents of society, acting on everyone's behalf in insuring the preservation of those records necessary in protecting the legal rights of each citizen and in preserving the historical record of human achievement, of cultural evolution, and of everyday life. We have a special role in society, and we are respected as ombudsmen acting in the public as well as each individual interest *Hickerson, 2001, p. 16, quoted in Bak, 2016, p. 375.*

In his discussion of the Trusted Digital Repositories Report (TDR) (OCLC-RLG, 2002), which considered issues related to the transition from the analog to the digital age for archives, Bak (2016) notes that although archives as cultural institutions are considered to be 'trusted agents of society', in the report, 'trust' remains undefined (p. 376). As Bak contends, such a 'brisk' treatment of the concept of trust by the report's authors 'belies the complexity of the concept of trust, which is understood to be social and contingent, in the Merriam Webster definition. The TDR report implies that trust exists in a simple binary with distrust: either an institution is trusted or it is distrusted; once an institution is trusted this trust extends to all of the divisions, systems and staff within the institution, and presumably over time as well. The TDR report suggests that trust in cultural institutions is a matter of mandate, while in the Merriam Webster trust is characterized as embedded within relationships' (2016, p. 277).

According to Bak, 'Populations that rely upon archives are not homogeneous in their view of archival institutions. Many Canadians, for example, may view their public archives as benign, trustworthy custodians of government records. Indigenous peoples often have a different view of these same public archives, seeing only one more organ of colonization' (2016, p. 377). Bak gives the example of the terms relating to information in the Indian Residential Schools Settlement Agreement (2006; see also Chapter 1), which required

> the government and churches who ran the residential school system [to] copy their relevant records and transfer these digital copies to the Truth and Reconciliation Commission of Canada (TRC), for use by the TRC in its truth telling, subsequently to be archived in the non-governmental National Centre for Truth and Reconciliation that was envisioned in the settlement agreement. The Indigenous negotiators of the Settlement Agreement evidently did not trust the federal government and church archives to preserve, arrange, describe and make available these records. Given that the TRC has found its work to be stymied by lack of cooperation from the government and some of the churches, and given that despite the clear terms of the Settlement Agreement, the TRC has had to take the government to court to gain access to relevant records, this lack of trust, in addition to being understandable, is seen to be good policy *TRC, 2012; Sinclair, 2014 (2016, p. 377).*

The proper trust that is required to establish productive relationships between researchers and participants, archives and community, is based on ongoing interaction and engagement, in addition to respect and mindfulness. For example, when it comes to issues of intellectual property, Engler et al. (2014, p. 52) discuss the importance of distinguishing between a cyber-cartographic atlas as a 'multimedia work' or 'a compilation consisting of multiple different works' where one solution is to refer 'the overall authors or author of the compilation' having 'a copyright in the atlas itself', while it is still possible to have simultaneous VGI contributions with their own copyright requirements, with the potential for a large number of copyright holders in one atlas project. This is an attempt to provide for more freedom in research processes where the authors recommend collaborative approaches that optimize cartographic interactivity, and contrast the previous era with strict licencing policies and the current more 'open' era, with a variety of examples reflecting freer access to information by the public, where VGI practice is ahead of VGI theory. This recommendation is in line with the rationality of care specification that protection not become excessive to the point of limiting freedom referred to above (Ruddick, 1989).

In another example of institutional ethics engaging in alleged overprotection, Iacono (2006) discusses institutional ethics processes related to the vulnerability of persons with intellectual disabilities, which have limited the participation of those persons in research relevant to them: 'This marginalizes and excludes entire groups of people who could benefit from partaking in, and helping find answers to, research questions proposed through various community- participatory projects (Fouché and Chubb, 2017, p. 27). And sometimes, the mere existence of a form and the need to sign it is enough to compromise community building: 'Procedures for obtaining consent were also identified by Cahill, Sultana and Pain (2007) as imposing unintentional harm for both participants (with or without disability) and researchers. It may well jeopardize months of dedication by the researchers to develop a balance in power when community members are required to sign a form. This can be seen as representing a legal contract and reaffirms power differentials (Fouché and Chubb, 2017, p. 28).

Brannelly and Boulton (2017) discuss an ethics of care approach as a two-pronged solution to issues that arise in the interaction between Indigenous and non-Indigenous researchers, and note how overprotection by researchers to avoid offending Indigenous communities can silence voices. Understanding this dynamic within an ethics of care framework can help to minimize these occurrences. McInerney (2016) invokes a 'hermeneutics of love' in his approach to community-orientated, participatory action research, and Ward and Barnes (2016) apply a feminist care ethics to social work practice with older people to create reflective spaces for deliberative engagement between older people and caregivers.

Reflecting on the legal and ethical issues relating to traditional knowledge (TK), Scassa et al. (2014) discuss the importance of informed consent; identifying the most appropriate ethical practices; community rights and fostering a culture of respect; the potential of Cyber-cartography for mapping TK; and the need for a new 'soft law' solutions to legal and ethical issues. This 'soft law' approach acknowledges the importance of the virtues of care in research relationships, and involves promoting them in an Inuit TK context. Cybercartographic atlas projects can extend over many years and involve ongoing research relationships. This has been the case with the RSLMMP, which involves research relationships rooted in earlier work, and a transdisciplinary approach to research based on intersecting with projects that also include enduring research relationships.

Soon after Fraser Taylor, the project's principal investigator, was informed that we were successful with the grant application, we were required to complete an ethics application, which required a more detailed examination of all of the activities and potential engagements with potential participants in the collaborative development of the prototype Residential Schools Land Memory Atlas Modules mentioned in Chapter 2. To reiterate, these would include:

i. six case study school modules to act as interactive geospatial portals to archival and other sources of information regarding these specific schools, where our initial research to create the prototype Shingwauk Industrial Home module was noted to have led to some productive ideas with respect to educational outreach, in addition to leading to the realization that extensive archival research is required in order to complete this module and construct the prototype modules for the other five case study schools;
ii. one prototype biographies map module with a series of map layers featuring work and travels of team participants;
iii. three orientation and support modules (mapping archival materials module, witness trauma education module, atlas orientation module) and
iv. one prototype project planning module.

At this beginning (or windup) phase of the project, we had not yet engaged in the literature review period, which gave rise to the decision to take a continuum approach, and we planned on devoting our first attention to exploring the archival records, historical publications and academic writing concerned with the Residential Schools theme (see Chapter 8). This component of the application was not subject to the research ethics process, which was geared toward safeguarding the rights of and 'participants' we might engage. However, throughout our research with this information, we nevertheless took a caring and reflective attitude toward the material, and considered possible educational and mapping activities that we could possibly engage participants in once the prototype atlas was sufficiently completed.

Since we had specified participatory mapping activities with participants for each of the six case study schools, we were responsible for spelling out the details of these activities in our ethics application; we did this to the best of our ability, considering that we planned to consult with our team, develop relationships and consult with others over time with respect to what we were actually going to do. Although it was challenging to describe a multidimensional project involving team members with intersecting projects and to describe events that could be subject to ethical approval, the process of preparing the initial application forced us to consider the feasibility of hosting six separate participatory mapping events across the country, where it takes considerable time and resources to implement just one activity adequately and in a way that is consistent with both institutional and participatory ethics. While we continued to work toward exploring participatory mapping possibilities for all six case studies, the most feasible school to begin with was Assiniboia Residential School, which was linked to the Remembering Assiniboia project referred to in Chapter 9. Andrew Woolford in collaboration with Ted Fontaine and the Assiniboia Survivors Group worked out a description of their planned activities, which included sharing circles, meal sharing, and a mandate to listen to, dialogue and work with the participating survivors toward their vision for how to go forward with their reconciliation work. Through a series of meetings, the group decided that it would like to host a reunion. I was fortunate to be included in the meetings, primarily as an observer; however, I was also able to describe the RSLMMP and present my vision for cybercartographically documenting the anticipated reunion events and activities (see Chapter 9).

Working over time with Andrew Woolford in connection with the Remembering Assiniboia Project has aided us in the process of negotiating institutional and participatory ethics. In addition to providing us with an ethics review model for participatory activities, based on the approved ethics application for the Remembering Assiniboia Project, Andrew discussed ideas around digitally mapping site-based tours led by survivors at the reunion with the group. In addition, via the iterative process approach to ongoing institutional ethics approval, Andrew submitted a revised ethics application for his project to include the projected reunion, and we were permitted to include a brief description of the RSLMMP and the plans to cybercartographically document the site-based storytelling tours in the application. The approved ethics application, which mentioned the intersection with the RSLMMP, was later included in the RSLMMP ethics application package to the ethics committee at Carleton, and approved handily. The institutional ethics process influenced decisions to focus on prototype development involving archival and historical materials; and, it forced us to sketch out a preliminary model of our participatory processes, in some cases before we had a chance to consult with those we had yet to meet; it also led us to identify the challenges and benefits of a transdisciplinary project involving intersections with other projects; with one of the greatest challenges being maintaining an emergent approach in an institutional ethics context. Many good discussions and related communications have been had with respect to both the institutional and participatory aspects of the RSLMMP. As John Allen contends in his relational space (see Chapter 8) approach (Allen, 2004), institutions are comprised by relations, which include opportunities to express personality and exert power. Institutional ethics regimes are both constructed and implemented by individuals — some engaged in their own CBR, others with vast experience with institutional ethics and research in many areas.

This chapter has presented an initial discussion of ethics and morality in relation to research, including a distinction that is often made between institutional and participatory approaches to ethics, ethics and archives, and a potential 'rationality of care' model (Ruddick, 1989, 2009) for an ethics of care approach, which could potentially provide a useful bridge between institutional and participatory approaches, and facilitate ethical approaches to archival information. In this regard, I have had some interesting and productive conversations with several people in the institutional ethics community, in addition to many others, around issues raised in this chapter, and plan to write a fuller follow-up account to the introduction provided in this chapter in a future publication following the launch of the Residential Schools Land Memory Atlas. This future work will comment on the manner in which a reflexive approach to research can help with the bridge-building project in ethics, and may well be included in the Project Journal map.

References

Allen, J., 2004. The whereabouts of power: politics, government and space. Geografiska Annaler 86B, 19–32.

Bak, G., 2016. Trusted by whom? TDRs, standards culture and the nature of trust. Archival Science 16, 373–402.

Bradley, M., 2007. Silenced for their own protection: how the IRB marginalizes those it feigns to protect. ACME: An International E-journal for Critical Geographies 6, 339–349.

Brannelly, T., Boulton, A., 2017. The ethics of care and transformational research practices in Aotearoa New Zealand. Qualitative Research 17 (3), 340–350.

Cahill, C., Sultana, F., Pain, R., 2007. Participatory Ethics: Politics, Practices, Institutions, ACME: An International E-Journal for Critical Geographies 6 (3), 304–318.

Cloke, P., 2002. Deliver us from evil? Prospects for living ethically and acting politically in human geography. Progress in Human Geography 26, 587–604.

Fouché, C.B., Chubb, L.A., 2017. Action researchers encountering ethical review: A literature synthesis on challenges and strategies. Educational Action Research 25 (1), 23–34.

Denzin, N., 2017. Critical qualitative inquiry. Qualitative Inquiry 23 (1), 8–16.

Dickens, L., Butcher, M., 2016. Going public? Re-thinking visibility, ethics and recognition through participatory research praxis. Royal Geographical Society (with the Institute of British Geographers) 41, 528–540.

Dyer, S., Demeritt, D., 2009. Un-ethical review? Why it is wrong to apply the medical model of research governance to human geography. Progress in Human Geography 33, 46–64.

Elwood, S., 2007. Negotiating participatory ethics in the midst of institutional ethics. ACME: An International E-Journal for Critical Geographies 6 (3), 329–338.

Engler, N.J., Scassa, T., Taylor, D.R.F., 2014. Cybercartography and volunteered geographic information. In: Taylor, D.R.F., Lauriault, T. (Eds.), Developments in the Theory and Practice of Cybercartography: Applications and Indigenous Mapping. Elsevier, Amsterdam (Chapter 4).

Goode, L., 2010. Cultural citizenship online: the Internet and digital culture. Citizenship Studies 14 (5), 527–542.

Glass, K.C., Kaufert, J., 2007. Research ethics review and Aboriginal community values: can the two be reconciled? Journal of Empirical Research on Human Research Ethics 2, 25–40.

Greenwood, D., 2007. Pragmatic action research. International Journal of Action Research 3 (1–2), 131–148.

Hickerson, H.T., 2001. Ten challenges for the archival profession. American Archivist 64 (1), 6–16.

Iacono, T., 2006. Ethical challenges and complexities of including people with intellectual disability as participants in research. Journal of Intellectual and Developmental Disability 31 (3), 173–179.

Lawson, V., 2007. Geographies of care and responsibility. Annals of the Association of American Geographers 97 (1), 1–11.

McInerney, R., 2016. A hermeneutics of love for community-based, participatory action research. Journal of Humanistic Psychology 56 (3), 263–285.

Massey, D., 2004. Geographies of responsibility. Geografiska Annaler 86B, 5–18.

Miller, F.G., Wertheimer, A., 2007. Facing up to paternalism in research ethics. Hastings Center Report 37, 24–34.

Mistry, J., Berardi, A., 2012. The challenges and opportunities of participatory video in geographical research: exploring collaboration with indigenous communities in the North Rupununi, Guyana. Area 44, 110–116.

Morton Ninomiya, M., Pollock, N., 2017. Reconciling community-based Indigenous research and academic practices: knowing principles is not always enough. Social Science & Medicine 172, 28–36.

OCLC-RLG, 2002. Trusted Digital Repositories: Attributes and Responsibilities. RLG, Mountain View, CA.

Ruddick, S., 1989. Maternal Thinking: Toward a Politics of Peace. Beacon Press, Boston.

Ruddick, S., 2009. Maternal Thinking: Philosophy, Politics, Practice. Demeter Press, Toronto.

Scassa, S., Lauriault, T., Taylor, F., 2014. Cybercartography and traditional knowledge: responding to legal and ethical challenges. In: Taylor, D.R.F., Lauriault, T. (Eds.), Developments in the Theory and Practice of Cybercartography: Applications and Indigenous Mapping. Elsevier, Amsterdam (Chapter 19).

Sinclair, M., 2014. What Do We Do about Th Legacy of Indian Residential Schools? Eleventh Annual SolKanee Lecture on Peace and Justice [speech]. Arthur V. Mauro Centre for Peace and Justice, University of Manitoba, Winnipeg. http://youtu.be/VYx99W7K40I.

Skelton, T., 2008. Research with children and young people: exploring the tensions between ethics, competence and participation. Children's Geographies 6, 21–36.

Strumińska-Kutra, M., 2016. Engaged scholarship: Steering between the risks of paternalism, opportunism, and paralysis. Organization 23 (6), 864–883.

Thrift, N., 2003. Practicing ethics. In: Pryke, M., Rose, G., Whatmore, S. (Eds.), Using Social Theory. Sage, London, pp. 105–121.

Truth and Reconciliation Commission (TRC), 2012. Truth and Reconciliation Commission of Canada Interim Report. The Commission, Winnipeg.

Valentine, G., 2005. Geography and ethics: moral geographies? Ethical commitment in research and teaching. Progress in Human Geography 29 (4), 483–487.

Ward, L., Barnes, M., 2016. Transforming practice with older people through an ethic of care. British Journal of Social Work 46, 906–922.

Wiles, R., 2015. What Are Qualitative Research Ethics. University of Southampton. https://www.youtube.com/watch?v=gHYg89Wg3-w.

Wood, L., 2017. The ethical implications of community-based research: a call to rethink current review board requirements. International Journal of Qualitative Method 16, 1–7.

Zeynep, M.Y., Fazli, M., Rahman, J., Farthing, R., 2016. Research ethics committees and participatory action research with young people: the politics of voice. Journal of Empirical Research on Human Research Ethics 11 (2), 122–128.

11

Broadening the cybercartographic research and education network: From Indian residential/boarding schools to Beltrami and back again

Stephanie Pyne*,[1], Tilly Laskey[2]

[1]Postdoctoral Research Fellow, Geomatics and Cartographic Research Centre (GCRC), Carleton University, Ottawa, ON, Canada; [2]Curator, Maine Historical Society, Portland, Maine, United States
*Corresponding author

O U T L I N E

11.1 Introduction

Taking a critical approach to tourism studies, Grimwood (2015) presents a conceptual approach to tourism in the context of a participant observation study in Canada's arctic that can encourage intercultural connections despite challenges. Mahrouse (2011) also reports on a participation observation case study — this time a 'Reality Tour' to South America — that warns of the risks that perpetuate colonialism and sociocultural, economic disparities. According to Mahrouse, increasing awareness of the ways tourism can contribute to the economic, cultural and environmental demise of the Global South has resulted in more people regarding 'conventional tourism as a gratuitous and crass form of exploitation and are opting for more socially responsible alternatives' (Mahrouse, 2011, p. 372). Mahrouse invokes the concept of mobility freedom to describe the economic privilege of tourists (mostly from the Global North). This socially conscious trend in tourism involves tourists being motivated to assuage themselves that their tourist dollars are helping the communities they visit, and at the same time to learn about and connect with 'the people' they visit — a motivation that is increasingly supported by commercial tour companies. While Mahrouse discusses the problematics of critical tourism within the context of the Global North-South dichotomy, her comments apply in other contexts ways that do not predefine the geopolitical context. Participant observation research is inherently reflexive since it involves documenting experience with personal notes, (sometimes) geographical coordinates and a reflective mentality, keeping in mind relationships with new peoples and lands encountered and the participant's evolving relationship with them and their own self.

While Mahrouse critiques commercial tour companies for failing to deliver on the authenticity they promise in 'promotional' messaging, Grimwood focuses on a case study in the Indigenous North, and critiques the eco-cultural tourist company Canoeroots for reifying the 'Canadian North as peripheral wilderness, landscape of loss and place to cross but not reside' (2015, 5). Seeking to transcend this, Grimwood introduces a series of "value-based metaphors" that may nurture cooperative spaces for just and sustainable Arctic tourism, [which] align with perspectives that identify "relations"as the fundamental basis of reality, knowledge and value (6)'. In contrast with Mahrouse, Grimwood appeals to the relational space (see Chapter 8) concept of 'moral terrain' as a vehicle to transcend colonial barriers to intercultural connections in his interpretation of the Thelon (River) field site:

'This geographic metaphor has been applied elsewhere for thinking across nature/culture and contextualizing ethics as everyday, corporeal and lived' (Figueroa and Waitt, 2008). According to Figueroa and Waitt (2008), moral terrains represent the web of values layered over places through discourses that establish normative practices and socioenvironmental belonging' (2015, 6–7).

In Grimwood's account, 'tourists' who have experience canoe camping over extended periods of time and paddling up and down rivers may have a greater connection with people whose ancestral history involves moving across the land at different times of the year than people who have not had a history of using their own physical energy in relation to the land. Similar to the 'tourists' referred to in Mahrouse's 'Feel Good' article, these people are economically 'privileged'; yet via their canoeing lifestyles, they have conditioned themselves to a lifestyle without the usual 'privileged comforts' — at least while on canoeing trips — which offers an opportunity for authentic connection with the people visited. To provide the reader with an idea of the tour, Grimwood provides the following description:

> The July 2010 route was staged with moderate current, no portages, several heritage sites, biome transitions and possible encounters with wildlife. Typical of the operator's other trips, the journey included nine clients (including myself), men and women, who convened just prior to boarding a chartered aircraft to the 'put-in'. Our descent involved paddling 7 hours each day in tandem canoes, travelling with all equipment and food, constructing new campsites each night and taking meals prepared by our guide. According to the operator, the trip was atypical in several ways. Clients were older (only one client was below the age of 60 years), but more skilled as canoeists, than most tour groups, and thick summer heat and moderate wind intensified the presence of the tundra's legendary insects. Moreover, while declining populations of caribou, wolf, muskoxen and grizzly bear within the Sanctuary had been observed, the operator noted our encounters with these mammals were particularly slight (2015,11).

'Emplacement' is another important concept invoked by Grimwood to convey that the tourism encounter in his case, which involved moving a considerable distance across space, was as simple as 'strangers getting to know one another' experiencing the same weather, terrain and challenges, with the 'locals' who were familiar with the route ultimately being appealed to as the authorities of the route:

> While the Thelon is promptly represented as spatial abstraction (e.g. 'wilderness' or 'frontier') whereby human inhabitation is deemed out-of-place, practices of canoeists and Inuit reflect the value in and of emplacement; that is, the condition of being in place, performing and encountering place or becoming emplaced […] People, things, ideas, meanings, histories and landscapes converged to enact place via strategically planned and/or spontaneous encounters (Bærenholdt and Granås, 2008). Within the Thelon, emplacement was expressed as practices of imagination/representation, practices of the body and practices of return (2015, 12).

Grimwood (2015) and Mahrouse (2011) provide examples of critical approaches to tourism studies as a backdrop for the discussion in this chapter of collaborative work to integrate cybercartographic research with teaching in the context of a Master's program in international tourism. This integrative work is part of an emerging international research and education network linked to the broader historical geography of Indian residential/boarding schools and involving disciplines such as art history, anthropology, geography and critical tourism studies. Similar to the strategy in Chapter 3 and elsewhere in the book, this chapter adopts the reflexive style of Bourdieu to examine the backgrounds and perspectives of the three originating actors in the emergence of this network: myself as a cybercartographer and member of GCRC at Carleton University in Ottawa, Canada; Tilly Laskey as an independent scholar specializing in American Indian art and culture and a curator at the Maine Historical Society, Portland USA; and Federica Burini, Associate Professor of Geography in the Department of Foreign Languages, Literatures and Cultures and the President of the Master's Degree Program in Planning and Management of Tourism Systems (PMTS) at the University of Bergamo, Italy. Burini's work with cartography acknowledges the pressing needs to develop forms of communication that optimize intercultural dialogue. Her research with the Diathesis Cartographic Laboratory aims to 'create a community mapping system able to communicate the social sense of place among communities living in a variety of territories' (Burini, 2012, 1), including rural Africa to urban Europe, primarily through a chorographic approach to cartography that is akin to Cybercartography.

The ongoing work discussed in this chapter provides a good example of ongoing transdisciplinary research that makes use of a cybercartographic approach at the intersection of theory and practice, research and education. Engaging with work in disciplines including anthropology, tourism studies, geography and cartography, this work also exemplifies the concepts of performativity, emergence and reflexivity. The chapter continues in Section 11.2 with a brief introduction to the people and relationships that generated the momentum for further research and education into Beltrami, a summary of a participatory mapping workshop held with Master's students in 2015 and 2016 that was linked to the Residential Schools Land Memory Mapping Project (RSLMMP), and subsequent teaching that involved the generation and use of cybercartographic mapping processes; then discusses the Beltrami Journey to America map in Section 11.3. Next, the chapter zooms in all the way with the story of Giacomo Costantino Beltrami from the material culture perspective of Tilly Laskey in Sections 11.4 and 11.5 (with screenshots from the participatory mapping referred to in Sections 11.2 and 11.3 to illustrate the intersectionality between Tilly's work, Cybercartography, research and education); while Sections 11.6 and 11.7 discuss two related examples of incorporating Cybercartography into teaching about the Residential Schools Legacy in a broader historical geographical context; and, Section 11.8 concludes with comments regarding the relationship between the residential/boarding schools legacy, and the emerging Beltrami Research Network.

11.2 Birth of a research and education network

This section provides a brief account of the emergence of the pilot participatory map module in the prototype Residential Schools Land Memory Atlas, the Exhibit Map, including related teaching, research and early stories of network development.

In March 2015, I accepted a guest professor position to teach half of a Master's level module on Intercultural Geography with a focus on Geographies of Tourism at the University of Bergamo, in Bergamo, Italy the following October. As I was making preparations for accommodations, I contacted my friend (and colleague through the Great Lakes Research Alliance for the Study of Aboriginal Arts and Cultures [GRASAC]), Adriana Greci Green, from Italy, to ask if she would be near Bergamo at that time. To my grateful surprise, this call turned out to be the beginning of some rewarding participatory mapping linked primarily to the technology and narrative dimensions of the RSLMMP.

Adriana was also a friend and colleague of Tilly Laskey who happened to be a material culture expert on Giacomo Costantino Beltrami, an atypical Italian lawyer who met and befriended Dakota and Anishinaabe peoples during his travels to North America in 1823. Tilly had researched the Beltrami Collection at the Museo di Scienze Naturali, a 5-minute walk from the university campus where I would be teaching.

Adriana connected me with Tilly, and before long I had an idea for a pilot participatory mapping exercise focussing on the Beltrami Exhibit in the Bergamo Museum, which intersected with research into 'mapping and touring' I had been doing in relation to planning for participatory mapping aspect of the RSLMMP (for example, Sprake, 2012). This idea tied in with the technological dimension of the RSLMMP, where the GCRC lab staff created a map module based on my specifications, which would later be populated by the students in the class; while the process of enacting the participatory mapping exercise with the students over a two-class workshop timeframe yielded observations and insights concerning the complexities associated with participatory mapping. I passed this idea by Federica Burini, the full-time professor for the course, and she enthusiastically accepted my proposal to engage the students in a participatory mapping exercise as part of my segment of the course. Not only was Federica from the same home town as Beltrami, but she too was aware of Beltrami's story and the museum collections, and had wanted to explore the idea of mapping with Beltrami further. Federica shared my decolonizing perspective toward the Beltrami collections and was interested in combining research with teaching in a way that incorporated mapping processes. The pilot participatory mapping exercises around the Exhibit Map Module allowed us to become more familiar with the logistical demands of on-the-ground participatory mapping and in terms of technology development.

It must be emphasized that none of this was planned, but was rather the result of a serendipitous encounter between myself and Adriana, which resulted in further connections through the GRASAC scholars network. The developments that have flowed from this encounter are consistent with the proposed RSLMMP objectives and outputs, yet at the same time, they reach in new directions, the most ambitious direction being the seeds for a future international mapping project. In addition, I was able to tell the students that they were contributing to the RSLMMP through their participatory mapping efforts, which included uploading photographic media and comments to the pilot Exhibit Module once our museum map tour was completed. This also required me to tell them about the project and about Residential Schools' historical geography as well. I was able to use the Residential School Map in the Lake Huron Treaty Atlas in my lectures and to tell students about how the RSLMMP was based on that previous iterative atlas development work. Now, we are learning from our experiences developing the Exhibit Module in our efforts to develop the Residential Schools Land Memory Atlas.

There have been many advantages to the participatory mapping exercise, both in terms of becoming more familiar with the logistical demands of on-the-ground participatory mapping and in terms of technology development. And, as it has turned out, Tilly and I have continued to work together each fall when I return to teach in Italy. To illustrate the output from these workshops, screenshots from the Exhibit Map and the Beltrami Journey to North America overlay are used in Section 11.5, which profiles a narrative by Tilly Laskey in relation to the material artefacts associated with Beltrami's journey (Figs. 11.2.1–11.2.3).

11.3 Mapping Beltrami's journey to America

In 2016, while I was preparing my 'Beltrami Lecture', I decided to supplement the Exhibit Map work with a map overlay of Beltrami's journey to North America, based on the account in his 1828 publication, *A pilgrimage in Europe and America leading to the discovery of the sources of the Mississippi and Bloody River with a description of the whole course of the former and of the Ohio*, and supporting documentation such as 'A Preview of the Beltrami Collection with a note on North American Ethnographic Material in Italian Museums' by Marino (1986); lecture notes that cover the ground of Tilly's essay in Section 11.5, which include images from Tilly's work; and, samples of student's photo essays from a related project during the course. Sample screenshots from this map, and the Exhibit Map referred to in Section 11.2, are included to illustrate points in the essay by Tilly Laskey in Section 11.5; but first, a little background on Tilly and the evolution of her interest in Beltrami's story (Fig. 11.3.1).

FIGURE 11.2.1 Screenshot from the Exhibit Map profiling an image of the students from the 2015 class standing in the middle of the museum during the participatory mapping phase involving interpretative photography.

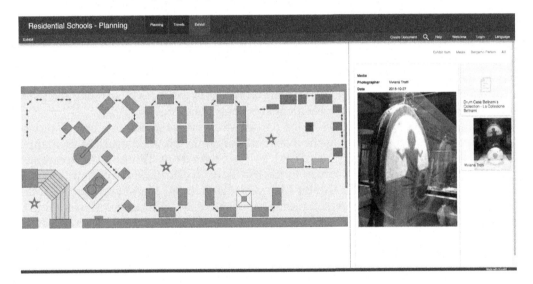

FIGURE 11.2.2 Screenshot from the Exhibit Map showing the current location of this drum in the Museo di Scienze Naturali in the dark blue box in the central window, with a photograph of the drum taken during the first class participatory mapping workshop in 2015. *Photograph taken by Viviana Trotti.*

11.4 Tilly Laskey and Beltrami

Tilly Laskey is a museum curator specializing in Indigenous art and culture whose academic training includes art history and anthropology. She received her MS in Museum Studies from the University of Colorado in 1999 and has curated at nationally recognized museums across the United States in Maine, Minnesota and South Dakota. Laskey was awarded a 2011 National Endowment for the Humanities Fellowship to study and document the Beltrami collections in Italy.

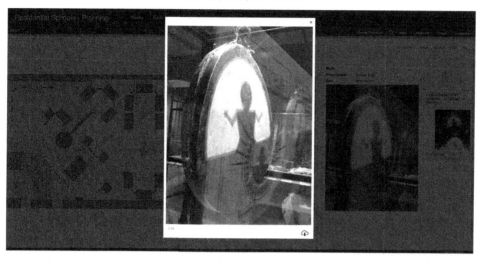

FIGURE 11.2.3 Screenshot of a darkened map display in the Exhibit profiling a photograph of the drum taken during the first class participatory mapping workshop in 2015. *Photograph taken by Viviana Trotti.*

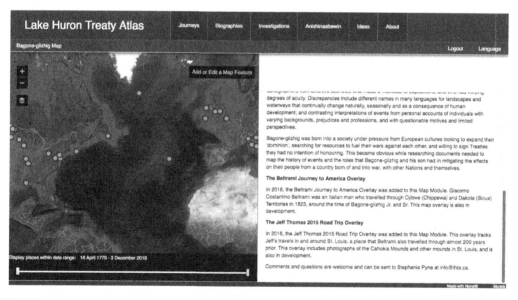

FIGURE 11.3.1 Screenshot from the Bagone-giizhig Map of the Lake Huron Treaty Atlas showing the points associated with Beltrami's life and journey to America in the central window, with introductory text describing the interrelated overlays in the right side panel.

Tilly first encountered Giacomo Costantino Beltrami in 2008, while she was curating the Bishop Whipple American Indian collection at the Science Museum of Minnesota. During a tour of the Whipple collection with attorney Fritz Knaak, she mentioned that the items were some of the oldest ethnographic collections from Minnesota, dating to about 1850. Mr. Knaak recalled that as an exchange student in Ancona, Italy, during the 1970s, he had been given a private tour of American Indian items in the Beltrami collections because — as with the items in the collection — he was from Minnesota. He also told Tilly that he thought the collections dated to the 1820s. She was incredulous, and her curiosity was piqued. A few months later, Tilly was on a flight to Italy to visit the two museums that house the Beltrami collections. She connected with the Museo Beltrami in Filottrano and with Marco Valle, the director of the Natural History Museum in Bergamo through Cesare Marino, an Italian who is also an American Indian Specialist working for the Smithsonian Institute and the foremost authority on the Beltrami collections.

Tilly kindly summarized the evolution of her interest in Beltrami in the following reflection:

It was during that trip in 2009 that I fell in love with the Beltrami collection, Giacomo Beltrami's larger-than-life personality, the caretakers of Beltrami's legacy, and Italy. I was anxious to study, digitize, and assess the pieces and to work with descendant and Indigenous source communities to understand more about the material culture pieces Beltrami collected. The collection is particularly important because it traverses the time when Native people were in complete control of the territory now known as Minnesota, and contains some of the first tourist-related art from that region. I received a National Endowment for the Humanities fellowship in 2011–12, allowing me to take a year-long sabbatical from my job at the Science Museum of Minnesota. I spent half of my time surveying similar collections relating to the Dakota, Anishinaabe, Menominee, and Cree tribes in America and Europe. The other half of my time was spent in Italy, examining and digitizing the Beltrami collections.

In addition to the material culture, I was interested in Beltrami's written accounts of his journey. I realized that scholars and casual historians had portrayed him as a buffoon and footnote in both Italy and Minnesota's history. The archives held at the Biblioteca Angelo Mai and my analysis of Beltrami's narrative indicate he had a keen intellect and foresaw conflict between Native people, trading companies, and the United States government decades prior to their occurrence — just as he foresaw issues in Italian unification. Beltrami was an important player in Italian social and cultural movement, Il Risorgimento, and it was his criticism of the Papal Government that caused him to flee from Italy to Europe and later to America. Beltrami's motives for leaving Italy and his failure in working toward the unification of Italian states are critical for understanding his need to explore and to do something "monumental". I realized this aspect of Beltrami's character had not been sufficiently explored, and required more analysis and research.

My NEH Fellowship allowed me to attend and present for the first time at the Native American and Indigenous Studies Association (NAISA) in May 2012, and in 2014 at the GRASAC's annual meeting in Brantford, Ontario, Canada. At these conferences, I began conversations with other scholars about the project, including Adriana Greci-Green who later connected me to Stephanie Pyne. Since the moment we began communicating, Stephanie has enthusiastically collaborated on this project by mapping my research into the layers of her work in Cybercartography; thus providing an opportunity to present research in a different conceptual way, while also providing an alternative method for source communities like the Dakota and Anishinaabe to connect land-based knowledge to a material culture item, or a historical point in time. It's very exciting and we are just in the beginning of our work.

Tilly's material culture account of Beltrami in the following Section 11.5 provides another example of 'emplacement' (Grimwood, 2015), in addition to providing some historical geographical context for this chapter, and a place to profile sample screenshots from the Exhibit Map Module participatory mapping with students in 2015 and 2016 (Section 11.2), and the Beltrami Journey to America map overlay produced in 2016 (Section 11.3).

11.5 Minnesota's first tourist? Giacomo Costantino Beltrami's 'transatlantic promenades' of 1823 (by Tilly Laskey): Introduction

Giacomo Costantino Beltrami (1779–1855) might have been the first tourist to visit the Dakota and Anishinaabe homelands, the region later known as Minnesota. Beltrami's claim to fame was his declaration that he found the source of the Mississippi River, and his penchant for carrying a red umbrella. He was a wealthy Italian explorer who became dually obsessed with Indian culture and mapping the source of the Mississippi River. While it is true that soldiers, priests, traders and diplomats certainly collected and preserved Minnesota-related heritage items from an early time, it is disputable how many visitors wandered into 'Indian Country' for completely personal reasons – or tourism – prior to 1823.

During his trip, Beltrami collected hundreds of items from Woodlands and Plains Nations. The collection that remains today, 110 Native items, paired with Beltrami's published narrative, are rare resources for learning more about Indigenous people and cultures during a period when they were firmly in the seat of power, and were living in what Michael Witgen describes as the 'intact, and unconquered Indian social world in the heartland of North America' (Witgen, 2012, 3). The items in the collection are mainly from the Dakota, Anishinaabeg and Cree people, with others possibly from Ho-Chunk and Menominee tribes. These heritage items are important because they were created at a time when culture was changing rapidly in Minnesota, with the construction of Fort St. Anthony (current-day Fort Snelling) and a rise in activity and competition between the Hudson's Bay Company and American Fur Company trading companies. These rare and early pieces illustrate negotiations between cultures, both within distinct Native tribes and with Europeans. The Beltrami items are a portal to understanding Minnesota before Treaties with the United States drastically altered tribal boundaries, prior to the influx of Euro-American settlers, and 39 years before complex cultural conflicts led to the Dakota/US War (Fig. 11.5.1).

Rather than collecting through middlemen, as many travellers did, Beltrami collected directly from Native people. He stated that he collected to preserve his 'Indian curiosities as a memorial and trophy of my labours in these my transatlantic promenades' (Beltrami, 1828, 403).

In 1823, a small production of souvenirs existed around the Fort areas and the trading posts along the Red River, where Beltrami likely purchased the nesting birchbark baskets, model canoe and decorated makuks in the collection. At Fort St. Anthony, Beltrami noted his commission of a miniature birchbark canoe, 'I have got them to make me a model, and to give me a specimen of their tar, which forms part of my little collection of Indian curiosities' (1828, 226) (Fig. 11.5.2)

While later generations of tourists desired nostalgic experiences and to gaze at Native people from a safe distance, Beltrami chose to charge into their homes. Like many travellers,

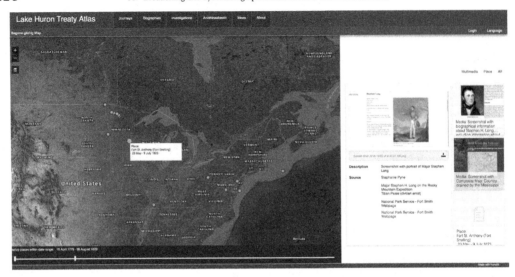

FIGURE 11.5.1 Screenshot from Beltrami Journey to America overlay in the Bagone-giizhig Map of the Lake Huron Treaties Atlas (referenced in Section 11.3) showing the location for Fort St. Anthony in the central window, and associated media related to Major Long in the right side panel.

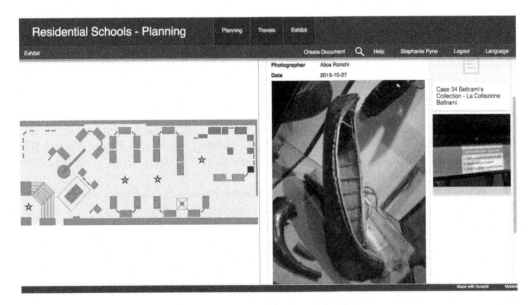

FIGURE 11.5.2 Screenshot from the Exhibit Map referred to in Section 11.2, showing the current location of the canoe in the Museo di Scienze Naturali in the central window, with a photograph of the canoe taken during the second class participatory mapping workshop in 2016. *Photograph taken by Alice Ronci.*

Beltrami was looking for exotica, proof of his journey and friendships with Indigenous people, to create a cabinet of curiosities in Italy, and as a mnemonic device to remember his journey.

11.5.1 Who was Giacomo Costantino Beltrami?

Beltrami was transformed by his journey, and three portraits illustrate this point: a circa 1818 ivory miniature portrait of Beltrami, the engraving which accompanied Beltrami's published narrative and a posthumous portrait by Enrico Scuri executed in 1857. Examining these portraits, the larger question arises of how an affluent Italian lawyer ended up in a birchbark canoe, wearing a captain's coat, in Leech Lake, Minnesota in 1823 (Fig. 11.5.1.1).

Beltrami served as a soldier, lawyer and magistrate during his lifetime. By all accounts, Beltrami was a larger-than-life character, always seeking adventure. At nearly six feet tall with corkscrew curly hair and fit from a lifetime of mountaineering, his imposing physical characteristics seemed to match his outspoken personality and bold behaviour, leading some to characterize Beltrami as a buffoon. While he may have been opinionated, impetuous, and overly romantic, Beltrami was no fool. He was a well-educated man who served in the Napoleonic military, and a cultured magistrate whose connections included the de' Medici family.

Beltrami began his odyssey at the age of 42 years, literally a man without a country. He wandered Europe for 1 year, surveying ancient sites of interest. With seemingly little forethought, he booked passage on a small sailboat in Liverpool, bound for America on 3 October 1822. The ship encountered numerous storms, was blown off course and Beltrami became sick — nearly dying according to his account. They landed in Philadelphia on 30 December 1822. A 'normal' crossing should have taken 21–29 days. Beltrami travelled immediately to Washington DC. He was interested in seeing America on his way to New Orleans, and afterwards

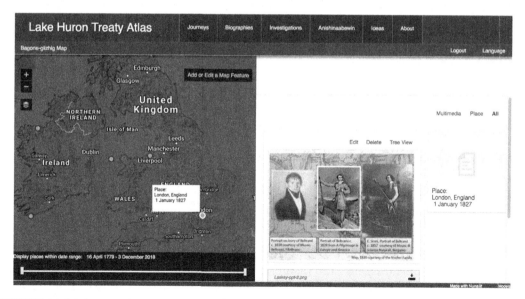

FIGURE 11.5.1.1   Screenshot from Beltrami Journey to America overlay in the Bagone-giizhig Map of the Lake Huron Treaties Atlas showing the location for Liverpool, England, from which Beltrami began his overseas trip in the central window, with a composite image in the right side panel displaying portraits of Beltrami pre and post journey.

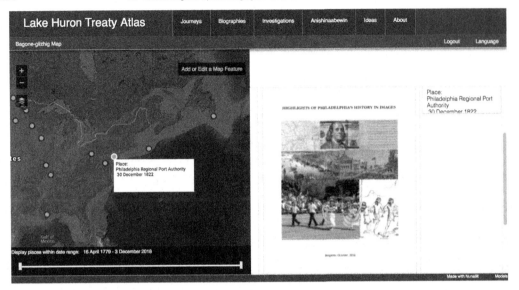

FIGURE 11.5.1.2 Screenshot from Beltrami Journey to America overlay in the Bagone-giizhig Map of the Lake Huron Treaties Atlas profiling the location for Philadelphia among other locations noted in Beltrami's writings in the central window, with a thumbnail of photo essay prepared by student, Polina Baskakova as part of the course requirements referred to in Section 11.3.

planned to travel to Mexico. He took a coach through Pennsylvania, and boarded a steamboat in Pittsburgh, which brought him 981 miles to the confluence of the Ohio and Mississippi Rivers, current-day Cairo, Illinois (Fig. 11.5.1.2).

Instead of heading south to New Orleans, Beltrami headed north on the Mississippi. Fellow passengers included Superintendent of Indian Affairs, Gen. William Clark, and Maj. Lawrence Taliaffero, the Indian Agent at Fort St. Anthony in current-day Minnesota. Clark's stories of his expedition with Meriwether Lewis (the Lewis and Clark Corps of Discovery Expedition, 1804–06) and an 'Indian room' in Clark's office in St. Louis fascinated Beltrami, who, at this point had yet to actually see a Native person. Beltrami was captivated by the scenery, cultures, and mostly by the Mississippi River. In his *Pilgrimage*, Beltrami explained his touristic reasons for changing his plans:

> A river of vast extent, of a majesty which is difficult to conceive; a country presenting extraordinary features at every step; a race of men entirely different from those of Europe; afford abundance of new and important subjects for philosophical meditation gratify the curiosity with the most agreeable surprise, and divert the afflicted mind from the subject of its regrets. I have felt every impression which so novel a scene is capable of producing (1828, 126).

Beltrami traveled upon the first steamboat to navigate the Upper Mississippi, the *Virginia*, in May 1823, with Indian Agent Maj. Taliaferro. The American landscape, rivers and people revitalized Beltrami. While traveling up the Mississippi River he reflected, 'Every object was as new to my imagination as to my eye' (1828, 151).

Beltrami began his Minnesota journey at Fort St. Anthony (established in 1819), present-day Fort Snelling. He stayed at the Fort for a total of 2 months, attended treaty negotiations with Maj. Taliaferro, and collected from Dakota and Anishinaabe people. In June 1823, Beltrami convinced General Josiah Snelling to allow him join the US mission that established the 49th parallel confirming the border between America and Canada. Along the way, Beltrami tracked the rivers, natural features and tribal territories. Beltrami and expedition leader Major Stephen Long disagreed during the journey, and in Pembina — two miles south of the 49th parallel and the border that Long established — they parted ways.

Beltrami left with three Anishinaabe guides to search for the source of the Mississippi River — his covert goal all along. Beltrami sought out the source of the Mississippi for his own quixotic reasons, saying, 'If I can survey the whole of its course, I will endeavour, as far as my attention and knowledge permit, to fill up this chasm in history and geography' (1828, 130). Never mind that Indigenous people in the area had their own understanding of these watersheds, and his Anishinaabe translator guided Beltrami to the lake, which he later erroneously deemed the true source of the Mississippi naming it Lake Giulia (Julia), after his friend, Giulia de' Medici Spada.

Although Beltrami lost his rights to the Euro-American 'discovery' of the sources of the Mississippi River to Henry Schoolcraft in 1832 (who was guided to Lake Itasca by Native people), Beltrami's legacy as a collector and documenter of American Indian culture far outlasts that honour.

Beltrami made cryptic notations in small travel notebooks during his trip, which are in the collections of the Biblioteca Angelo Mai, Bergamo. These notes formed the basis for his 1824 publication in French, later translated and published in English as *A pilgrimage in Europe and America leading to the discovery of the sources of the Mississippi and Bloody River with a description of the whole course of the former and of the Ohio*, in 1828.

Factors contributing to Beltrami's reputation for eccentricity included his red umbrella. After leaving the Long Expedition, Beltrami's group was ambushed by Dakota men, leaving one of his guides wounded. The guides tried to convince Beltrami to walk a shorter route to Red Lake with them, but he refused to leave the canoe and his collections. Alone and unable to master paddling solo, Beltrami dragged the canoe after him and propped his umbrella in the stern—both, he reasoned, to make onlookers curious before they shot at him and to keep his collections dry.

Although this story reeks of legend, Beltrami's red umbrella exists, and is part of the collections of the Museo di Scienze Naturali in Bergamo. The red umbrella apparently saved Beltrami on numerous occasions. Upon returning to Fort St. Anthony, Dakota men told him that, [he] 'had acted judiciously in making myself known to them by means of my umbrella signal, as I should otherwise have experienced a shower of balls as well as arrows' (1828, 484).

Diplomats or translators had moderated Beltrami's first interactions with Natives, but after he left the Long Expedition, his survival and success hinged on Indigenous people. He was generously accepted into their communities, sheltered, fed and even clothed. Part of Beltrami's physical transformation included his clothing and his encounters with the Pillager Indians of Leech Lake, Minnesota, a woman he calls 'Woascita', and her father, Pokeskononepe, also known as Cloudy Weather or Big Cloud. Two pieces of clothing, Beltrami's coat — often referred to as a Captains coat — and his birchbark top hat, exemplify the transcultural experiences occurring between Native people and those of European descent.

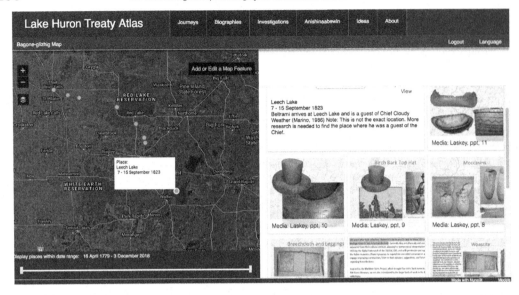

FIGURE 11.5.1.3 Screenshot from Beltrami Journey to America overlay in the Bagone-giizhig Map of the Lake Huron Treaties Atlas profiling the location for Leech Lake, which Beltrami visited from September 7 to 15, 1823 in the central window, with thumbnails of related material artefacts from Laskey PowerPoint presentation slides (2014) in the right side panel.

Describing his outfit and his interaction with Woascita, Beltrami states:

> My head was covered with the bark of a tree, formed into the shape of a hat and sewed with threads of bark; and shoes, a coat, and pantaloons, such as are used by Canadians in the Indian territories, and formed of original [moose] skins sewed together by thread made of the muscles of that animal, completed the grotesque appearance of my person. I am indebted for my new wardrobe to the fair Woascita, who had compassion on the nakedness to which the thorns and brambles of the forest had reduced me. ... The gift therefore is valuable in itself, and as such I shall preserve it with care, but still more as a memorial of regard and friendship (1828, 481) (Fig. 11.5.1.3).

If accurate, Beltrami's artistic attribution to Woascita is one of the earliest European accounts linking objects to a specific and named female Anishinaabe artist in the Northwest Territories. Beltrami's new outfit transformed him temporarily, and upon reaching Fort St. Anthony after being away for four and a half months, he said: 'On the arrival of the flotilla all the officers hastened down to enquire about me. They were answered by the supposed dead man himself. While replying to their kind questions I divested myself of the skin covering which I had on, in the disguise of an Indian' (1828, 482).

Captain's coats are a hybrid genre, based on a European cloth military cut, but are made with indigenously based materials and artistic sensibilities. David Penney explains the coats as, 'negotiations from European producer to Native consumer through the fur trade, then from Native producer back to European American consumer via trade in souvenir market arts' (Penney, 2007) (Fig. 11.5.1.4).

Beltrami was in the Leech Lake area (see Fig. 11.5.1.3) for approximately 15 days. While not impossible, it is unlikely Woascita was able to create and embellish five pieces of clothing, including the painting of hides and creating porcupine quill and moose hair embroidery. It

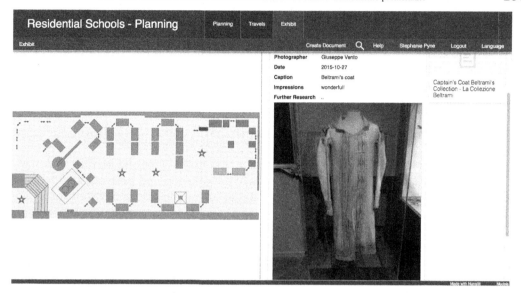

FIGURE 11.5.1.4   Screenshot from the Exhibit Map referred to in Section 11.3, showing the current location of the captain's coat in the Museo di Science Naturale in the central window, with a photograph of the captain's coat taken during the first class participatory mapping workshop in 2015. *Photograph taken by Giuseppe Vento.*

is, however, probable that two pair of moccasins were specifically made for Beltrami, as evidenced by their size, approximating a US size 12 shoe. Both pairs exhibit signs of use. Beltrami was appalled at Native peoples' lack of daily head coverings, noting, 'They wear a covering round the loins, all the rest of the body, even the head, is naked whether it rains, hails or freezes' (1824, 143).

Somewhere between Pembina and Leech Lake, Beltrami procured a birchbark top hat. Like his coat, the hat illustrates that cultural negotiation functions two ways. It is unclear whether Beltrami commissioned his top hat or if some enterprising Native person riffed the European design into this wonderful birchbark creation.

11.5.2 Minnesota's first tourist? Giacomo Costantino Beltrami's 'transatlantic promenades' of 1823 (by Tilly Laskey): Conclusion

Beltrami's Pilgrimage is a classic Enlightenment work, where he set out to classify, know, explore and order the world, documenting it for future visitors. Although *Pilgrimage* contains difficult 19th-century language and is vehemently against organized religions, Beltrami's writing is tempered with humanistic and feminist perspectives, and a postmodern viewpoint of cultures. Beltrami noted distinctions between Native Nations, named individual people, and respected Indigenously named places by recording them in Ojibwe and Dakota languages. He scrutinized gender roles, discussed women's work, their appearance, clothing and activities more so than most 20th-century ethnologists did.

Beltrami's *Pilgrimage* was not well received by Americans. He expected to be celebrated, but instead he was heavily criticized. *Pilgrimage* refuted ideas of Indians as dangerous people and illustrated a place full of vibrant and sophisticated civilizations, which opposed American propaganda extolling a disappearing Indian and empty frontier.

Beltrami returned to Italy 16 years after his American adventure, but the affronts to his honour and claims that he fabricated his journey plagued him to his dying day. He spent his last years sequestered in one room of his palazzo, requesting to be buried in an unmarked grave. Against his wishes the people of Filottrano interned him in a tomb, of all places, inside the local Catholic church.

11.6 Mapping the Mississippi: 2018 class workshop

Inspired by my presentation of Laskey's account of Beltrami's life and journey to America, the 2018 Master's class and I devoted two classes to a workshop and presentations revolving around Beltrami's original journey, and expanding to include locations in seven states along the Mississippi River. This exercise continued the precedent of engaging students in mapping workshops, this time with the Mississippi Tourism Mapping Workshop, which was also intended to explore the extent of Native American tourism initiatives along the Mississippi River. To encourage participation, each student adopted a website to investigate from the list provided on a national tourism website, and contributed a one-slide PowerPoint critical summary for a group presentation to the class. I mapped the locations for the venues referred to in the sites as a new overlay in the Bagone-Giizhig Map, which also included the Beltrami overlay. There is potential to upload the students' slides to the points on the Mississippi River overly, although this has not yet been done. In addition, further research remains to be completed on tourist venues, sites and activities not included on the national site; and on universities, colleges, cultural centres and even boarding schools (Fig. 11.6.1).

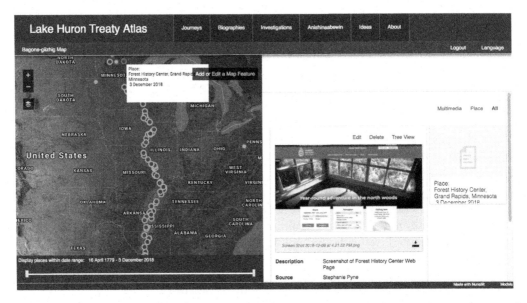

FIGURE 11.6.1   Screenshot from the Bagone-giizhig Map of the Lake Huron Treaty Atlas showing the points for touristic centres along the Mississippi River (dark purple dots), which superimpose many of the points on Beltrami's Journey to America (light purple dots) in the central map window, with the point for the forest history centre chosen and associated media displayed in the right side panel.

11.7 Transposing Beltrami's Journey to America onto the Residential Schools Map

This map work follows the template established in Chapter 3, which involved a similar method with respect to transposing the Jeff Thomas, 2015 road trip onto the Residential Schools Map.

Again, focussing on place as an organizing factor in an experimental mapping approach at the intersection of written narrative and more conventional forms of mapping, I began to transpose the Minnesota leg of Beltrami's journey from the Bagone-Giizhig Map onto the Residential Schools Map. This allowed me to see the boarding schools that had been established in the region Beltrami travelled through. It intrigued me to think that Beltrami may have met Dakota and Anishinaabe relatives of those who later attended boarding school. The action of transposing Beltrami's story onto the Residential Schools Map is an example of broadening the historical geography of boarding schools to include related knowledge and perspectives (Fig. 11.7.1).

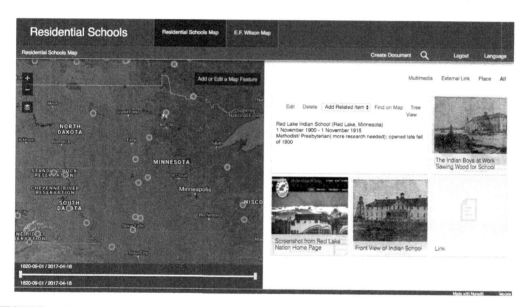

FIGURE 11.7.1   Screenshot from the Residential Schools Map showing the points associated with the portion of Beltrami's Journey going primarily through North Dakota and Minnesota (blue triangles), superimposed onto the points for Indian Boarding Schools in those states and surrounding regions in the central window, with thumbnails showing the school in 1904 and an image from the Red Lake Nation website in the right side panel.

11.8 Discussion and conclusion: Research, teaching and the emergence of the Beltrami Research network project

The interest in learning about Beltrami's life and journey to America is linked to the broader historical geography of the Indian Residential/Boarding schools legacy. Beltrami visited with the people whose descendants would later be corralled into schools who the US and Canadian governments attempted to assimilate into the dominant White culture (see Chapters 5 and 6). Engaging in the mapping exercise of transposing a segment of Beltrami's journey onto the Residential Schools Map provides a quick snapshot of colonial activities following Beltrami's tour. Engaging in this process is a way of including the perspective of 'Minnesota's first tourist' in the story of the prehistory and emergence of boarding schools in the United States. As I was engaging in this transposition, I wondered if any of the descendants of the people Beltrami met went to Canadian residential schools, and how many, if any went to the Red Lake Indian School. The answers to these questions could be relevant to Tilly Laskey's quest to find the descendants of those who created the Beltrami Collections.

Over the past 4 years, cybercartographic mapping exercises involving Beltrami's story have been carried out in conjunction with lectures that include a talk on critical tourism studies, with a focus on the articles by Grimwood (2015) and Mahrouse (2011) mentioned above, which discuss socially just approaches to tourism that involve the establishment of personal connections and common experiences. In this regard, I characterized Grimwood's relational space concept of 'emplacement', as being 'as simple as "strangers getting to know one another" experiencing the same weather, terrain and challenges, with the "locals" who were familiar with the route ultimately being appealed to as the authorities of the route'. In our course we began to develop a method of 'cybercartographic emplacement' through the interactivity in each assignment or learning activity. For example, during the first year of the participatory mapping at the museum, I spent some time talking about Anishinaabe world view, and shared knowledge about the four medicines, including sage and tobacco. Before our trip to the museum, we acknowledged the Ancestors related to the material objects we were about to go and view; and before entering the museum, we each put down tobacco outside the doors, again to acknowledge and honour the Ancestors.

The story of how the RSLMMP became involved with the Beltrami story is another example of emergence in cybercartographic research, including the emergence of an idea for broadening research to integrate it with teaching. This collaborative approach to teaching and research has involved Cybercartography in the conception and development of an idea for international research project with potential partners in North America and Europe. As Laskey has contended, Beltrami was an unconventional character for his times. For example, he documented his encounters with and impressions of Dakota and Anishinaabe peoples in a far more favourable and realistic light than his contemporaries were in the habit of doing. Since first learning of Beltrami's character and story from Tilly Laskey in 2015 and soon after connecting with Federica Burini who shares my interest in combining mapping, research with teaching, we began discussing possibilities for developing research/teaching exchange and internship opportunities linked to the emerging Beltrami Research and Education Network.

The emerging Beltrami Research and Education Network project is one example of new research that began near the beginning of the Residential Schools Land Memory Mapping Project (RSLMMP), which in turn emerged from work on the cybercartographic Lake Huron Treaty Atlas (Pyne, 2013; Pyne and Taylor, 2015). The impetus for this project lies at the intersection of the ongoing work of (1) those taking a relational, decolonizing approach to historical material artefacts, including Tilly Laskey, (2) those representing institutional concerns, such as the Museo di Scienze Naturali in Bergamo, (3) academics including Anton Treuer (Bemidji University) and Alan Corbiere (York University), (4) Dakota and Anishinaabe descendants of those who gave or sold artefacts to Beltrami, (5) representatives of American institutions such as the Smithsonian, including Cesare Marino, (6) the international tourism program (Planning and Management of Tourism Systems) at the University of Bergamo, (7) those concerned with applying cartographic networking (and related) solutions to interdisciplinary research, such as the Geomatics and Cartographic Research Centre (GCRC, Carleton University), and (8) GRASAC, Carleton University, which includes digital records associated with the collection in Bergamo, which Laskey worked with in 2011–12.

This emerging project broadly follows the historical geographical logic of Beltrami, who kept a journal and made a map of his river journey from Philadelphia to Pembina, North Dakota and beyond. It is envisaged to begin with a cybercartographic project to create a Beltrami Atlas comprised of an initial set of pilot maps, including a research network map. This map would be reflexive in nature, reflecting aspects of the various stages of individual projects related to areas including, material culture, repatriation, and community economic development via tourism. The inspiration to include tourism in the scope of the project is in part due to what is known about Beltrami, one of the first 'tourists' into Dakota and Anishinaabe territories, in addition to being the result of ongoing development in work with colleagues from the International Tourism Program at the University of Bergamo. There is potential in this regard to include Native communities in tourism development, including research tourism. There is also potential to incorporate mapping strategies in what could be a rather diverse, distributed set of projects linked to the narrative logic of Beltrami's journey, which included travels up the Mississippi and Missouri Rivers. With the underlying and ongoing spirit and motivation of the project lying in the ethos of good relationships exhibited by Beltrami in his encounters with Dakota and Anishinaabe people, this project aims to contribute to education, intercultural reconciliation and community development.

This chapter has provided another example of 'relationship' and network building that has been made possible through cybercartographic practice. In this case intersecting interests include intercultural reconciliation, museology and conservation studies, tourism studies, Indigenous studies, and historical geography. The process of bringing these interests together has involved mapping processes, theory and practice and research and education. In the future, we would like to combine cybercartographic mapping processes initiated thus far with the chorographic approach practised by Burini — a critical approach looking at maps in terms of being reflexive semiotic instruments. In an iterative manner, we are continuing to slowly grow the Beltrami Research and Education Network, with the broad aim of contributing to intercontinental intercultural reconciliation. Engaging in reflexive and holistic processes, we continue to engage in research and education that reflects upon one another and contributes to society.

References

Beltrami, J.C., 1828. A Pilgrimage in Europe and America Leading to the Discovery of the Sources of the Mississippi and Bloody River With a Description of the Whole Course of the Former and of the Ohio, vols. I and II. Hunt and Clark, London.

Burini, F., 2012. Community Mapping for Intercultural Dialogue. EspacesTemps.net. https://www.espacestemps.net/en/articles/community-mapping-for-intercultural-dialogue/.

Figueroa, R.M., Waitt, G., 2008. Cracks in the mirror: (Un)covering the moral terrains of environmental justice at Uluru-Kata Tjuta National Park. Ethics, Place & Environment 11 (3), 327–349.

Grimwood, B., 2015. Advancing tourism's moral morphology: relational metaphors for just and sustainable arctic tourism. Tourist Studies 15 (1), 3–26.

Mahrouse, G., 2011. Feel-good tourism: an ethical option for socially-conscious Westerners? ACME: An International E-Journal for Critical Geographies 10 (3), 372–391.

Marino, C., 1986. A Preview of the Beltrami Collection with a Note on North American Ethnographic Material in Italian Museums.

Penney, D., 2007. "Captain's coats" in three centuries of Woodlands Indian art: a collection of essays. In: King, J.C.H., Feest, C.F. (Eds.), European Review of Native American Studies Monographs. ZKF Publishers, Altenstadt no. 3.

Pyne, S., 2013. Sound of the Drum, Energy of the Dance: Making the Lake Huron Treaty Atlas the Anishinaabe Way. Unpublished PhD Dissertation. Carleton University, Ottawa.

Pyne, S., Taylor, D.R.F., 2015. Cybercartography, transitional justice and the Residential Schools Legacy. Geomatica 69 (1), 173–187.

Sprake, J., 2012. Learning Through Touring: Mobilising Learners and Touring Technologies to Creatively Explore the Built Environment. Sense Publishers, Rotterdam, Boston.

Witgen, M., 2012. An Infinity of Nations: How the Native New World Shaped Early North America. University of Pennsylvania Press, Philadelphia.

12

Conclusion: building awareness to bridge relationships

Stephanie Pyne*,[1], D.R. Fraser Taylor[2]

[1]Postdoctoral Research Fellow, Geomatics and Cartographic Research Centre (GCRC), Carleton University, Ottawa, ON, Canada; [2]Chancellor's Distinguished Research Professor of International Affairs, Geography and Environmental Studies, and Director, Geomatics and Cartographic Research Centre (GCRC), Carleton University, Ottawa, ON, Canada

*Corresponding author

OUTLINE

12.1 Introductory words

At the launch of the Cybercartographic Atlas of Indigenous Perspectives and Knowledge (Great Lakes — St. Lawrence Region), we offered Grandfather Commanda tobacco for his prayers for mapping our ongoing work in Cybercartography and reconciliation, which led to the development of the Lake Huron Treaty Atlas, and our current focus on residential schools (Pyne, 2013). These prayers helped shape the two-pronged Anishinaabe and critical academic approach to gathering, mapping and sharing stories related to the ongoing relationships between Turtle Island's first peoples and others (Johnson et al., 2006; Pyne, 2013). This approach can be understood in terms of reflexivity; talk, templates and tradition; implicit and explicit forms of cartography; and intersectionality with a common commitment to social, spatial and other forms of justice reflecting an ethics of care. It features values such as inclusivity, transparency and a commitment to ontogenesis or emergence, which is reflected in critical approaches to both cartography and archival studies (Kitchin and Dodge, 2007; Kitchin, 2008; Turnbull, 2007; Upward et al., 2011). When it comes to engaging with mapping in a way that involves Indigenous peoples, the challenges of "remapping" include maintaining an ongoing attentiveness to avoiding misappropriations of knowledge, understandings and perspectives: 'The problem that faces Indigenous peoples worldwide is to find a way to incorporate Western [geospatial technologies] and cartographic multimedia while minimizing the mistranslations, recolonizations, and assimilations of conventional technoscience' (Pearce and Louis, 2007, p. 123). A primary critical concern is whether or not cartography is capable of meaningfully conveying such things as experience, Indigenous perspectives and knowledge, and critical academic approaches to the status quo (Johnson et al., 2006; Johnson and Murton, 2007; Palmer, 2009, 2012; Palmer et al., 2009; Turnbull, 2007; Pyne and Taylor, 2012).

A number of new issues are arising in this digital age, which has brought with it both opportunities and challenges for democratization and social justice. With a new capacity for the creation, storage and sharing of a vast amount of information, the term 'archives' has become more and more ubiquitous, and is increasingly being 'practiced' in the spirit of justice with an emphasis on the value of inclusion. As the 'preeminent goal of transitional justice' (Crocker, 2015), thick conceptions of reconciliation also require effective public deliberation in order to be achieved. This involves 'a person-to-person relation characterized by the African concept of ubuntu: forgiveness, mercy (rather than justice), a shared moral vision, mutual healing, and social harmony' (Crocker, 2015, p. 5). Noting the difficulties associated with achieving maximum reconciliation and the desire to surpass minimal reconciliation, Crocker offers three intermediary degrees of reconciliation, which offer some hope for getting past 'the depth of hostility between past opponents, as well as both moral and practical objections to coercing mutuality, contrition, or the granting of forgiveness': 'mutual security', 'deliberative reciprocity' and 'historical reconciliation' (Crocker, 2015, p. 5).

The chapters in this book can be seen to contribute to both the 'deliberative reciprocity' and 'historical' phases of reconciliation, where the 'deliberative reciprocity' phase of reconciliation involves 'exchange of proposals, reasons, and criticism where the parties try to solve a concrete problem, make a policy choice together, or establish a rule under which they agree to live'(Crocker, 2015, p. 6); and 'historical reconciliation' involves 'people reconciling with their common past via multiethnic (or multinational) and multidisciplinary historical investigation concerning the causes, nature, and consequences of past wrongdoing. Former enemies can settle accounts with the past and to some extent with each other in and through forging at least the general outline of a common narrative about what happened and why' (Crocker, 2015, p. 7).

In the case of reconciliation in the context of the Residential Schools Legacy in Canada, the Truth and Reconciliation Commission of Canada has paid some attention to the goals of truth by including previously excluded perspectives, in addition to dealing with compensation. Each of these areas has problematic aspects, which are being addressed in situ by a variety of intersecting approaches reflecting common themes and interests (see Chapter 1) (Cooper-Bolam, 2014, 2018; Reimer and Bombay, 2010). It would seem that — although contested in many respects — the processes that have led to the work of Canada's Truth and Reconciliation Commission on Indian Residential Schools and the publication of its final report have contributed to progress at various nodes of David Crocker's 'reconciliation continuum' (see Chapter 1). This includes work reported on in this volume, which contributes to both 'democratic deliberation' and 'historical reconciliation' (Crocker, 2015).

The work to develop the Residential Schools Land Memory Atlas under the Residential Schools Land Memory Mapping Project (RSLMMP) discussed throughout much of this volume, includes this volume, and is based on the concept of 'spatializing history', which emerged in the production of the Lake Huron Treaty Atlas (Pyne, 2013, 2014; Pyne and Taylor, 2015). This framing responds to Brenna Bhandar's implicit cartographic call to 'spatialize' history in a nonlinear, nonteleological way, [which] 'could open up possibilities for political change and transformation' (Bhandar, 2004, p. 831). Spatializing history involves both forward and backward looking elements (Dwyer, 1999) in addition to an inclusive approach to 'space'. This is reminiscent of Johnson et al. (2006), who have allowed the Hawaiian practice of looking backwards to face the future to guide their work:

> The concepts of 'past' and 'future' are explained by Hawaiians using bodily directions, the front of the body faces the 'past' while the back faces 'future'. Hawaiians 'face' their 'future' with their backs because the future is unknown. On the other hand, 'past' is knowable; it can be 'seen' in front of each of us, shaping our character and consciousness. Hawaiians believe that knowing who they are, genealogically, and where they came from, geographically and metaphysically, makes them capable of making more to move in the future. By allowing this concept to guide our work we are focussing our attention first on Western cartography's history, including its role in creating and perpetuating European colonialism (Harley, 1992, p. 532) and second on the history of Indigenous cartographies (Johnson et al., 2006, p. 82).

Both 'spatializing history' and the Hawaiian approach referred to above are circular, holistic approaches involving nonlinear approaches to time and space. The ongoing collaborative project to create the Residential Schools Land Memory Atlas 'looks in all directions', with aim of spatializing history by creating geonarratives that:

1. participate in the critical cartographic movement;
2. assume a relational approach to space that emphasizes performance and process;
3. are multidimensional;
4. give rise to emergent knowledge (i.e. are nonteleological);
5. bring together past and present;
6. emphasize context;
7. involve a holistic view of development and
8. make knowledge accessible for people in a way that allows for personal discovery.

The cartographic dimension of the Atlas contributes to its ability to spatialize history by virtue of its concern with appealing to place as an organizing factor for information (Pyne and Taylor, 2015; Taylor, 2003, 2005). The critical cartographic approach contributes further by dissolving the binary between object and action, between maps as objects and maps as processes. In the case of the Residential/Boarding Schools legacy, successful reconciliation involves 'revisiting' the historical geography of Residential Schools relationships from a variety of perspectives, and identifying links between the stories of today, tomorrow and yesterday. The project to create the Residential Schools Land Memory Atlas and this volume provide a way to gather these perspectives and present them in a way that questions the epistemological and ontological assumptions associated with modernism, including — but not limited to — the colonial 'world view'.

As a virtual geospatial public outreach and education interface, the Atlas is being designed and developed to shed light on a variety of themes relating to Residential Schools and space. In addition to the augural Residential Schools and E.F. Wilson Maps, which emerged through previous work on the Lake Huron Treaty Atlas, a series of new maps are planned for the Residential Schools Land Memory Atlas (see especially Chapters 2, 3, 4, 8, 9 and 11, which provide some examples). After the launch of the Lake Huron Treaty Atlas and its outreach phase (2012–14), the Residential Schools Map took on a life of its own in terms of usage and relevance to reconciliation processes that are part of the Truth and Reconciliation Commission's mandate (see Chapter 2). Since this time, the Residential Schools Land Memory Atlas has been developing in an iterative manner along a variety of dimensions (Brauen et al., 2011; Pyne and Taylor, 2012; Pyne, 2013). Iterative processes give rise to emergent knowledge (Turnbull, 2007; Kitchin and Dodge, 2007; Pyne and Taylor, 2012) and represent a nonlinear, cyclical view of development where past elements can be incorporated, new things can be brought forward and certain elements can be left behind, with the possibility of reintegrating them into the development process after any number of project iterations (Cowen and Shenton, 1996; Pieterse, 2009; Brauen et al., 2011; Pyne, 2013). Collaborative relationships are central aspects of the types of iterative processes that give rise to the Atlas, and iterative processes occur along at least four interrelated dimensions: conceptual; financial; technological and narrative — which correspond the people-technology-content analytical triangle employed by Dr. Ali Arya in his approach to digital media (personal communication throughout the teaching of the doctoral core course in digital media, Carleton University, Fall, 2017). The iterative interplay between theory (or concepts) and practice is a hallmark of the cybercartographic atlas-making framework, which guides the atlas project (Brauen et al., 2011; Pyne, 2013, 2014, 2019; Taylor and Lauriault, 2014; Taylor, 2005).

The central purpose of both Cybercartography and the collaborative project to design and develop the cybercartographic Residential Schools Land Memory Atlas continues the work of the Lake Huron Treaty Atlas: to contribute to enhanced understanding and improvement in relationships between people and between people and the land. In the case of Cybercartography, it is the broader goal of achieving a new awareness of social political and economic processes through cartographic narrative; and, in the case of this specific atlas project, it is the more focused goal of enhancing awareness and mutual understanding in Residential Schools reconciliation processes. This consistency in fundamental goals has made Cybercartography an ideal theoretical and methodological framework for the Residential Schools Land Memory Atlas, which has involved the emergence of critical, holistic and reflexive mapping practices.

Cybercartography is a set of concepts and tools that provides an effective atlas building framework for approaching complex social, political and economic phenomena that include reconciliation processes (Taylor, 1997, 2003, 2005, 2009, 2013; Taylor and Caquard, 2006; Taylor and Pyne, 2010; Pyne and Taylor, 2012). Cybercartography participates in the critical turn in cartography in terms of theory and application, and represents a processual (Kitchin and Dodge, 2007; Turnbull, 2007) approach to mapping and the politics of space that welcomes and works well with other forms of knowledge. The Cybercartographic Atlas Framework makes mapping practices possible that are both responsive to and part of reconciliation processes along a variety of dimensions. The approach to maps as practices is consistent with the 'ontogenetic' conception of mapping practices put forward by Kitchin and Dodge (2007) who 'propose a radical departure in ontological thinking concerning maps: a shift from ontology (how things are) to ontogenesis (how things become), or from the nature of maps to the practices of mapping [...] they are not ontologically secure representations, but, rather, a set of unfolding practices' (Kitchin, 2008, p. 213); and is consistent with critical thinking in archival studies (see Chapter 8).

Understanding the evolution and existence of the Residential Schools Land Memory Atlas with reference to 'a set of unfolding practices' is not only the most apt way to describe and explain the Atlas, it is also an example of spatializing history (Bhandar, 2004; Brown, 2001). Adopting the Bourdieu's (1992) reflexive high-speed tour approach in documenting some of these practices serves to 'put them on the map', open to further reflection and dialogue in the future. In addition, both the cybercartographic atlas-making framework and the atlas project it supports emphasize interactivity and broad community participation (Pyne, 2019; Taylor and Pyne, 2010; Pyne and Taylor, 2012, 2015). They are transdisciplinary and holistic in nature, with an emphasis on storytelling, knowledge sharing and enhancing awareness of different perspectives. The Inuktitut name of the software developed to create atlas modules — 'Nunaliit', which means 'settlement', 'community' or 'habitat' — illustrates the community orientation of the project. This name was given to the cybercartographic framework to emphasize the community-based approach that was driving the development of the software in different domains: (1) open specification approaches; (2) modularity; (3) 'live' data; (4) geospatial storytelling and (5) audiovisual mapping (Caquard et al., 2009).

12.2 Reflexivity, intersections and implicit and explicit approaches to cartography

The transdisciplinary approach to developing the Residential Schools Land Memory Atlas includes work with individuals and organizations who are often engaged in their own projects, which nevertheless intersect with the Atlas project in a variety of ways, some of which are discussed in this volume. These intersections involve reflexive relationships between theory and practice; map and text (Wood and Fels, 2008), map and art and teaching and learning. When it comes to the reflexive relationship between teaching and learning, the examples briefly described in Chapter 11 involved students learning about cybercartographic mapping and then contributing to the generation of cybercartographic maps, at the same time beginning to learn about the worldviews and ways of the first peoples on Turtle Island and the colonial legacy of their relationship with settler populations, including the Residential Schools Legacy.

Writing reflexively by reflectively telling stories about the iterative processes in the making of the Residential Schools Land Memory Atlas can be considered as a form of map performance that enriches people's understandings of the Atlas. Some of these stories are already included in the Project Journal and Travels Maps (see Chapter 2), which begins to document a diverse array of processes involved in the map-making that are also a part of the maps themselves. The contributions to this book that are not directly related to discussions of the Atlas (Chapters 5, 6 and 7) are nevertheless related in a reflexive way to the Atlas project and exhibit reflexivity in their presentation. For example, the mapping initiatives of the Carlisle Indian School Digital Resource Center (CISDRC) discussed in Chapter 5 provide excellent examples for us in the ongoing development of the Atlas, in addition to providing the beginning to a potential ongoing research (for reconciliation) relationship. Chapter 3 provides a good example of an approach to Ellingson's (2009) 'multigenre crystallization' with its reflexive geonarrative approach that begins to synthesize cartography and art by introducing the concepts of implicit and explicit cartography. For example, the creation, discussion and incorporation into the chapter of the pilot map of the essay by Jeff Thomas, 'I Have A Right to Be Heard', which makes more explicit the implicit cartography. Rather than there being a binary relationship between art and cartography, they can be considered as different phases of the same process along a continuum that bends back upon itself, reflexively. Our collaborative work in Cybercartography to develop the Residential Schools Land Memory Atlas has allowed both phases to manifest in the form of implicit cartographic writing about the work and explicit cybercartographic mapping. This is consistent with critical approaches to cartography that emphasize its performative or processual nature, which includes a non-binary approach (del Casino and Hannah, 2006); Turnbull's (2007) thinking on 'hodology', which considers trail-making in different dimensions — from neural mapping to digital mapping, and many forms of 'mapping' in between; and the relational concepts of epi-map and para-map discussed by Denis Wood and Martin Fels in *The Natures of Maps: Cartographic Constructions of the Natural World* (2008).

The discussion of 'the settler-colonial mesh' in Chapter 6 provides another example of implicit cartography in its concern with the spatiality of the residential schools legacy in North America, with phrases such as 'the space of destructive assimilative education' and 'the space of Indigenous boarding schools'; in addition to its use of a key spatially oriented concept: 'the settler-colonial mesh', which is a multiscalar and multidimensional concept that responds to the complexity of 'settler colonial practices of assimilative education over time'. While the chapter focuses on 'assimilative boarding schools', it is also interested in 'broader processes of settler colonial domination', reflecting the significance of broader historical geographical issues going beyond the particular stories associated with residential schools. Also reflecting a focus that goes beyond residential schools, Chapter 7 considers issues related to social and spatial justice (an implicit cartographic concept) to '[establish] lines of ideological, operational, and eventually museological, continuity between 19th century workhouses in England and Ireland and Indian residential/boarding schools in North America'. Trina Cooper-Bolam's approach is consistent with the talk, templates, tradition approach discussed further in this book and acknowledges the importance of in-person interactions, relationships and reciprocities, and of experiential knowledge and memory, derived in part, from the affective and embodied phenomena that only arises from visiting and dwelling in sites of research. While Chapter 10 begins to explore an ethics of care approach to transdisciplinary research involving a broad notion of 'community', which is inherently reflexive insofar as it requires positionality, and iterative processes that involve ongoing consultation, negotiation and revision based on consensus involving plenty of talk, the development of templates and the emergence of new traditions. In this regard, despite the fact that institutional ethics processes reflect a certain degree of principalism and consequentialism, which often manifest in written guidelines and reporting procedures, and have a tendency toward deontological ethics, they also include interpersonal dialogue between ethics committee members and others involved in institutional research ethics administration and 'the researcher'. These relational encounters help to situate and provide context for ethics committee decisions concerning the progress of research projects, at the same time opening a window for an ethics of care approach in institutional ethics processes. More effort to document and reflect on the relational dimension of institutional ethics process will go a long way in bridging the divide between institutional and participatory approaches noted by Elwood (2007), in addition to paying attention to Indigenous knowledge and practices regarding 'good relations' (see Chapter 1), and to issues relating to what some refer to as 'decolonizing our minds'. For example, the Circle of All Nations grounding ideology is embedded in the concept of Gina-waydaganuc, which affirms that everything is connected and therefore utmost respect and responsibility constitute the grounding principles in relationship (Thumbadoo, 2017).

12.3 Talk, templates and tradition

The many iterative processes involved in the atlas-making process provide the best examples of the performative aspect of the Atlas project. Many of these processes involve the kind of 'talk' referred to in the 'talk-templates-tradition'. A performative or processual approach to map and atlas-making is central to this project (Caquard et al., 2009; Pyne and Taylor, 2012; Pyne, 2013). For example, the process of writing about the Residential Schools Land Memory Atlas is part of the broader Atlas design and development process, in addition to being 'about' the Atlas (Wood and Fels, 2008). From this perspective, the atlas is an ongoing process oriented toward contributing to the conditions for sufficient intercultural mutual understanding. The atlas production process involves a variety of interpersonal interactions, knowledge exchanges and actions related to better understanding the nature of the Residential Schools Legacy through atlas design, development and use. The line between the Atlas as a material object and the design and development processes involved in its making are blurred, the dichotomy is diminished. Insofar as the Atlas is being designed to allow for ongoing critical input and contributions, the map user can also become the map-maker. In this respect, designing, developing and using the atlas are all intertwined (Brauen et al., 2011).

Chapters 8 and 9 provide insight into the in-development Residential Schools Land Memory Mapping Atlas with reflexive accounts of the processes involved in the emergence of the Residential Schools Land Memory Atlas. Chapter 9 discusses an interesting approach to community-based research involving volunteered geographic information generated by survivors of Assiniboia Indian Residential School in Winnipeg, Manitoba, Canada at their 2017 reunion; and Chapter 8 charts aspects of roughly the first half of the iterative development Residential Schools Land Memory Atlas, including work with research assistants from universities across Canada. This chapter sheds light on the role of Cybercartography as both a research and a presentation framework at the intersection of community-based approaches to archival studies and cartography. Combined with the concept of emergence in the context of project design and development is not only research but also extensive talking between team members, which has resulted in a series of templates with the aim of contributing to both the broader reconciliation tradition, and the more specific cybercartographic tradition.

12.4 Development for reconciliation in a transitional justice context

Although Crocker (2015) has identified 'reconciliation' as the preeminent goal of transitional justice, the other seven goals of truth, compensation, providing a public platform for victims, accountability and punishment, rule of law; compensation to victims, institutional reform, long-term development and public deliberation are also essential. Achieving transitional justice involves working toward these goals, which are mutually interdependent and actually become means for each other. For example, inclusive public deliberation contributes to effective long-term development.

There is vast agreement concerning the broader reach and relevance of contemporary cartography to social and other forms of justice. Not only has it become a reflexive practice with historical and critical cartographers 'bending back' to look critically at our cartographic past; but cartography has also reached out to the general public, dramatically 'democratizing' participation in map and atlas-making activities and use over the past 20 years (Pyne, 2013, 2019). A similar broadening has occurred in approaches to development. For example, the previous emphasis on maximization of self-interest as an end of economic development has been largely replaced with a host of individual well-being indicators reflecting a diverse range of individual rights (Goulet, 1971; Sen, 1999; Haq ul, 1995). Going beyond yet a concern for individual well-being and agency, a relational approach to development aims at healthy, productive, creative and fair relationships. Cybercartography provides a good example of a contemporary approach to cartography that takes a relational approach to development.

An important theme that ties the Atlas project to Cybercartography is its approach to development. For example, its emergent, iterative approach taken to atlas development is linked to its ability — through collaborative participation — to foster human development through shared experiences, and teaching and learning through transdisciplinary interactions. These processes are all supported by a theoretical and practical framework that provides a certain degree of support and stability, while allowing considerable room for creative flow and the 'natural' emergence of geonarratives or interactive multimedia 'mapped stories'. The iterative processes involved in the creation of the Atlas maps are rooted in a nonlinear, cyclical view of development (Cowen and Shenton, 1996; Nederveen Pieterse, 2009): incorporating past elements, bringing new elements forward, and leaving certain elements behind, with the possibility of reintegrating them into the development process after any number of project iterations. This is consistent with a preeminent Indigenous value of 'not wasting'. Atlas development occurs along many dimensions through a variety of cybercartographic mapping practices that reflect Taylor's 'development from within' approach (Taylor and McKenzie, 1992; Pyne, 2013, 2019) and Carmen's (2000) get your feet in the mud approach. This transformative approach to development emphasizes nonhierarchical, holistic relationships and involves meeting people in their knowledge spaces, finding common ground and participating in emergent mapping practices. These practices in turn feed back into further developments in the cybercartographic atlas framework, reinforcing the mutually interdependent relationship that exists between the framework and the Atlas (Pyne, 2013).

Mirroring Elwood's (2007) concerns with institutional research ethics, Carmen (2000) critiques to 'abstract formal' approaches to development, which he claims are 'outside-in' approaches that cannot help but promote dependency and inhibit true human flourishing. Carmen paints a dismal picture of development from the outside in:

> The excluded [are not] well-served with development intervention packages which are 'not theirs', nor with projectile projects — targeted 'at' them from the outside. Those interventions and those capital-intensive outside expert-led and outside-funded projects, be they of GO or NGO ilk have been shown for five long disheartening development decades, to have served — and very efficiently so — the tastes and 'capabilities' for self-preservation, self-aggrandizement, self-promotion and institution-building ('empire-building') of the interventionists and of the self-appointed assorted modernizers, extensionists, civilizers and handsomely remunerated trainers, capacity builders, empowerers and developers, and their respective organizations themselves, often without increasing by one iota — sometimes quite the reverse — the capacity for sustained and sustainable autonomous human agency of those being intervened the 'target' populations and so-called 'beneficiaries' (Carmen, 2000, 1021–22).

Carmen describes himself as someone who 'spent half a lifetime looking from the outside in, i.e. from the engine rooms of knowledge and power (in academia and in a large multinational)', until he decided to 'step in the mud' (2000, 1020). Discussing the effect of a simple change in perspective — looking from the 'inside out', instead of from the 'outside in' — Carmen echoes the view that development discourse in terms of functionings and capability is 'light years removed from the soulless and culturally disembedded shibbolets of economic development transfer', (2000, 1021). However, he cautions these concepts are about 'what ought to ' and 'what ought to be done', rather than about how. According to Carmen, Sen's (1992, 1999) capability concept is limited because it is presented in 'static, nonprocess language' and it is 'a concept relatively closer to an individualistic mindset'. The danger of this mindset is that it promotes an us-them attitude, discouraging the development of 'autonomous human agency', at the same time imposing a 'foreign intellectual framework' in the name of furthering such agency.

In contrast, Carmen presents a 'growth and learning' conception of capability stating 'capability can be enhanced by (1) human learning and (2) cooperation and solidarity with others' (2000, 1023) in an equitable environment of peers. Carmen insists this dynamic, process-oriented, 'growth and learning' model of capability can itself be a powerful approach to development thinking. Consistent with his distinction between an 'outside in' and an 'inside out' perspective, Carmen contrasts the 'culture of power' with 'power of culture', where the 'culture of power' is an 'isolationist attitude' that does not 'need the other' and the 'power of culture' is powerful 'because of the factor of solidarity (cooperation with others …) and because it has the dynamic, creative capacity to learn' (2000, 1023). For Carmen, true human autonomous agency can only be achieved by learning in cooperation with others. Finally, one of Carmen's main contributions to development thinking is his main point that the teacher does not transmit knowledge; rather, people learn through their interactions with 'the object', which is some form of common output or product produced by a group.

In order to truly overcome a paternalistic approach to development, it is necessary to adopt a broader relational conception of development that emphasizes: (1). nondivisible social goods (Gore, 1997); (2). emphasis on process language; (3). nonhegemonic relationships; (4). interactive 'object'-oriented teaching and learning; (5). learning in solidarity with others; (6). the 'power of culture' and (7). an 'inside-out'perspective and approach. In addition, it is necessary to acknowledge the postcolonial context of many development initiatives, and the corresponding need for these initiatives to contribute to intercultural reconciliation as 'a widely accepted objective and guiding principle in attempts to deal with the aftermath of painfully repressive regimes around the world' (Bhandar, 2004, p. 834).

Similar to Carmen's inside out approach is the 'development from within' approach (Taylor and McKenzie 1992), which emphasizes inclusion and participation, and is contrasted with 'development from below' approaches. In addition to other critiques, 'considerable skepticism was expressed by some Indigenous planners who said that development from below and concomitant ideas, such as agropolitan development, were just one more example of theories and prescriptions which are developed in the North applied to the South' (Taylor and Mckenzie, 1992, p. 234). The Cybercartography of today is rooted in this 'development from within approach'. It reflects a relational approach to development; and, it acknowledges the need for reconciliation through an emergent, context-sensitive processual approach to mapping that is consistent with the relational development features identified above (Pyne, 2013).

Relational development requires nonhegemonic relationships, interactive 'object'-oriented teaching and learning, learning in solidarity with others; the 'power of culture', and, an 'inside-out' perspective and approach. In addition, it is necessary to acknowledge the postcolonial context of many development initiatives, and the corresponding need for these initiatives to contribute to intercultural reconciliation (Bhandar, 2004). The potential for the Residential Schools Land Memory Atlas to integrate education and research provides an example of 'object'-oriented teaching and learning along the sociocultural political development dimension (Carmen, 2000; Pyne, 2013, 2019). Since its inception with the initial work to create the Treaties Module, in 2007, the Residential Schools Land Memory Atlas has been developing as the outcome of a variety of collaborative relationships. This speaks not only to the ability of the Atlas, including the Residential Schools Map, to contribute to learning in solidarity with others but also to its ability to involve the 'power of culture' and an 'inside-out' perspective in map development processes.

12.5 Concluding words

This book's contributions are broadly consistent with qualitative inquiry in its overriding concern with social and related forms of justice (Denzin and Lincoln, 2005; Denzin et al., 2008) and in its commitment to participating in the broad trend toward the democratization of research, which involves aspects such as intersectionality and inclusion (Rescher, 2001 Pyne, 2013). In this spirit, the book includes chapters discussing the work and thought of members of the RSLMMP research community. This is important because — despite disciplinary differences — many common themes exist in issues discussed by the contributors. The *Introduction to the Handbook of Qualitative Research* (Denzin and Lincoln, 2005) provides a useful description of qualitative methods that identifies eight historical moments:

> [T]he traditional (1900–50); the modernist, or golden age (1950–70); blurred genres (1970–86); the crisis of representation (1986–90); the postmodern, a period of experimental and new ethnographies (1990–95); postexperimental inquiry (1995–00); the methodologically contested present (2000–04); and the fractured

future, which is now (2005). The future, the eighth moment, confronts the methodological backlash associated with the evidence-based social movement. It is concerned with moral discourse, with the development of sacred textualities. The eighth moment asks that social sciences and the humanities become sites for critical conversations about democracy, race, gender, class, nation-states, globalization, freedom and community (Denzin and Lincoln, 2005, p. 2—3).

Although the temporal periods referred to in Denzin's account of the eight historical moments of qualitative research differ from the far older Seven Fires Prophecy (See Chapter 1), which reaches back at least 600 years, it is important to remember that the Seven Fires Prophecy is not necessarily linked solely to a linear approach to time. From a Seven Fires Prophecy perspective, we interpret Denzin's eight historical moments as an evolution in thought with the potential of contributing to the 'lighting of the eighth fire' discussed in Chapter 1.

The two related turns in thinking that have occurred in this historical evolution are the 'spatial turn', with an emphasis on performance, process and relationships); and, the 'interpretative turn', with its emphasis on narrative and reflexivity (Denzin and Lincoln, 2005): 'In both inductive analytic […] and more artistic approaches to qualitative research, researchers [have] abandoned claims of objectivity in favour of focussing on the situated researcher and the social construction of meaning' (Ellingson, 2009, p. 1—2). Subjective bias, as it is understood in a conventional science sense, does not exist according to the view of 'research' that is emerging in this eighth moment.

The interpretive view is moving toward a holistic perspective where the binary distinction between subjectivity and objectivity does not exist. This is not to say that the individual's perspective does not matter, however, quite the contrary. Instead we have the 'situated researcher', a concept more akin to Indigenous viewpoints, who can be conceived of as an individual with a unique perspective and gifts with the capacity for vision, and a responsibility to develop that vision and those gifts for themselves and their communities.

Whether implicit or explicit, the approaches to 'mapping' for reconciliation in the context of the Residential Schools Legacy discussed through this volume reflect a rich or thick approach to mapping that is reflexive in many ways. In addition to sharing knowledge about the systemic legacy (for example, Jeff Thomas' discussion of ethnocide in Chapter 3 Andrew Woolford's discussion of genocide in Chapter 6), stories are also shared about efforts to contribute to healing and reconciliation that document intersecting initiatives, and providing for the emergence of new strategies.

This book provides an example of Cybercartography as an umbrella framework in terms of being a hub for 'development' that includes transdisciplinary participation with collaborators who contribute to atlas 'products', while at the same time pursuing their own projects. In this regard, Engler et al. (2014) have noted 'how [Cybercartography is] a technology, a mode of production, and a unifying framework [that] is ideal for the management, dissemination, and visualization of VGI' (44). In addition, it begins to document various ways to expand the narrative potential of Cybercartography, in part by expanding the scope and kind of 'research partners' and through contributions to the conceptual dimension with observations and reflections on concepts such as reflexivity, implicit and explicit cartography, and development.

As the result of many good conversations, templates have been developed in the form of pilot maps and mapping strategies, and an emerging Residential Schools mapping tradition, there remain many things yet to do and a variety of ongoing challenges. For example, in a pleasantly reflexive way, writing and mapping processes involved in the collaborative creation of Chapter 3 have led to the activation of a new collaborative research relationship with the Legacy of Hope Foundation with respect to 'mapping' its Where Are the Children Website, which involves intersections between the research and teaching, and new ideas related to education and outreach. Following the launch of the Residential Schools Land Memory Atlas, a publication including more details on this and the other endeavours introduced in this book is in order.

In this book, we have focused on bringing together a number of intersecting perspectives, which share a common aim of contributing to reconciliation in a primarily Residential/ Boarding Schools context in particular, yet extend to other contexts, including ontological and epistemological spaces. While it includes details concerning the historical geography of school systems, attitudes and occurrences, it also delves reflexively into educational strategies involving Cybercartography for working toward reconciliation to acknowledge and address the wrongs that have been committed under this system, the resilience of the school Survivors, and a host of related issues. The motto 'building awareness to bridge relationships' (Pyne, 2013), which was coined to describe the Lake Huron Treaty Atlas project, applies equally to the RSLMMP, which is beginning to develop some critical mass in this endeavour.

References

Bhandar, B., 2004. Anxious reconciliation(s): unsettling foundations and spatializing history. Environment and Planning D: Society and Space 22, 831−845.

Bourdieu, P., 1992. The practice of reflexive sociology. In: Bourdieu, P., Wacquant, L. (Eds.), An Invitation to Reflexive Sociology. University of Chicago Press, Chicago, pp. 217−253.

Brauen, G., Pyne, S., Hayes, A., Fiset, J.P., Taylor, D.R.F., 2011. Transdisciplinary participation using an open source cybercartographic toolkit: the atlas of the Lake Huron treaty relationship process. Geomatica 65 (1), 27−45.

Brown, W., 2001. Politics Out of History. Princeton University Press, Princeton, NJ.

Caquard, S., Pyne, S., Igloliorte, H., Mierins, K., Hayes, A., Taylor, D.R.F., 2009. A "living" atlas for geospatial storytelling: the cybercartographic atlas of indigenous perspectives and knowledge of the Great Lakes region. Cartographica 44 (2), 83−100.

Carmen, R., 2000. Prima mangiare poi filosofare. Journal of International Development 12, 1019−1030.

Cooper-Bolam, T., 2014. Healing Heritage: New Approaches to Commemorating Canada's Indian Residential School System. Master's Thesis. Carleton University, Ottawa. https://curve.carleton.ca/search?s=cooper+bolam.

Cooper-Bolam, T., 2018. On the call for a residential schools national monument. Journal of Canadian Studies 52 (1), 57−81.

Cowen, M.P., Shenton, R.W., 1996. Doctrines of Development. Routledge, London; New York.

Crocker, D., 2015. Obstacles to reconciliation in Peru: an ethical analysis. Unpublished English translation of Obtsaculos para la reconciliacion en el Peru. In: Giusti, M., Gutiérrez, G., Salmón, E. (Eds.), La Verdad nos Hace Libres. Sobre las Relaciones entre Filosofía, derechos Humanos. Religión y Universidad, Fondo Editorial de la Pontificia Universidad Católica del Perú, Lima, Peru.

Del Casino, V.J., Hanna, S.P., 2006. 'Beyond the 'binaries': a methodological intervention for interrogating maps as representational practices. ACME: An International E-Journal for Critical Geographies 4 (1), 34−56.

Denzin, N.K., Lincoln, Y.S., 2005. The SAGE Handbook of Qualitative Research, third ed. Sage Publications, Thousand Oaks.

Denzin, N.K., Lincoln, Y.S., Tuhiwai Smith, L., 2008. Handbook of Critical and Indigenous Methodologies. Sage, Thousand Oaks, California.

Dwyer, S., 1999. Reconciliation for Realists. Ethics and International Affairs 13, 81–98.

Ellingson, L., 2009. Engaging Crystallization in Qualitative Research. Sage, Thousand Oaks.

Elwood, S., 2007. Negotiating participatory ethics in the midst of institutional ethics. ACME: An International E-Journal for Critical Geographies 6 (3), 329–338.

Engler, N., Scassa, T., Taylor, D.R.F., 2014. Cybercartography and volunteered geographic information. In: Taylor, D.R.F., Lauriault, T. (Eds.), Developments in the Theory and Practice of Cybercartography: Applications and Indigenous Mapping. Elsevier, Amsterdam (Chapter 4).

Gore, C., 1997. Irreducible Social Goods and the Informational Basis of Amartya Sen's Capability Approach. Journal of International Development 9 (2), 235–250.

Goulet, D., 1971. The cruel choice: a new concept on the theory of development. The Western Political Quarterly 24 (3), 594–596.

Haq ul, M., 1995. Reflections on Human Development. Oxford University Press, New York.

Harley, J.B., 1992. Rereading the maps of the Columbian encounter. Annals of the Association of American Geographers 82 (3), 522–542.

Johnson, J., Louis, R., Pramono, H., 2006. Facing the future: encouraging critical cartographic literacies in indigenous communities. ACME: An International E-Journal for Critical Geographies 4 (1), 80–98.

Johnson, J., Murton, B., 2007. Re/placing native science: indigenous voices in contemporary constructions of nature. Geographical Research 45 (2), 121–129.

Kitchin, R., Dodge, M., 2007. Rethinking maps. Progress in Human Geography 31 (3), 331–344.

Kitchin, R., 2008. The practices of mapping. Cartographica 43 (3), 211–216.

Pieterse, N., 2009. Critical holism and the tao of development. In: Nederveen Pieterse, J. (Ed.), Development Theory: Deconstructions/Reconstructions. Sage, London.

Palmer, M., 2009. Engaging with indigital geographic information networks. Futures 41, 33–40.

Palmer, M., 2012. Theorizing in digital geographic information networks. Cartographica, Special Issue on Indigenous Cartography and Counter Mapping 47 (2), 80–91.

Palmer, M., Elmore, D., Watson, M., Kloese, K., Palmer, K., 2009. Xoa:dau to Maunkaui: integrating indigenous knowledge into an undergraduate earth systems science course. Journal of Geoscience Education 57 (2), 137–144.

Pearce, M.W., Louis, R., 2007. Mapping Indigenous depth of place. American Indian Culture & Research Journal 32 (3), 107–126.

Pyne, S., 2013. Sound of the Drum, Energy of the Dance: Making the Lake Huron Treaty Atlas the Anishinaabe Way. Unpublished PhD Dissertation. Carleton University, Ottawa. https://curve.carleton.ca/392a68ac-086c-4470-976d-101d4e96f9f7.

Pyne, S., 2014. The role of experience in the iterative development of the Lake Huron treaty atlas. In: Taylor, D.R.F., Lauriault, T. (Eds.), Developments in the Theory and Practice of Cybercartography: Applications and Indigenous Mapping. Elsevier, Amsterdam (Chapter 17).

Pyne, S., 2019. Cybercartography and the critical cartography clan. In: Taylor, F., Anonby, E., Murasugi, K. (Eds.), Further Developments in the Theory and Practice of Cybercartography: International Dimensions and Language Mapping. Elsevier, London. Forthcoming).

Pyne, S., Taylor, D.R.F., 2012. Mapping indigenous perspectives in the making of the cybercartographic atlas of the Lake Huron treaty relationship process. Cartographica, Special Issue on Indigenous Cartography and Counter Mapping 47 (2), 92–104.

Pyne, S., Taylor, D.R.F., 2015. Cybercartography, transitional justice and the residential schools legacy. Geomatica 69 (1), 173–187.

Reimer, G., Bombay, A., 2010. The Indian Residential Schools Settlement Agreement's Common Experience Payment and Healing, first ed. Aboriginal Healing Foundation, Ottawa, Ont.

Rescher, N., 2001. Philosophical Reasoning. Blackwell Publishers, Oxford.

Sen, A., 1992. Inequality Re-examined. Harvard University Press, Cambridge, Mass.

Sen, A., 1999. Development as Freedom. Oxford University Press, Oxford.

Taylor, D.R.F., 1997. Maps and mapping in the information era. In: Ottoson, L. (Ed.), Keynote Address to the 18th ICA Conference, Stockholm, vol. 1, pp. 1–10. Proceedings.

Taylor, D.R.F., 2003. The concept of cybercartography. In: Peterson, M.P. (Ed.), Maps and the Internet. Elsevier, Amsterdam, pp. 405–420.

Taylor, D.R.F., 2005. Cybercartography: theory and practice. In: Modern Cartography Series, vol. 4. Elsevier, Amsterdam.

Taylor, D.R.F., 2009. Maps, mapping and society: some new directions. In: Proceedings of Global Map Forum. Geographical Survey Institute, Tskuba, Japan, pp. 32–35.

Taylor, D.R.F., 2013. Fifty years of cartography. The Cartographic Journal 50 (2).

Taylor, D.R.F., McKenzie, F., 1992. Development from within: Survival in Rural Africa. Routledge, London, New York.

Special issue on cybercartography. In: Taylor, D.R.F., Caquard, S. (Eds.), Cartographica 41 (1), 1–5.

Taylor, D.R.F., Pyne, S., 2010. The history and development of the theory and practice Cybercartography. International Journal of Digital Earth 3 (1), 1–14.

Taylor, D.R.F., Lauriault, T. (Eds.), 2014. Developments in the Theory and Practice of Cybercartography: Applications and Indigenous Mapping. Elsevier, Amsterdam.

Thumbadoo, R.V., 2017. Ginawaydaganuc and the Circle of All Nations: The Remarkable Environmental Legacy of Elder William Commanda. Way. Unpublished PhD Dissertation. Carleton University, Ottawa. https://curve.carleton.ca/system/files/etd/aa4e3cbb-5b83-464d-8286-a901fcd77b06/etd_pdf/2accd64a4b4deca5aa4dd344fdea830d/thumbadoo-ginawaydaganucandthecircleofallnationsthe.pdf.

Turnbull, D., 2007. Maps, narratives, and trails: performativity, hodology, and distributed knowledges in complex adaptive systems: an approach to emergent mapping. Geographical Research 45, 140–149.

Upward, F., McKemmish, S., Reed, S., 2011. Archivists and changing social information spaces: a continuum approach to recordkeeping and archiving in online cultures. Archivaria 72, 197–237.

Wood, D., Fels, J., 2008. The Natures of Maps: Cartographic Constructions of the Natural World. University of Chicago Press, Chicago.

Index